CHEILOSTOMATOUS BRYOZOA
PART 2

HIPPOTHOOIDEA - CELLEPOROIDEA

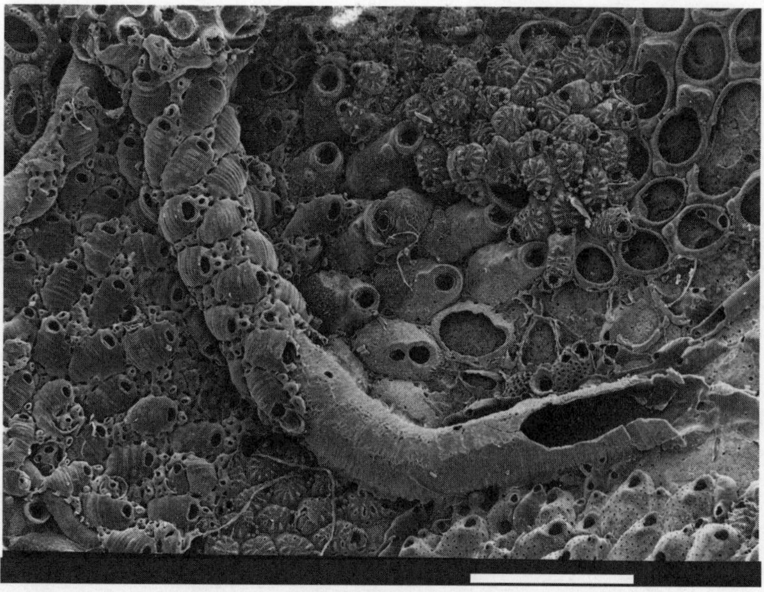

Frontispiece
A cheilostomate community on the inner surface of a bivalve shell.
The species present are: *Chorizopora brongniartii, Escharina johnstoni, Trypostega venusta, Crassimarginatella solidula, Neolagenipora collaris, Hippothoa flagellum, Callopora discreta, Schizotheca fissa, Collarina balzaci* and *Puellina praecox.*
Scale bar = 1mm.

Synopses of the British Fauna (New Series)
Edited by R. S. K. Barnes and J. H. Crothers
No. 14 (Second Edition)

CHEILOSTOMATOUS BRYOZOA
PART 2

HIPPOTHOOIDEA - CELLEPOROIDEA

Notes for the identification of British species

P. J. HAYWARD*
and
J. S. RYLAND

School of Biological Sciences, University of Wales Swansea, Singleton Park, Swansea, SA2 8PP
* and *Research Associate, The Natural History Museum, London SW7 5BD*

1999
Published for
The Linnean Society of London
and
The Estuarine and Coastal Sciences Association
by
Field Studies Council
Shrewsbury

Synopses of the British Fauna (New Series)

Fellows of the Linnean Society of London, and members of the Estuarine and Coastal Sciences Association, may obtain copies at a discount by calling at the Linnean Society's rooms in Burlington House, Piccadilly. London

or, through the post, from
FSC Publications, Preston Montford, Montford Bridge, Shrewsbury SY4 1HW

non-members may order copies from :-

FSC Publications, Preston Montford, Montford Bridge, Shrewsbury SY4 1HW
or
Backhuys Publishers, Postbus 321, 2300 AH Leiden,
The Netherlands

ISBN 1 85153 263 3

Cover Illustration: *Escharella abyssicola* (Norman)

Printed in Great Britain by Henry Ling Ltd at The Dorset Press, Dorchester.

Cheilostomatous Bryozoa
Part 2. Hippothooidea - Celleporoidea

P. J. Hayward and J. S. Ryland

Contents

Foreword

The original edition of this book was written by the same authors and was published, under the title of *British Ascophoran Bryozoans,* in 1977. It was the second of four *Synopses* (the others being no. 10, 1977; nos. 33 and 34, 1985) which formed the first comprehensive account of the British marine Bryozoa since the classic monograph of T. Hincks (1880a). Together, they described 267 species. Their publication, and the continual enthusiasm of the authors for bryozoans, have stimulated unprecedented advances in our knowledge and understanding of the phylum, and of the British fauna in particular.

One consequence of this increased understanding is that the original versions of the two *Synopses* treating the Cheilostomatida (nos. 10 and 14) are now out of date and we have decided to replace them with new, fully-revised volumes. The titles have had to be changed when it became apparent that the distinction between the traditional suborders "Anasca" and "Ascophora" had no basis in phylogeny. The second editions of the two volumes each treat a sequence of superfamilies. Part I (*Synopsis* No. 10, Second Edition 1998) described those classically assembled under the Anasca, together with the Cribriomorpha, whilst Part 2 covers the rest of the old suborder Ascophora. The key to genera includes all British cheilostomatous Bryozoa.

We welcome these second editions, and congratulate the authors on their scholarship and dedication to their phylum.

R. S. K. Barnes
Cambridge University
and
Estuarine & Coastal Sciences Association

J. H. Crothers
Field Studies Council
and
Linnean Society of London

Introduction

Bryozoans are aquatic, sessile, colonial animals. The phylum includes a minority of freshwater and estuarine species but is otherwise predominantly marine. They are surprisingly common animals, forming a conspicuous part of the hard substratum epifauna on most rocky shores, and often becoming especially abundant in shallow sub-littoral habitats. Bryozoan colonies are usually small, individually covering less than a few square centimetres of substratum, but may be so numerous in some habitats that they occupy considerable areas of suitable rock or shell substrata. On moderately exposed rocky shores, the blades of kelps and the fronds of lower shore fucoid algae may be covered with broad sheets of bryozoans. Some species develop relatively large, erect, three-dimensional colonies, rigid or bushy, which themselves provide habitats for numerous encrusting or motile invertebrates. The British bryozoan fauna includes about 300 species. The world fauna is presently unknown; the oft-quoted 4000 species is now seen to be entirely unrealistic. Possibly 1000 new species have been described worldwide in the second half of the twentieth century, and the rate of description of new species shows no sign of slackening. This is partly attributable to modern taxonomic precision, but also to the fact that bryozoan taxonomists have only recently discovered the astonishing diversity of the phylum in coral reef habitats, or in coarse ground continental shelf environments. It is probable that the true total number of living species of Bryozoa is more than twice the estimated 4000.

Bryozoans will be familiar to students and specialists of marine biology, but they may not be immediately recognised for what they are by the beginner. Each colony originates as a single founding zooid, the ancestrula, which results when a sexually produced larva settles and metamorphoses upon its preferred substratum. The colony grows through a process of budding, producing replicated series of genetically and usually morphologically identical units, autozooids, to develop branching chains, encrusting sheets, mounds or nodules. Many develop erect, branching forms, flexible or rigid, which may be two- or three-dimensional. As many as 30 species may be found between the tides on a moderately exposed, fully marine, rocky shore. On the middle shore small white patches of the cheilostomates *Haplopoma graniferum* (p. 314),*Cauloramphus spiniferum* and *Cribrilina cryptooecium* may be abundant on the undersides of flat stones, in pools or damp, shaded areas. Towards mean low water mark, stones and boulders will bear thicker, broader encrustations of *Cryptosula pallasiana* (p. 194) and *Umbonula littoralis* (p. 104), both a deep orange colour, and pinkish sheets of *Schizoporella unicornis* (p. 220). This same suite of species will often be abundant on larger kelp holdfasts, together with *Escharoides coccinea* (p.114), which also develops broad, orange sheets, and reddish nodular colonies of

Celleporina hassallii (p. 324). The encrusting fauna of a kelp holdfast may well include a dozen bryozoan species, and both its stipe, and the small red algae typically growing on it, often bear tufts of small, erect-growing species such as the cyclostomates *Crisia* and *Crisidia*, and the cheilostomates *Scrupocellaria*, *Aetea* and *Scruparia*. These latter may also occur on rock surfaces, with shaded aspects, growing among small red algae, or developing a bryozoan turf. However, the most conspicuous intertidal rocky shore bryozoans are undoubtedly the algal encrusting species, most notably *Membranipora membranacea*, which develops extensive lacy sheets on kelp blades, and the stellate, bristly colonies of *Electra pilosa*, which can occur on kelps but is especially common on *Fucus serratus*. The *Fucus* community also includes a number of sheet-forming ctenostomates. *Flustrellidra hispida, Alcyonidium gelatinosum* and *A. hirsutum* are the major components, but it is certain that the latter, at least, comprises a complex of genetically distinct, sympatric species and the *Fucus* community is potentially more diverse ecologically and taxonomically than is presently apparent.

Bryozoan populations become increasingly diverse, and frequently abundant, on hard ground sub-littoral habitats, and probably achieve a peak of taxonomic diversity on offshore shell gravels. The British fauna described in the four *Synopses* (no. 10, 1977; no. 14, 1979; nos. 33 and 34, 1985) totalled 267 species; these four volumes together constituted the first comprehensive account of the British marine Bryozoa since the classic monograph of T. Hincks (1880). During the past two decades taxonomic, morphological and ecological research on the Bryozoa has expanded tremendously, both in Britain and overseas, yielding an impressive new literature. This new flowering has had natural, and sometimes profound, consequences for the taxonomy of the phylum, and the classification has undergone some revolutionary changes. First editions of the two *Synopses* treating the Cheilostomatida (nos. 10 and 14) are now quite out of date and there was a need to replace both with new, revised volumes. The number of species included has increased too, partly as a result of recent systematic revision, but also through the recognition of undescribed species in the synonymies of otherwise familiar taxa, and the discovery of several entirely new species in recent years. A few species which were excluded from the first two editions as being largely unknown have since been found to be more widely distributed, and perhaps more common, than was previously apparent and are included in these more comprehensive surveys.

The format of these new *Synopses* is similar to that of the first editions, with chapters on general structure, reproduction, ecology, predators and practical methods. However, although the higher classification of the Cheilostomatida is still uncertain, it is clear that the traditional distinction between the Anasca and the Ascophora has no basis in phylogeny. Consequently these two volumes each treat a sequence of superfamilies; no. 10 encompasses those classically assembled under the old suborder Anasca, together with the Cribriomorpha, and no. 14 covers the rest of the old suborder Ascophora. The account of General Structure introduces the variety of morphological expression throughout the Bryozoa, and is followed by a more detailed survey of the Cheilostomatida. The key to genera includes all British cheilostomatous Bryozoa.

General Structure

Bryozoan colonies consist of replicated series of **zooids**, and grow by continuous, non-sexual budding from a single, sexually-produced founding zooid, the **ancestrula**. The colony may consist entirely of identical **autozooids**, or may include variously specialised polymorphs, termed **heterozooids**. Autozooids (Fig. 1) are functionally independent, each being equipped with a feeding **lophophore**, gut and reproductive organs, but the energetic demands of heterozooids must be supplied by the autozooids. Quite obviously, all of the zooids of a colony are genetically identical; all are in organic continuity, and may be integrated to a degree which contributes considerable individuality to the colony itself. A majority of bryozoan colonies develop as flat, encrusting sheets, in which the autozooid has a more or less box-shaped outline. The surface of the box applied to the substratum is morphologically dorsal, and the upper surface consequently ventral. The posterior end of an autozooid is the first part of the autozooid to differentiate from a developing bud, and the opposite anterior end is that furthest from the colony origin. As with many colonial animals, orientation of single zooids can become confusing and a more useful notation has been developed. Thus, the surface attached to the substratum is termed **basal**, and the upper surface **frontal**; the end of the autozooid closest to the colony origin is referred to as **proximal**, and the opposite end, generally corresponding with the growing tip or edge of the colony, is **distal**. In certain cylindrical autozooids, especially among the Ctenostomatida (Fig. 1A) and the Cyclostomatida (Fig. 1D), it may still be difficult to determine frontal and basal surfaces without reference to internal anatomy, but, with these few exceptions, even the most modified zooids may be orientated using this terminology.

Each autozooid has an eversible, bell-shaped **lophophore** of ciliated tentacles. The mouth is situated at the vertex of the bell and a short, U-shaped gut descends into the body cavity of the autozooid. This consists of a short, fat **pharynx**, a slender **oesophagus**, and a three-part stomach comprising a **cardia**, a voluminous **caecum** at the loop of the gut, and an ascending **pylorus**. A slender intestine passes to a short **rectum**, which opens at a dorsally situated anus, which is found at the base of the lophophore, outside the tentacle bell. In some families of bryozoans a conspicuous **gizzard** (Fig. 1A), with dark-coloured crushing teeth, is located between the oesophagus and the stomach. A nerve ganglion is present between the mouth and the anus, with nerves radiating to the gut, musculature and other parts of the zooid. The lophophore and gut, and their associated muscles, are collectively termed the **polypide**, while the rest of the zooid, the outer cuticle and skeleton, and the peritoneum lining the body cavity, are referred to as the **cystid**. In many species the polypide may regress, break down and be replaced one or more times during the lifespan of the autozooid. Polypide regression may occur in response to environmental deterioration, or as part of the reproductive cycle of the autozooid, or simply as a consequence of the normal physiological cycle of growth. The products of polypide breakdown are partly resorbed; unresorbed residues are compacted to form residual **brown bodies**. These may be enclosed by the developing gut of the new polypide, and eventually ejected, or may accumulate in the cystid as evidence of previous growth/regression cycles. Stained preparations of bryozoan autozooids reveal a network of tissue strands within the body cavity, most densely developed around the gut, but radiating through the zooid with branches extending to the communicating mural pores which link adjacent zooids. This network is referred to as the **funiculus**; it is important in the transport of metabolic products, especially in supplying the energy requirements of non-feeding polymorphic zooids, or shifting food reserves to regions of the colony where autozooids are no longer able to feed. In the Cyclostomatida, however, the funiculus is poorly developed and appears not to link adjacent autozooids.

The withdrawn lophophore is enclosed within a **tentacle sheath**, formed from an introverted portion of flexible outer body wall. The **orifice** through which lophophore and tentacle sheath pass on eversion may be closed by muscular contraction, or by a hinged flap, the **operculum**. Eversion of the lophophore is achieved through the operation of a simple hydrostatic system. In cylindrical ctenostomates, such as *Bowerbankia* (Fig. 1A), the flexible, uncalcified body wall is pulled inwards by the contraction of small **transverse parietal muscles**. As internal hydrostatic pressure rises above that of the surrounding seawater, **longitudinal parietal muscles** relax and the lophophore everts smoothly. The tentacle sheath also everts, lifting the expanded tentacle bell well above the level of the orifice. Lophophore retraction is generally very swift, powered by long **retractor muscles** which arise from close to the proximal end of the autozooid and insert around the base of the bell. Retraction of the introvert is completed by the longitudinal parietal muscles. In box-shaped, encrusting ctenostomates, and in those cheilostomates (Fig. 1B) which have a flexible frontal wall, the transverse parietal muscles simply depress the frontal surface of the autozooid, but lophophore eversion otherwise proceeds in exactly the same manner as in *Bowerbankia*. The frontal surface of ascophorine cheilostomates (Fig. 1C) is calcified and thus inflexible; parietal muscles now insert on a flexible membrane, or the floor of a compensating sac, lying beneath the calcified frontal shield. Contraction of the muscles raises internal hydrostatic pressure and seawater flows into the space or sac beneath the frontal shield, compensating for the reduced volume of the cystid, and lophophore eversion proceeds as in the other groups. It is a far more complex system in the Cyclostomatida. Cyclostomate zooids (Fig. 1D) are typically elongate and tubular, with a tiny terminal membrane. Body wall deformation can thus have little effect on internal pressures, so in the Cyclostomatida lophophore eversion operates through the redistribution of internal hydrostatic pressures between two non-communicating body cavities.

Zooid polymorphism is characteristic of all taxonomic groups within the Bryozoa but is especially well expressed in the Cheilostomatida. The form, function and homology of the often bewildering variety of heterozooids seen

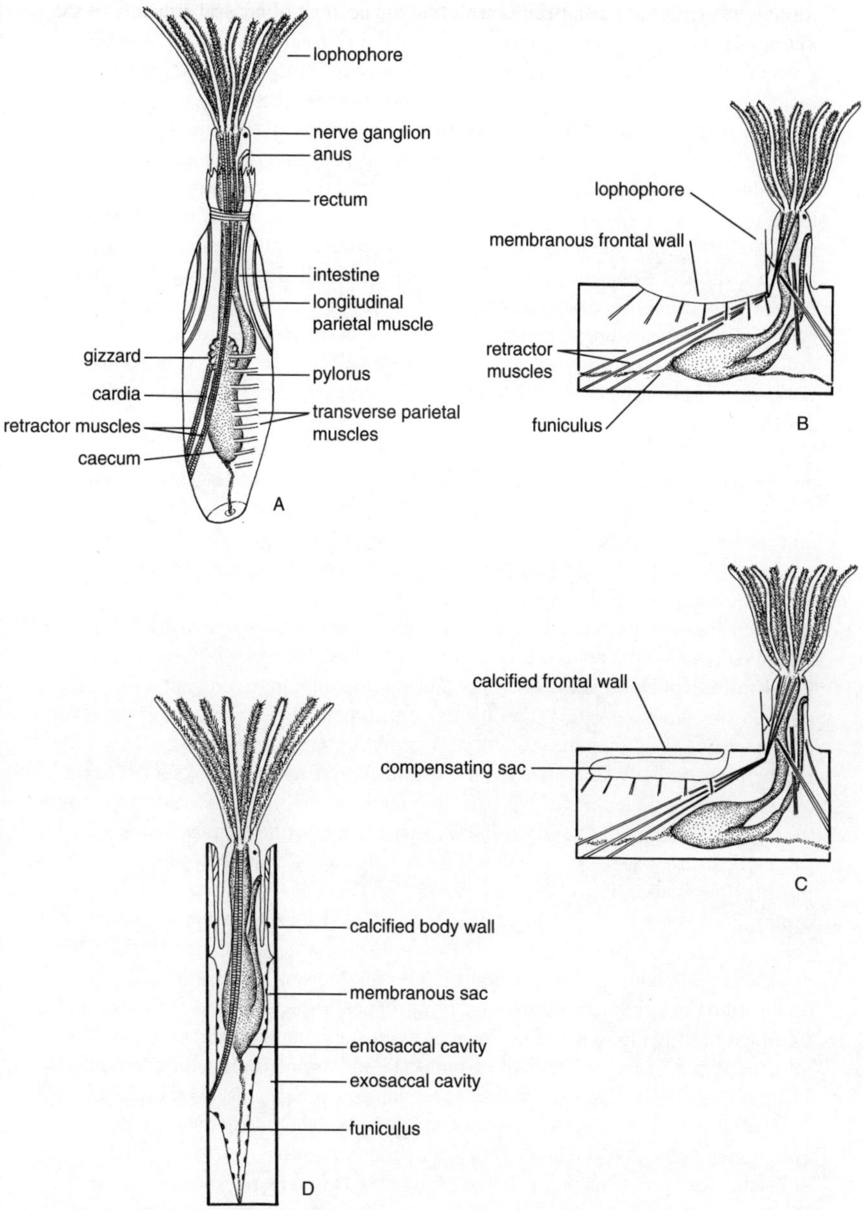

Fig. 1. Autozooid structure in four groups of Bryozoa.
A, A stoloniferous ctenostomate (*Bowerbankia*). B, A neocheilostomatine.
C, An ascophorine. D, A cyclostomate.

among this Order are discussed at length in the account of general structure of the cheilostomates.

Classification of the Bryozoa

Living bryozoans are classified into three classes. Of these, the PHYLACTOLAEMATA is wholly freshwater in distribution. The characters of the other two may be summarised as follows.

Class STENOLAEMATA: zooids cylindrical, body wall calcified. Lophophore eversion achieved by redistribution of internal coelomic pressures. Limited polymorphism. Marine. Four orders, only one of which includes living species.

Order Cyclostomatida (Fig. 1D): orifice of zooid circular, terminal, closed by muscular contraction. Sexual reproduction involving polyembryony; embryos brooded in conspicuous swollen gonozooids.

Class GYMNOLAEMATA: zooids cylindrical or squat. Body wall calcified or uncalcified. Lophophore eversion involving deformation of part of the body wall. Zooid polymorphism well developed.

Order Ctenostomatida (Fig. 1A): zooids cylindrical, flask-shaped or flattened; encrusting, erect or creeping. Orifice primitively terminal, closed by muscular contraction, appearing puckered or squared. Body wall uncalcified, membranous or gelatinous. Heterozooids limited to stolons, spines and kenozooids; no avicularia. Embryos brooded within the parent zooid. Mostly marine; a few freshwater species.

Order Cheilostomatida (Fig. 1B, C): zooids calcified, typically box-shaped. Orifice frontal or subterminal, closed by a hinged flap, the operculum. Heterozooids commonly present, often polymorphic within a species. Embryos brooded in specialised brood chambers (ovicells), in membranous internal ovisacs, in modified gonozooids, within the tentacle sheath, or frontally beneath a lattice of overarched spines. Almost exclusively marine. Constitutes the majority of living bryozoans.

Key to the Higher Taxa of Marine Bryozoa

1. Non-calcified bryozoans (Fig. 2D); colonies forming fleshy or gelatinous incrustations, stoloniferous networks, or dense, erect tufts. Zooids frequently semi-transparent; orifice terminal, or frontal in encrusting species, without an opercular flap. Chitinous spines may be present, other heterozooids represented only by stolons.
   .... CTENOSTOMATIDA (Synopsis No. 33)

   Calcified bryozoans. Colony erect, creeping or encrusting. At least the vertical walls calcified; colony often consisting of a rigid, calcareous formation in which the position of each zooid is indicated only by the orifice.
   .... **2**

Fig. 2. Bryozoan zooids showing the characteristics of the major groups.
A, *Tubulipora phalangea* (Cyclostomatida): encrusting colony showing ancestrula and two gonozooids. B, *Crisia ramosa* (Cyclostomatida): part of an erect, branching colony. C, *Diplosolen obelia* (Cyclostomatida): part of an encrusting colony. D, *Bowerbankia gracilis* (Ctenostomatida): part of a stolon with a cluster of autozooids. E, *Electra pilosa* (Cheilostomatida, Malacostegina): two autozooids showing proximal gymnocyst and extensive frontal membrane through which the polypide is visible. F, *Amphiblestrum flemingii* (Cheilostomatida, Neocheilostomatina): three autozooids showing frontal membrane underlain by a broad cryptocyst. G, *Cribrilina cryptooecium* (Cheilostomatida, Acanthostegomorpha): two autozooids showing frontal shields formed from fused marginal spines, with pores marking the sutures between them. H, *Escharoides coccinea* (Cheilostomatida, Umbonulomorpha): two autozooids showing calcified frontal shield with marginal pores, adventitious avicularia and an ovicell. I, *Schizoporella unicornis* (Cheilostomatida, Lepraliomorpha): two autozooids showing calcified frontal shield with scattered frontal pores, adventitious avicularia and an ovicell.

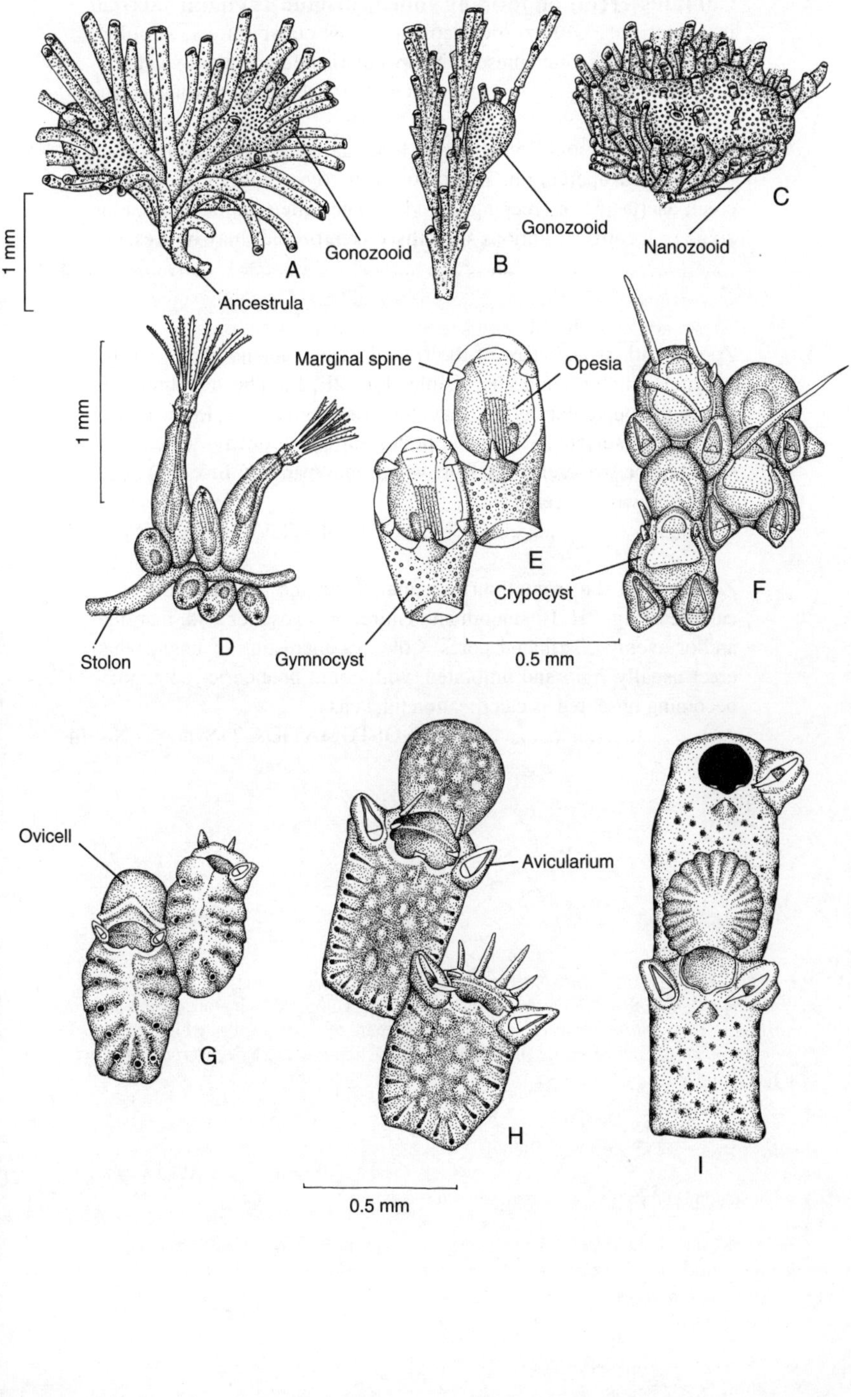

1 mm
Gonozooid
Ancestrula
A
Gonozooid
B
Nanozooid
C
1 mm
Marginal spine
Opesia
Stolon
D
Gymnocyst
E
0.5 mm
Crypocyst
F
Ovicell
Avicularium
G
H
I
0.5 mm

2. Colonies erect, or forming lobed, discoid or hemispherical incrustations. Gonozooids, appearing as conspicuous, stippled, swollen regions, often present. No avicularia, vibracula or ovicells.
.... CYCLOSTOMATIDA (Synopsis No. 34)

Orifice of the zooid frontal or sub-terminal, almost invariably closed by a hinged operculum. Zooids often partly membranous. Colony erect, encrusting or creeping. Slender spines often present around the zooid or its orifice. Various specialised heterozooids may be present.
.... **3**

3. Zooids with at least part of the frontal surface membranous, so that internal structures may be visible (Fig. 2E, F). The membrane, in which the operculum is set, may comprise a greater or lesser part of the frontal surface; it may be underlain by a concave calcareous shield or be overarched by ribs, which may partially fuse (Fig. 2G), but it is always present.
.... CHEILOSTOMATIDA 1 (Synopsis No. 10)

Zooids with the frontal surface apart from the orifice completely calcified (Fig. 2H, I); smooth and entire, or rugose, or with marginal and/or evenly distributed pores. Colonies encrusting or erect; when erect usually rigid and unjointed, with zooid boundaries sometimes becoming obscured as calcification thickens.
.... CHEILOSTOMATIDA 2 (Synopsis No. 14)

General Structure of the Cheilostomatida

Colony form has been a conservative feature throughout the evolution of the Bryozoa, with a few basic types occurring repeatedly in every major radiation. The simplest, and undoubtedly the most primitive, colony type seen among the cheilostomates, is the uniserial encrusting chain characteristic of *Pyripora catenularia* (Fig. 3). Similar colony forms were prevalent among the Jurassic ctenostomates (Todd, 1994), and are seen in the earliest known cheilostomates, such as the Jurassic *Pyriporopsis portlandensis* Pohowsky (Taylor, 1981). Uniserial encrusting colonies have been termed **runners** (McKinney & Jackson, 1989), and are especially suited to crowded substrata, spreading widely and simply outrunning competitors and predators. Indeed, they are generally poor competitors; but although parts of such colonies may be overgrown or destroyed by predators, their spreading habit increases the chances that part of the colony will survive. A majority of cheilostomate colonies develop encrusting sheets, with autozooids in multiserial, contiguous rows. Size varies greatly; in some species colonies may achieve a maximum size of as few as a dozen zooids, others develop extensive sheets over tens of cm^2, and the largest may comprise several million zooids. In the smallest sheets, or **spot** colonies (Bishop, 1989) which appear adapted to ephemeral or small, unstable substrata (Fig. 3), survival is short, individuals reproducing at a small size before succumbing through loss of substratum, and several generations may occur within a single season. Species developing very large encrusting sheets (Fig. 3) may not necessarily be better equipped to cope with competition or predation, but have different strategies to enhance the prospects of genotype survival. In the epiphytic *Membranipora membranacea*, for example, rapid directional growth, large size and early reproduction confer important benefits; the colony occupies the maximum area on the perennial part of the kelp frond, ensuring at least partial survival when the annual blade is shed, and essentially outgrows the appetites of its nudibranch predators. However, not all sheet-encrusting cheilostomates are so defenceless. Generally, species developing extensive sheets on permanent hard substrata are often successful in competition for space, frequently overgrowing other bryozoan species. They are often long-lived, with fast-acting repair and regeneration responses, and their competitive advantages may be increased by multilaminar growth and resulting high relief. Additionally, although parts of such sheets may be destroyed by physical damage or abrasion, isolated portions of the colony are able to repair, regenerate and regrow, giving rise to extremely long-lived clonal populations. The most persistent encrusting species may prove to be those in which growth through frontal budding is as vigorous as that achieved through peripheral budding, leading to the development of massive domed or nodular colonies (Fig. 3). Colonies of *Turbicellepora avicularis* (p. 336), for example, may achieve volumes of several cm^3, while similarly growing species in shallow tropical habitats form solid, multilaminar growths with volumes an order of magnitude larger than this.

Erect growth may be viewed as another response to competitive pressure for space or food which has been adopted in many cheilostomate families. Some sheet encrusting species, for example among *Schizomavella*, may be only lightly attached to the substratum and have the ability to grow upwards from it, developing brittle, unilaminar sheets whose form and extent are determined largely by the physical and biological constraints of their habitat. Such sheets may fold back upon themselves, or adjacent lobes may grow together to form rigid bilaminar plates. However, more regularly ordered bilaminar growth has evolved in many cheilostomate families, with colony form varying from flat or folded, but basically indeterminate, bilaminar plates, to regularly branching, rigidly ordered, arborescent shapes in which branch width and thickness, and frequency and angle of bifurcation vary relatively little within a population (Fig. 3). *Palmicellaria elegans* (p. 354), for example, develops a slender two-dimensional form, branching regularly by equal dichotomy, with autozooids in back-to-back pairs, each pair rotated through 90° relative to its predecessor. Colony form in *Palmicellaria* seems to be constant through the life of the individual, and longevity is perhaps related to the ability of the animal to thicken and strengthen its basal regions as it grows. In *Porella compressa* (p. 158), new sheets of autozooids, and new branches, develop on the lower regions of growing colonies, which thicken massively, and develop richly branched three-dimensional forms as they grow larger. The most complex and highly ordered colonies are those of the lace corals (Phidoloporidae), which develop three-dimensional shapes, typically folded and enrolled, from reticulate bilaminar sheets (Fig. 3). The defined frontal surface of the colony consists of feeding autozooids and adventitious polymorphs; the basal surface is built of sealed, non-feeding kenozooids. The mesh size, and the number of autozooid series bordering the lacunae, vary little within a species. The architecture of these highly structured colonies implies a considerable degree of morphological and functional integration within the colony.

Erect, branching unilaminate colonies are especially characteristic of the cellularine Cheilostomatida (Fig. 3). Autozooids are arranged in alternating linear series, most often just two to four series but sometimes many more, all facing a defined frontal surface. These lightly calcified, twiggy or bushy colonies appear as disordered clumps out of water, but display a highly ordered form in their natural habitat. Branches have a regular, little varying shape, developing open fans or spirals around a major axis, with frequency of branching, interval between branches and angle of dichotomy more or less constant within a species. In most

Fig. 3. Colony form in cheilostomate bryozoans.

A, A spot colony, *Phylactella labrosa.* B, A uniserial runner, *Pyripora catenularia.* C, *Membranipora membranacea,* a sheet colony. D, The fenestrate colony of a phidoloporid. E, the distinctive, jointed, rounded-sectioned colony of *Cellaria.* F, *Palmicellaria elegans,* an erect twiggy colony with biserial arrangement of autozooids. G, A nodular colony characteristic of *Turbicellepora.* H, A jointed cellularine, *Tricellaria.*

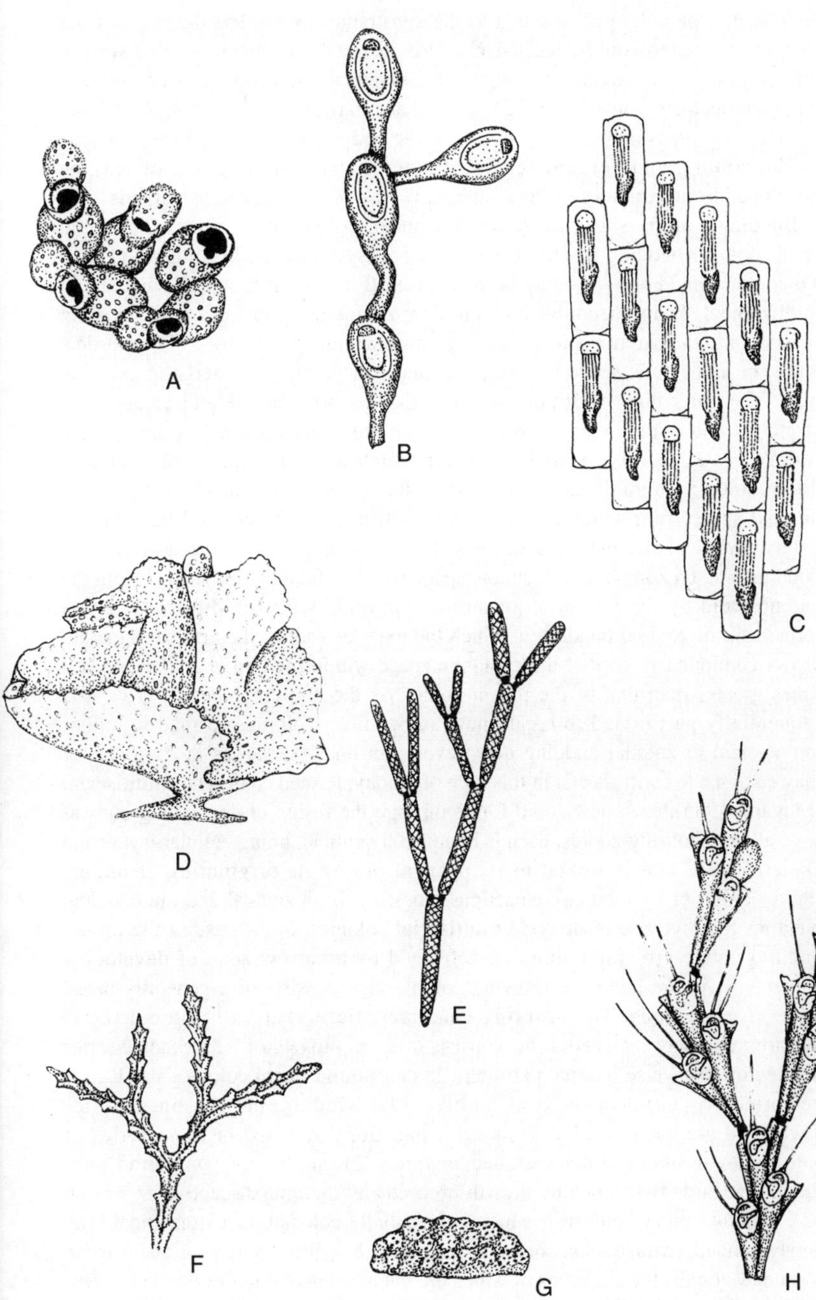
A
B
C
D
E
F
G
H

cellularines the colony is attached to the substratum by rootlets developed from stoloniform heterozooids, termed **rhizoids**. Rooted attachment is also seen in other types of erect colony covering a wide systematic spectrum. The colony may be cylindrical and jointed (Fig. 3E), or rigid and narrowly arborescent. Rootlets in these colony types may develop from specific budding sites on the lower surfaces of the colony, and perhaps represent single polymorphic units, or may be developed from expansions of the outer body wall of one or more autozooids.

Budding pattern is a primary determinant of colony form in bryozoans, and cheilostomate budding has been described and systematised by Lidgard (1985). Three main budding types may be recognised (Fig. 4): intrazooidal, zooidal and multizooidal. In **intrazooidal** budding new zooids develop from a pre-existing space within the parent zooid, either a chamber within or at the base of the vertical walls, or a coelomic space between a frontal cuticle and an underlying calcified wall. The outer body wall bounding the space – both cuticle and epidermis – stretches and expands by a process termed **intussusception**, the new zooid forming as a steadily expanding blister which at a particular size begins to differentiate a polypide and calcify its walls, at the same time forming a new internal space from which the next zooid will develop. In **zooidal** budding the exterior wall of the parent zooid begins intussusceptive growth at a defined budding site, creating a new coelomic space which is then partitioned off from the parent zooid by the formation of an interior wall. **Multizooidal** budding is a refinement of zooidal budding in which the exterior wall of the originating zooid grows continuously so that the coelomic space within the bud is usually several times greater than that of the parent zooid. As the bud continues to grow it is sequentially partitioned to create new autozooids. Colonies formed by either intrazooidal or zooidal budding may develop in linear series, as in *Pyripora*, or may coalesce to form sheets. In this type of colony, termed **coalescent multiserial** (Lidgard, 1985), each new zooid forms through the fusion of developing buds of several neighbouring zooids; each is bounded by cuticle, lining cellular layers and skeletal wall, and is linked to its progenitors by its originating chambers (intrazooidal) or by tubular connections (zooidal). Both zooidal and multizooidal budding can give rise to **discrete multiserial** colonies. In the case of the former budding mode growing colonies are fringed by a narrow zone of developing autozooids; in the latter the growing colony edge consists of a generally broad zone of giant buds. The unifying characteristic is that within the discrete multiserial colony each zooid row consists of lineal progenitor – descendant series divided by transverse internal partitions. In **compound lineal** colonies sheets may be formed by either zooidal or multizooidal budding, but the lineal series developed are each divided by interior partitions into two or more series of autozooids. In the most complex budding/growth form, termed **compound non-lineal** (Lidgard, 1985), colony growth proceeds by the intussusceptive expansion of a cuticular outer body wall which is essentially colonial, or extrazooidal. The newly defined, extrazooidal, coelomic space may be partitioned in various ways to form new zooids, the skeletons of which are entirely internal in origin.

The nature and formation of the skeletal components of the body wall have always been important in cheilostomate classification. Anascan cheilostomates

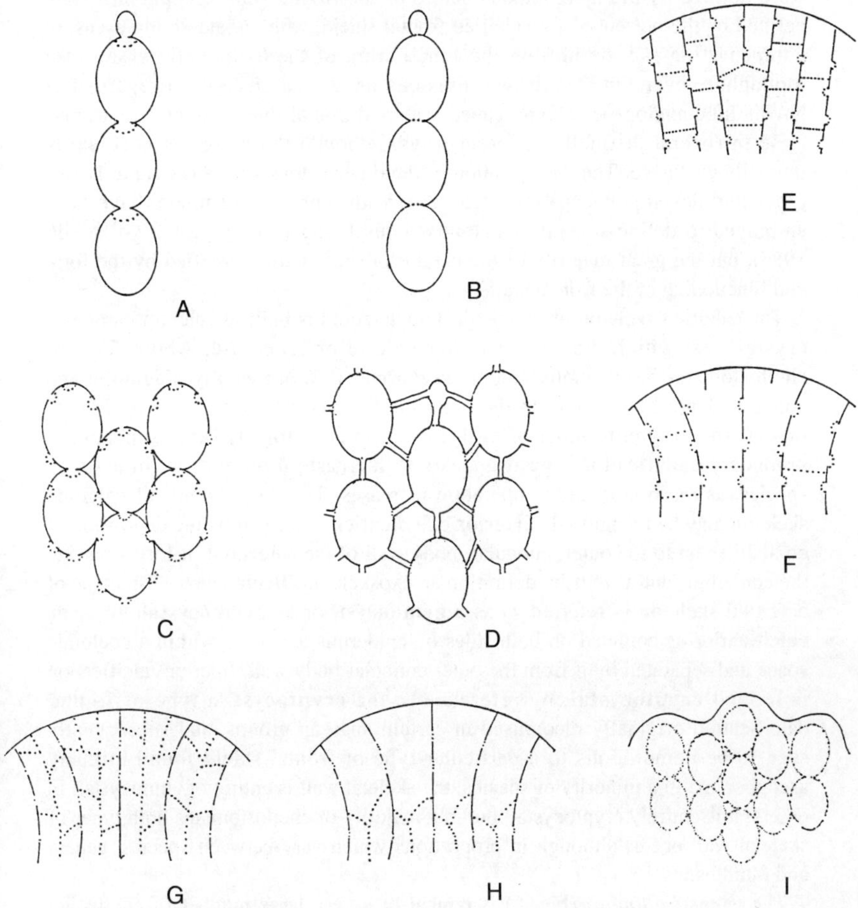

Fig. 4. Budding patterns in cheilostomate bryozoans.
Solid lines indicate exterior, cuticle-bounded calcification; dots indicate interior calcification. A, Intrazooidal: uniserial. B, Zooidal: uniserial. C, Intrazooidal: coalescent multiserial. D, Zooidal: coalescent multiserial. E, Zooidal: discrete multiserial. F, Multizooidal: discrete multiserial. G, Zooidal: compound lineal. H, Multizooidal: compound lineal. I, Zooidal: compound non-lineal. [After Lidgard, 1985]

were defined by the membranous nature of the frontal wall; ascophorans were defined by the presence of a calcified frontal shield, with an underlying ascus, or compensating sac, permitting the functioning of the hydrostatic system for lophophore eversion. These two divisions have long since been regarded as having little phylogenetic significance; calcified frontal shields, in particular, may arise in several distinctly different ways, although these are not necessarily mutually exclusive. The classification of the order is thus still rather fluid. Larval types and development patterns, together with reproductive mode, have been employed to define several, apparently natural, superfamily groups (d'Hondt, 1985), but the great majority of the cheilostomates is still classified by the form and functioning of the frontal wall.

The calcified skeleton of the cheilostomate zooid is built of calcium carbonate crystallites, which may be of either calcite or aragonite. About 50% of cheilostomates have wholly calcitic skeletons, 10% are wholly aragonitic and about 40% are bimineralic. In the latter the two crystal forms occur in discrete layers; the aragonite most often forms in later ontogeny, as a superficial, secondary calcification, and it appears to increase in proportion, in a single species, as mean seawater temperature increases. Two major kinds of calcified skeleton may be recognised. **Exterior calcification** is formed by the deposition of crystallites on to the outer, cuticular, body wall of the autozooid. It forms outside the epidermis, and is thus by definition an exoskeleton (Banta, 1969). This type of calcified skeleton is referred to as a **gymnocyst**, or as **gymnocystal. Interior calcification** is bounded on both sides by epidermis, enclosed within a coelomic space and separated by it from the outer cuticular body wall. Interior calcification is termed **cryptocystal**, by reference to the **cryptocyst**, a type of frontal calcification originally recognised in certain anascan groups, and subsequently seen to be homologous to a particular type of frontal shield found in many ascophorans. In a minority of species, the skeletal wall is entirely gymnocystal; in others, it is entirely cryptocystal, but in a majority of cheilostomates both types of skeletal wall occur, although in proportions which vary between species, genera and families.

The genus *Callopora* (Fig. 5) is typical of a very large number of encrusting neocheilostomate genera, and can be used to demonstrate the range of autozooid morphology seen in the group. It forms encrusting sheets, with new autozooids developing, through intrazooidal budding, from conspicuous chambers visible at the base of the vertical walls of peripheral autozooids. The vertical walls of the autozooids curve inwards, and proximally are continuous with a smooth frontal gymnocyst. Much of the frontal surface is a transparent membrane, with the operculum set within it at the distal end, through which the polypide can be clearly seen. The membrane is bordered by spines, set on a distinct ridge, the **mural rim**, marking the boundary between the gymnocystal calcification of the frontal and lateral walls, and a narrow border of granular cryptocystal calcification underlying the membrane (Fig. 6). The cryptocyst defines a space, the **opesia**, a useful descriptive term for the residual area of uncalcified frontal surface. The proportions of frontal cryptocyst and gymnocyst vary throughout the Calloporidae. In *Amphiblestrum*, for example, the gymnocyst is considerably

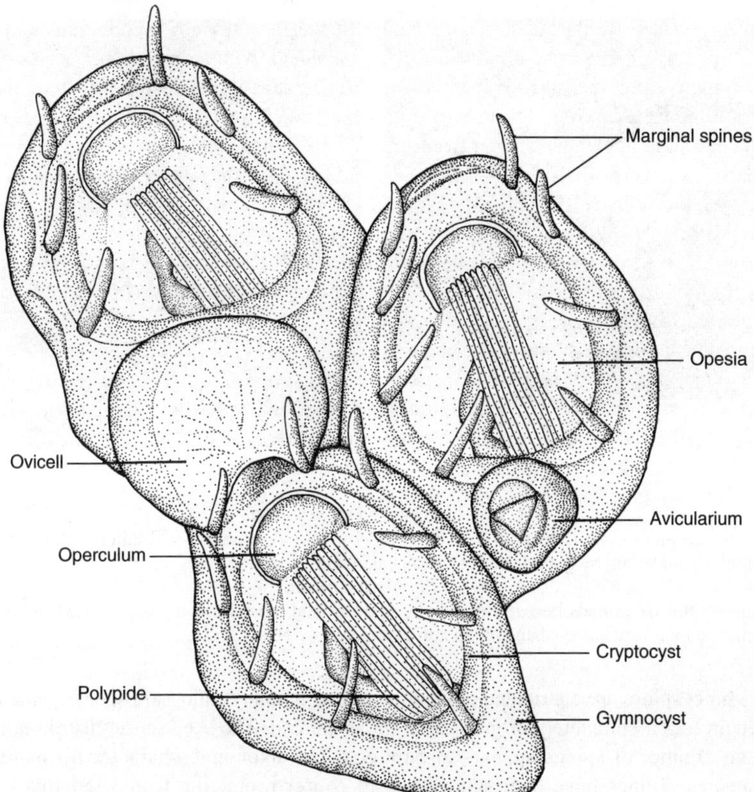

Fig. 5. Autozooid structure in *Callopora lineata*

reduced and the cryptocyst comes to occupy a much larger area of the frontal surface, with a correspondingly smaller, and characteristically trifoliate opesia. In the Microporoidea (Fig. 6B) the frontal gymnocyst has disappeared completely and the frontal membrane, visible as a glistening sheet in living specimens, is entirely underlain by a granular cryptocyst. The opesia is now coincident with the orifice of the autozooid, and the operculum forms a close-fitting lid above it. The vertical walls of microporoidean species are still bounded by cuticle, continuous with the membranous frontal wall, and thus by definition exterior calcification. Boundaries between zooids therefore represent two compressed vertical walls with a doubled layer of cuticle between. Species of *Membranipora* and *Flustra*, for the most part, lack frontal calcification entirely, apart from minimal areas of gymnocyst or cryptocyst in the proximal corners of the autozooids. However, the lateral walls in these genera are lightly calcified, external gymnocyst, with a doubled cuticular layer between each autozooid series.

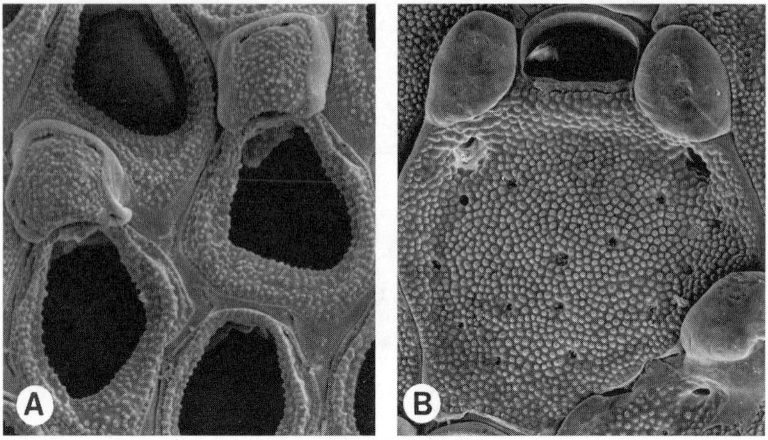

Fig. 6. A, *Alderina solidula*,
with the membranous frontal wall bordered by smooth gymnocystal calcification and partially underlain by granular cryptocystal calcification, x60.
B, *Micropora normani*:
the cryptocyst extends beneath the whole of the frontal membrane, the edges of which are marked by a very narrow band of smooth gymnocyst, x150.

Erect spines are distributed around the opesia in many calloporid species, either encircling it completely or grouped at the distal end. The occurrence, distribution and number of spines appears to be a constant taxonomic character for many species. Spines have been considered to confer protection from predators or physical damage, or to modify surface water currents, although there is no good evidence for any supposed function and a significant number of species lack them altogether. Nonetheless, there seems to be a trend through cheilostomate evolution towards a reduction of the exposed membranous area of the frontal surface, and the evolution of larger, or more numerous, or closely linked spines perhaps in part reflects this trend.

In some calloporids, for example, the spines arch medially and fuse along the midline of the autozooid, forming a convex, basket-like shield above the frontal membrane. This morphology foreshadows that of another distinctive cheilostomate group, the Acanthostegomorpha (= Cribriomorpha), which underwent an immense radiation in the Cretaceous, declined sharply thereafter, but still constitutes a significant proportion of the living cheilostomates. Cribriomorph autozooids have gymnocystal frontal calcification which may be very reduced, as in *Puellina*, or may comprise up to half the total frontal length, as in *Figularia*. There is a small cryptocystal rim in some species, but the most important feature of the cribriomorphs is a rigid, reticulate frontal shield, formed from flattened, closely fused spines. The calcification of each spine, or **costa**, is essentially gymnocystal, and the limits of each are marked by lines of small pores, or larger lacunae between the points of fusion.

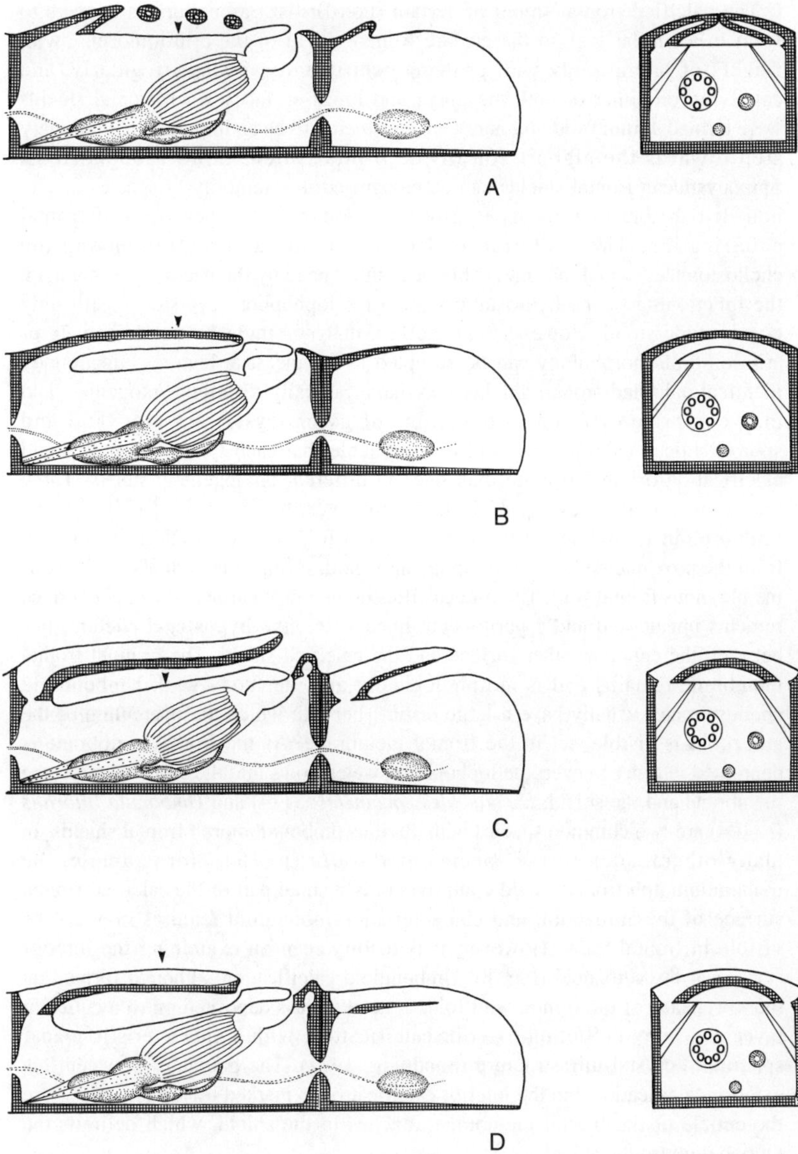

Fig. 7. Four types of frontal shield in ascophorine cheilostomates: longitudinal sections of autozooids, each with a developing bud, and fully-developed autozooids in transverse section. A, Cribriomorph shield in *Cribrilina* (Acanthostegomorpha). B, Gymnocystidean shield in *Hippothoa* (Hippothoomorpha). C, Umbonuloid shield in *Escharoides* (Umbonulomorpha). D, Cryptocystidean shield in *Schizoporella* (Lepraliomorpha).

The calcified frontal shield of certain (non-British) ascophorines appears to form in a similar way to the costate frontal shield of the cribriomorphs, with fingers of lateral body wall growing centripetally, fusing irregularly, and calcifying on either or both the upper and lower surfaces. Such frontal shields were termed **spinocystidean**, and were considered to be an important evolutionary step towards the elaboration of the frontal shield in all ascophorines. Spinocystidean frontal shields can be recognised in a minority of species (again, non-British) but they most often occur together with other types of frontal calcification. The evolution of frontal shield calcification among the cheilostomates, which has inevitably been influenced by the necessity to maintain the functioning of the hydrostatic system for lophophore eversion is still only poorly understood. However, it is certain that no single neocheilostomate or cribriomorph morphology can be accepted as an ancestral feature. Functionally identical calcified frontal shields may have radically different ontogenies, and may comprise different proportions of gymnocystal, cryptocystal and spinocystidean calcification, and it is probable that each type of frontal shield calcification originated more than once in different phylogenetic stocks. Three major structural types of shield may, however, be recognised (Fig. 7). An **umbonuloid** frontal shield is formed when a fold of body wall grows distally from the proximal end of a developing autozooid, forming an arch above the still-membranous frontal wall. The fold calcifies on its lower surface; the upper surface remains uncalcified and a peritoneum-lined space, the **hypostegal coelom**, lies between the cuticular outer surface and the calcified shield. The original frontal membrane remains and is morphologically an outer body wall. Umbonuloid cheilostomates usually have a large orifice, beneath which the operculum of the autozooid is visible, set in the frontal membrane. As the frontal membrane is depressed in order to evert the lophophore, water flows into the space between the membrane and the shield. *Escharoides coccinea* (p. 114) and *Umbonula littoralis* (p. 104) are two common species with obvious umbonulomorph frontal shields. In many others, such as the species of *Porella* (p. 154), for example, the umbonulomorph frontal shield comprises only a small part of the calcified frontal surface of the autozooid, and characteristic umbonuloid features may not be visible in frontal view. However, it is readily seen on examining the interior surface of the autozooid (Fig. 8). Umbonuloid calcification, although formed on the lower face of the frontal wall fold, is nevertheless deposited on to a cuticular layer and is by definition exterior calcification, with a characteristic planar spherulitic crystal ultrastructure (Sandberg, 1977). The boundary between this exterior calcification and the interior calcification is marked by a ring scar, where the cuticle of the frontal membrane attaches to the shield, which delimits the umbonuloid frontal shield.

In a minority of cheilostomates, the frontal shield forms simply through calcification of the outer body wall; it is by definition exterior, or gymnocystal, calcification and consequently this type of shield is termed a **gymnocystidean** shield. The frontal wall of the autozooid is unable to play any role in lophophore eversion and a new structure, the **ascus**, has evolved for the purpose. The ascus forms through the invagination of an area of frontal wall proximal to the orifice of

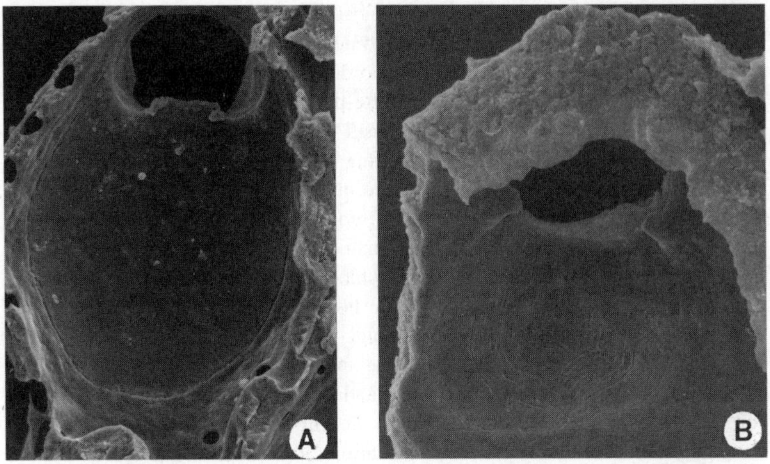

Fig. 8. The frontal shield in two umbonulomorphs, viewed from the lower surface to show umbonuloid ring scars.
A, *Porella concinna*, x145 B, *Neolagenipora collaris*, x210

the autozooid; its development coincides with the formation of the operculum in a developing autozooid. It is a hollow sac, lined with cuticle and enclosing a space which is morphologically exterior to the autozooid; it is not homologous to the frontal wall of other cheilostomates but is an entirely new structure. Gymnocystidean cheilostomates are relatively few in number, but include several common and geographically widespread genera, such as *Hippothoa* (p. 84) and *Chorizopora* (p. 100). In **cryptocystidean** cheilostomates, the frontal shield is entirely of interior, or cryptocystal, calcification. The original, membranous frontal body wall remains, above a frontal hypostegal coelom, and is again excluded from participating in lophophore eversion. An ascus is thus also present in cryptocystidean cheilostomates. A very large proportion of living Cheilostomatida appears to have cryptocystidean frontal shields. In some the flexible outer membrane, or the presence of secondary calcification on the outer frontal surface, both indicative of a hypostegal coelom, imply the nature of the frontal shield. However, SEM examination is often necessary to establish the ontogeny of the frontal shield and the reinvestigation of the skeletal structure of even the most familiar cheilostomate species in recent years has provided some surprising results, the taxonomic consequences of which have yet to be fully assessed. The most interesting revelation is that a large number of species seem to have compound shields. In most cases these consist of a cryptocystidean shield with a distal umbonuloid portion, but other combinations of interior and exterior calcification also occur (Gordon, 1988).

The frontal calcification of many ascophorans is punctured by pores. These are mostly uncalcified patches plugged with tissue, and are termed **pseudopores**. However, the organic continuity of a cheilostomate colony is maintained by tissue connections linked by complex communication pores (Fig. 9). The **basal pore chambers**, or **dietellae**, which give rise to new buds in intrazooidal budding, are separated from the originating autozooid by a perforated plate of interior calcification termed a **septulum**, or **rosette plate**. The pores of the septulum are each blocked by a single dumbbell-shaped cell; aggregations of cells develop around each end of the dumbbell cells during zooid ontogeny and it is to these cell masses that strands of funiculus and fine nervous tissue lead. Metabolites and nervous impulses can only be transmitted between zooids at these cellular aggregations. Transverse walls between autozooids may bear similar septula, or small, simple pores which are presumably blocked also by special cells. However, lateral walls of autozooids may enclose homologues of the diatellae, termed **mural pore-chambers**. These begin as aggregations of special cells at defined uncalcified areas in the lateral walls of the autozooid. Rosette plates form around these cells, sealing the uncalcified area, and as the adjacent autozooid develops the cuticular layers between it and its neighbour break down establishing communication between the two. The interior calcification of the rosette plate is laid down by the most proximal of the two autozooids; thus, mural pore chambers and basal pore chambers in the distal half of an autozooid result from partitioning the inner space of that zooid, while those in the proximal half are products of the coelomic space of the previous astogenetic generation. In umbonulomorph and cryptocystidean cheilostomates the hypostegal coelom is linked to the main coelom of the autozooid via frontal septula, usually located in large marginal pores, termed **areolae**.

Zooidal polymorphism occurs in all bryozoan orders but is especially characteristic, and most marked, in the Cheilostomatida. Polymorphism may be limited to the lophophore, some species developing modified male lophophores during the reproductive season, with few, very long tentacles which serve simply to release sperm. More commonly, zooid polymorphs, or **heterozooids**, have a skeletal morphology quite distinct from that of the autozooids. In most heterozooids homologous structures can be seen which demonstrate their nature as polymorphic zooids. In some, however, it is only the outer cuticular wall, lining peritoneum and inner coelomic space, separated from neighbouring zooids by one or more septula that define the structure as a zooid polymorph. Spines, for example, may be simple, hollow or solid, outgrowths of the body wall of an autozooid, or they may be coelomic chambers, cuticle-bounded, and with a lining peritoneum and a communicating septulum, and are thus regarded as zooid polymorphs. **Kenozooids** are a frequently occurring type of polymorph. These are rectilinear or irregular chambers, smaller or larger than autozooids, which lack both orifice and operculum but which may otherwise resemble an autozooid. Generally they are found in disordered areas of sheet-encrusting colonies, where regular budding sequences are disrupted by irregularities or obstacles in the substratum. In linear encrusting colonies they usually represent apparently aborted autozooids, unable to complete growth and development through lack of space. Kenozooids retain their blastogenic potential, however, and are able to produce buds which develop into new, fully functioning autozooids if there is sufficient space. In highly organised erect colonies, such as those of the lace corals

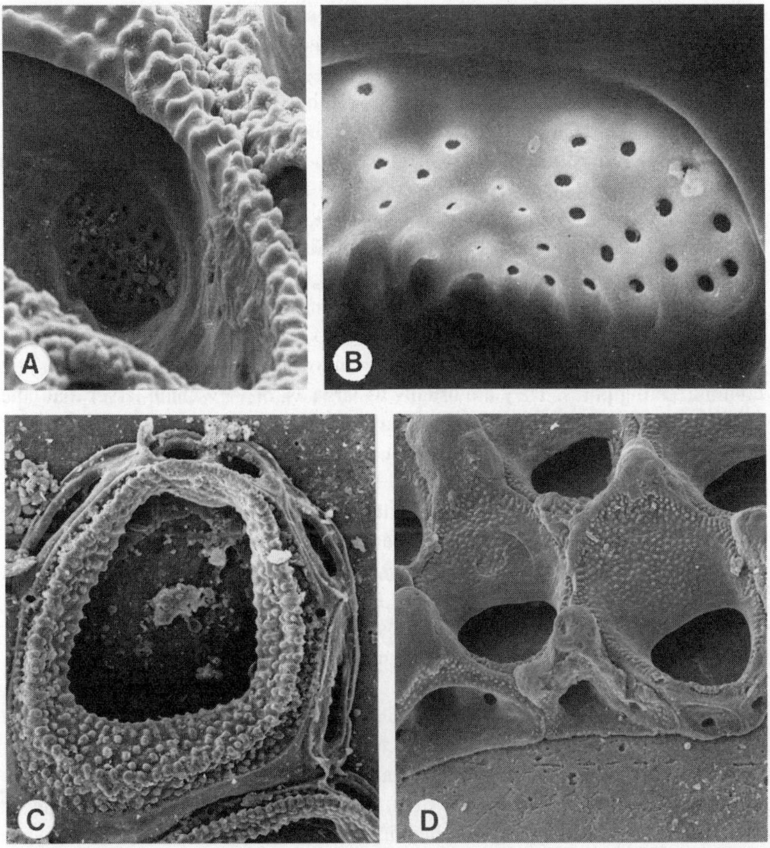

Fig. 9. A, B, *Conopeum reticulum*:
A, The septulum in a lateral wall, x 260; B, Detail of the septulum in the distal terminal wall, x850. C, *Alderina imbellis*: autozooid at a colony margin showing large basal pore chambers, x270. D, *Ammatophora nodulosa*: autozooids at the colony margin, showing mural pore chambers, x100.

(Phidoloporidae), kenozooids are specialised structures, budding from defined sites to form the abfrontal layer of the colony, and thus contributing importantly to the architecture and structure of the colony.

The most familiar, and characteristic, cheilostomate polymorph is the **avicularium**. These are polymorphs in which the zooid operculum is modified as a heavily sclerotised **mandible**, the orifice rim forms a thickened **rostrum**, and the opercular musculature is greatly developed to form mandibular occlusor and divaricator muscles. With a few exceptions, avicularia lack functioning polypides and the body cavity is entirely occupied by muscles. Avicularia vary greatly in size; they may be as large as, or very much larger than autozooids, or so small as to be seen only with difficulty using average magnification. Three types may be recognised: **vicarious avicularia** replace autozooids in normal budding sequences and are linked to all neighbouring zooids, autozooids or other heterozooids, by communicating pores. They are usually as large as, or very much larger than, the autozooids. **Interzooidal avicularia** may be wedged in between autozooids, or even appear to replace an autozooid, but are budded from, and only communicate with, a single autozooid and are usually smaller than it. **Adventitious avicularia** are each borne by a single originating autozooid, budded from a frontal septulum. They may be almost as large as the bearing autozooid, or very much smaller. Avicularia vary in form almost endlessly throughout the cheilostomates (Fig. 10). They may be stalked (**pedunculate**) and mobile, sessile, or columnar; the mandible may be elongate, triangular, spoon-shaped, elliptical or semicircular, and is often conspicuously hooked, or equipped with prominent marginal teeth. In some the mandible forms a whip-like seta, and in one particular type, the **vibraculum**, an additional gyrator muscle enables the seta to be moved in a wide flailing action. Avicularia are widely used in cheilostomate taxonomy. Their presence, morphology and situation relative to the autozooid are used to define both species and genera. Some taxa are characterised by a single type of avicularium present at a particular budding site; in others two, three or more types of avicularia may be present in a single colony, the position of some varying considerably from one part of the colony to another. Yet, for almost all kinds of avicularia, no function has been recognised. The bird's head avicularia of *Bugula*, and the tiny frontal avicularia of some ascophorines, appear to serve as a deterrent against particular micropredators, or to discourage surface fouling of the colony by tube-building amphipods or polychaetes (Winston, 1984, 1986). But for the great majority of avicularia, including the most extraordinarily modified types, function remains a matter for speculation.

Fig. 10. Some cheilostomate heterozooids.

A, Vicarious avicularium of *Flustra foliacea*. B, Pedunculate adventitious avicularium of *Bugula avicularia*. C, D, Pedunculate adventitious avicularia of *Epistomia bursaria*. E, Sessile adventitious avicularium of *Callopora lineata*. F, Vibraculum of *Scrupocellaria scruposa*. G, Adventitious avicularium of *Schizoporella dunkeri*. H, Setiform interzooidal avicularium of *Escharina hyndmanni*. I, Adventitious avicularium of *Microporella ciliata*. J, Adventitious avicularium and mandible of *Smittoidea reticulata*. K, Vicarious avicularium and mandible of *Turbicellepora avicularis*.

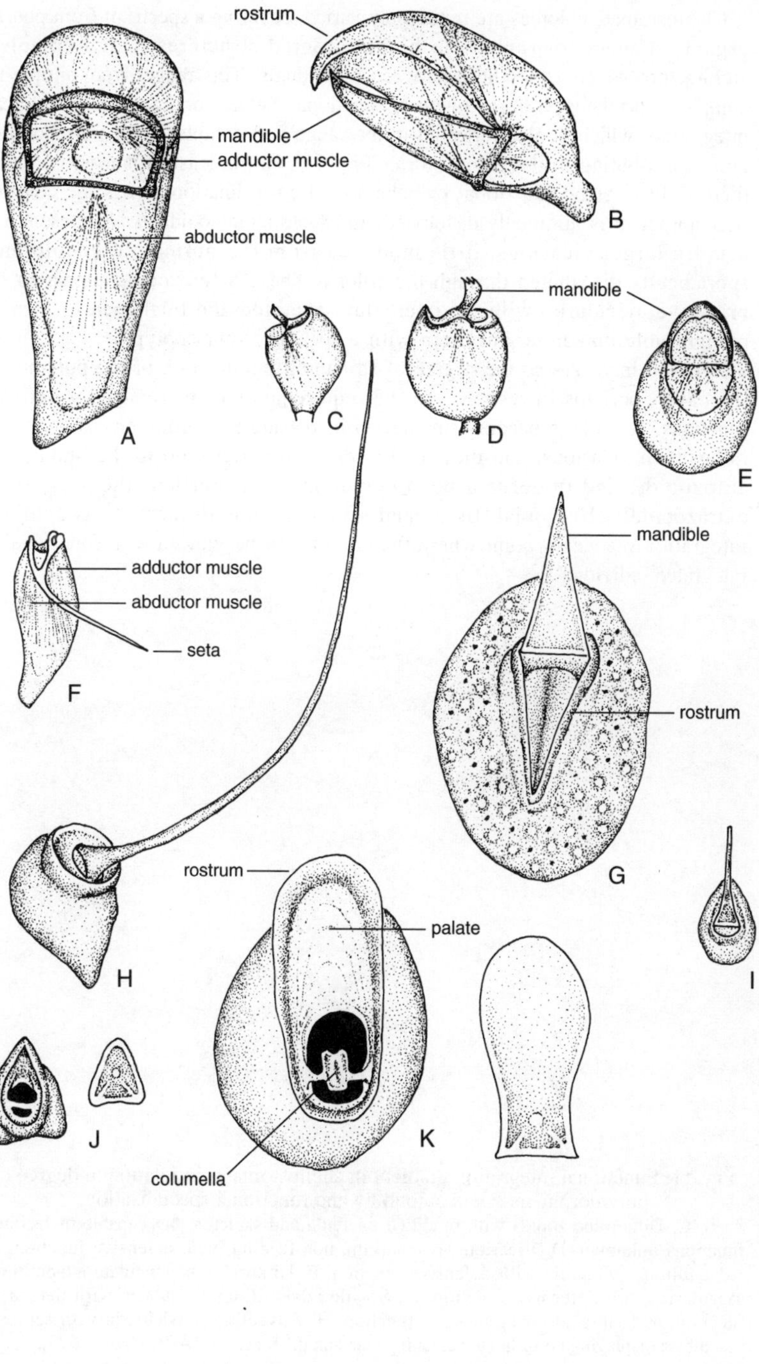
rostrum
mandible
adductor muscle
abductor muscle
B
A
mandible
C
D
E
adductor muscle
abductor muscle
seta
F
mandible
rostrum
G
rostrum
palate
H
I
J
K
columella

Cheilostomate colonies are immensely varied, covering a spectrum from poorly organised sheets or runners to highly ordered structures with a complex architecture which varies little between individuals. This range, from simple to complex, correlates with a gradient of functional, behavioural and morphological integration, which is also reflected in the range and complexity of polymorphic units contributing to the colony form (Fig. 11). In the least organised colonies there is little or no functional or behavioural co-ordination across the colony; polymorphism is minimally developed, and such heterozooids as are present are usually large, vicarious, little modified from the normal autozooid and sporadically distributed through the colony. These heterozooids appear to be primitive avicularia, with the operculum enlarged and thickened to form a recognisable mandible, but often with a still functional polypide. As colony integration increases so the degree of structural modification of the polymorph increases, perhaps in response to the requirements of more strictly defined functions. Avicularia become smaller, an increasing proportion are adventitious rather than vicarious, and they tend to occur in strict ratio to the number of autozooids, and in defined positions upon them. Further, the amount of extrazooidal, or colonial, tissue and skeleton tends to increase as colonial integration reaches a point where the colony can be viewed as a functionally integrated individual.

Fig. 11. Functional integration gradient in cheilostomates in relation to degree of polymorphism, zooid autonomy and functional specialisation.

A, B, C, Dimorphic zooids with modified opercula and skeleton, but capable of feeding; functions unknown. D, Vicarious avicularium, non-feeding, with defensive function. E, Adventitious avicularia with defensive function. F, Dimorphic pedunculate adventitious avicularia, with defensive function. G, Sessile adventitious avicularia with defensive function, and vibracula with cleaning function. H, Adventitious avicularia with setiform mandibles displaying co-ordinated cleaning and support functions [After Cook, 1979].

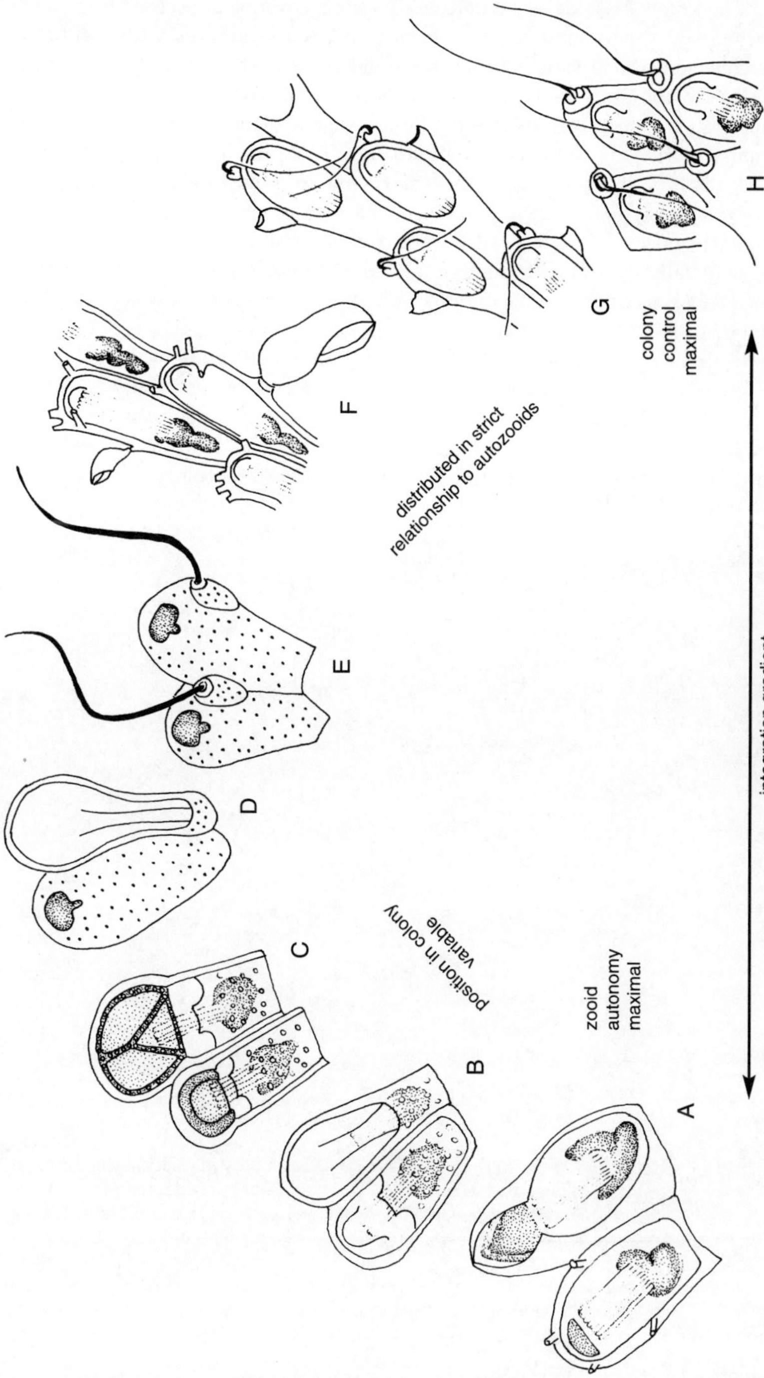
A
B
C
D
E
F
G
H
zooid autonomy maximal
position in colony variable
distributed in strict relationship to autozooids
colony control maximal
integration gradient

Reproduction and Life Histories

All bryozoan colonies are hermaphroditic, producing both male and female gametes. Gonads are ductless, differentiating from peritoneum in the proximal half of the zooid in the breeding season. Autozooids may be dioecious; or monoecious, and protandrous or protogynous. Sex ratios, reproductive strategy and life cycles vary widely throughout the phylum and are not easily generalised. In many cheilostomates ovaries and testes occur together in each autozooid but mature at different times, with protandry being the usual rule. In others, perhaps many others, an autozooid may first develop a testis, becoming thus an **androzooid** (Dyrynda & Ryland, 1982). Following maturation and release of the sperm, the polypide degenerates and is replaced, and an ovary develops, the zooid being then a **gynozooid**. A minority of dioecious species, such as *Epistomia bursaria*, develops only one polypide per autozooid, and androzooids and gynozooids are thus determined in a fixed ratio during early ontogeny (Dyrynda & King, 1982). The males and females of some dioecious species may be morphologically dissimilar. Males may be smaller than other zooids, with smaller orifices, and modified lophophores with few tentacles. Females may be smaller or larger than autozooids; they may be enlarged and specialised for internal brooding of embryos, or reduced and serving merely to produce ova and support a brooding ovicell. However, presence of an ovicell does not necessarily indicate sex; once the reproductive period is over ovicelled zooids may regenerate initially as androzooids (Dyrynda & Ryland, 1982). In *Celleporella hyalina* (p. 94), and related species, the colony consists at first of a sheet of non-sexual, feeding autozooids which supply the energetic requirements for reproduction; both androzooids and ovicelled gynozooids develop subsequently through frontal budding, and the ratios of these to the autozooids will depend partly upon the feeding success of the autozooids (Hunter & Hughes, 1993a).

Ripe eggs are discharged into the coelom of the zooid and released to the exterior through a dorsally situated **coelomopore**, at the base of the lophophore. In cheilostomates with planktotrophic larvae the coelomopore is usually modified as a short, fat tube, termed the **intertentacular organ**. Sperm are shed through the pores at the tips of the tentacles; in some species this occurs through all tentacles, in others just the dorsal, or adneural pair. Fertilisation has been observed on few occasions, and its timing, and the site where it occurs, seem to differ widely. Released sperm are entrained by the feeding currents of everted lophophores and can be seen aggregated along the tentacles. In oviparous bryozoans possessing intertentacular organs sperm have been observed aggregated within the tube and, in some at least, fertilisation appears to be effected as the eggs are shed. In others, such as *Membranipora* from the Pacific coast of North America, fertilisation – in the sense of insemination – occurs much earlier, whilst the eggs are still developing. In brooders such as *Umbonula littoralis* the embryos develop in the body cavity and sperm must also enter the maternal zooid, presumably through the coelomopore. In ovicelled brooding species the lophophore has been seen to bend proximally as the egg streams through the coelomopore into the lumen of the ovicell, and it has been widely assumed that fertilisation was most likely to occur at this time. However, there are now several unequivocal reports of sperm nuclei embedded in developing, even ovarian, oocytes of ovicellate brooders and it seems that sperm transfer and insemination may occur precociously, although syngamy (fusion of male and female pronuclei), followed by the appearance of the fertilisation membrane, occurs only after the egg has passed through the coelomopore (Ryland & Bishop, 1993). This appears true for oviparous species and both ovicellate and non-ovicellate (including ctenostomate) brooders. Genetic cross-fertilisation has been assumed to be the rule in cheilostomates, but there is mounting evidence for occasional self-fertilisation, although most often correlated with reduced larval viability (Hunter & Hughes, 1993b).

Two major larval types are seen in the Cheilostomatida. The Membraniporoidea shed numerous small eggs directly into the sea. These develop into shelled, planktotrophic larvae, termed **cyphonautes**, which feed and grow during several weeks or months spent in the plankton. This appears to be the primitive bryozoan larval state, and a majority of species produces bulky lecithotrophic larvae which generally settle within a few hours of release (Fig. 12). Five main morphological variants have been described (Zimmer & Woollacott, 1977), but all are fundamentally alike in bearing a tuft of apical cilia, a posterior tuft of plume cilia, a well-marked girdle, or **corona**, of short cilia, and are consequently referred to as **coronate larvae**. Behavioural patterns of cyphonautes larvae are poorly known; they appear to be rheophilic, and perhaps show a limited selectivity for substrata. However, coronate larvae are obliging experimental subjects and more is known of their behaviour. In coastal species light is an important stimulus to larval release, and many cheilostomates shed larvae during the first few hours of daylight. The high light intensity of mid-day inhibits larval release in these species (Cancino, Hughes & Orellana, 1994). The larvae are phototactic and have a swimming period of just a few hours; early morning release thus permits them to locate and settle on their preferred substratum within the brief period available to them. In contrast, some species in which light appears unimportant to larval release produce larvae with a longer swimming period, able to delay settlement for 12 hours or more without loss of viability. Larvae are brooded in a variety of structures. In the Aeteoidea embryos develop in membranous sacs attached at the distal end of the autozooid. In the Scruparioidea dimorphic gynozooids brood several embryos simultaneously in a distal coelomic cavity enclosed by a bivalved ovicell. In a minority of ascophorine cheilostomates the gynozooid is an inflated zooid, termed a **gonozooid**, as in *Reptadeonella* (p. 184), or an ovicelled dwarf zooid, as in *Celleporella* (p. 92). In a few species, the common intertidal *Umbonula littoralis* (p. 104), for example, single larvae develop within the body cavity of otherwise unmodified autozooids. However, throughout the Cheilostomatida the most common brooding chamber is the **ovicell**, a globular structure situated at the distal end of the maternal zooid (Fig. 13). The ovicell

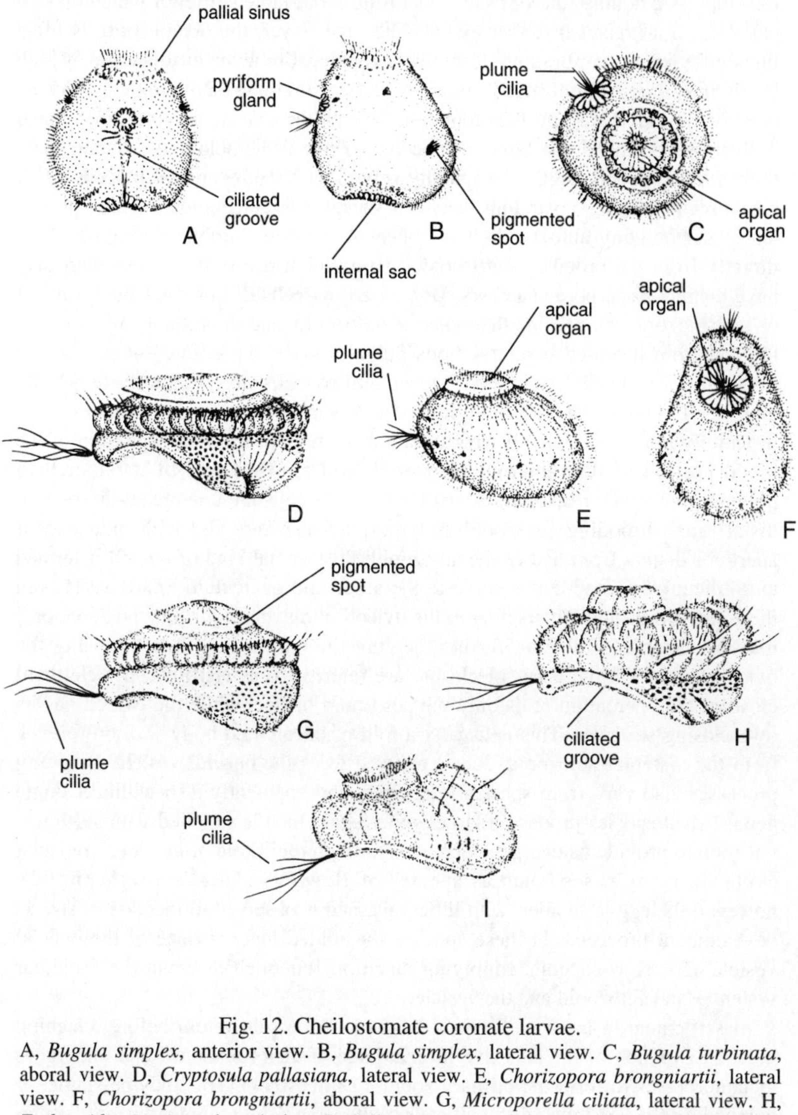

Fig. 12. Cheilostomate coronate larvae.
A, *Bugula simplex*, anterior view. B, *Bugula simplex*, lateral view. C, *Bugula turbinata*, aboral view. D, *Cryptosula pallasiana*, lateral view. E, *Chorizopora brongniartii*, lateral view. F, *Chorizopora brongniartii*, aboral view. G, *Microporella ciliata*, lateral view. H, *Escharoides coccinea*, lateral view. I, *Cellepora pumicosa*, lateral view. A, C, E, F, after Calvet, 1900: D, G, H, I, after Barrois, 1877]

develops as a double fold of tissue, enclosing a flat lumen, both of which may be calcified to a greater or lesser extent. The outer layer, the **ectooecium**, is often uncalcified and membranous frontally, exposing the inner **entooecium**, which itself may have uncalcified spots, or larger lacunae. Both layers may be completely calcified, and the frontal surface of the ovicell may be ornamented with ridges or knobs (**umbones**). Ovicells are not all homologous structures and their developmental origins are actually very diverse. **Independent ovicells** either rest directly on the substratum, between succeeding autozooids, or may project above surrounding autozooids, and appear to be polymorphic modules budded directly from the brooding autozooid. Those which rest on the substratum may have distinct basal pore chambers. **Dependent ovicells** rest on the frontal surface of the autozooid succeeding the maternal autozooid, and in probably all cases are budded from it either as adventitious polymorphs or as an integral part of the distal autozooid (Nielsen, 1981). Dependent ovicells may be partly or wholly embedded in the distal autozooid; however, in some species embryos are brooded in entirely immersed ovicells which seem to be formed from an expansion of the distal portion of the maternal autozooid, and are thus part of it rather than polymorphic structures. In the genus *Cellaria*, ovicells are completely immersed, distal to the brooding autozooid, polymorphic in nature and with independent apertures distant from that of the autozooid; this special kind of ovicell is termed **endotoichal**. In some cheilostomates the autozooid operculum jointly seals both the zooid orifice and the opening of the ovicell; it may have just one position, or it may be capable of closing further, sealing the autozooid while exposing the ovicell aperture. Both types of closure are referred to as **cleithral**. In **acleithral** closures the operculum seals only the autozooid orifice, while the ovicell has its own closing structure. This is usually a fold of uncalcified body wall originating from the maternal autozooid, and is referred to as the **ooecial vesicle**. Brooding processes also vary from species to species, and apparently also within a single genus. Most species produce a large, yolky egg which is supplied with sufficient nutrition to provide its energetic demands through the whole of the developmental cycle. In many cases, such as species of *Bugula* and *Celleporella* (p. 92), however, the egg is smaller, with little yolk, and is observed to increase in size as development proceeds. In these species, the folded inner surface of the ooecial vesicle acts as a placenta, supplying nutrition funnelled to it via the funicular system of the autozooid and the vesicle.

At settlement, a larva will appear to explore the substratum before attaching itself by eversion of the adhesive **internal sac**. It then rounds off and undergoes metamorphosis to form the primary zooid, or **ancestrula**. This may be single, or twinned; or the larva may differentiate simultaneously into three or six ancestrular zooids. The ancestrula is usually smaller than later autozooids and sometimes morphologically distinct from them as well. In some encrusting malacostegines, for example, the ancestrula bears more spines than later zooids, while in many ascophorines it has a flexible, malacostegine-like frontal membrane, and is termed **tatiform**. Budding commences often before the ancestrula has differentiated a functional polypide and the elaboration of **ephebic** (i.e., mature, or adult) morphological characteristics continues through few to many generations of

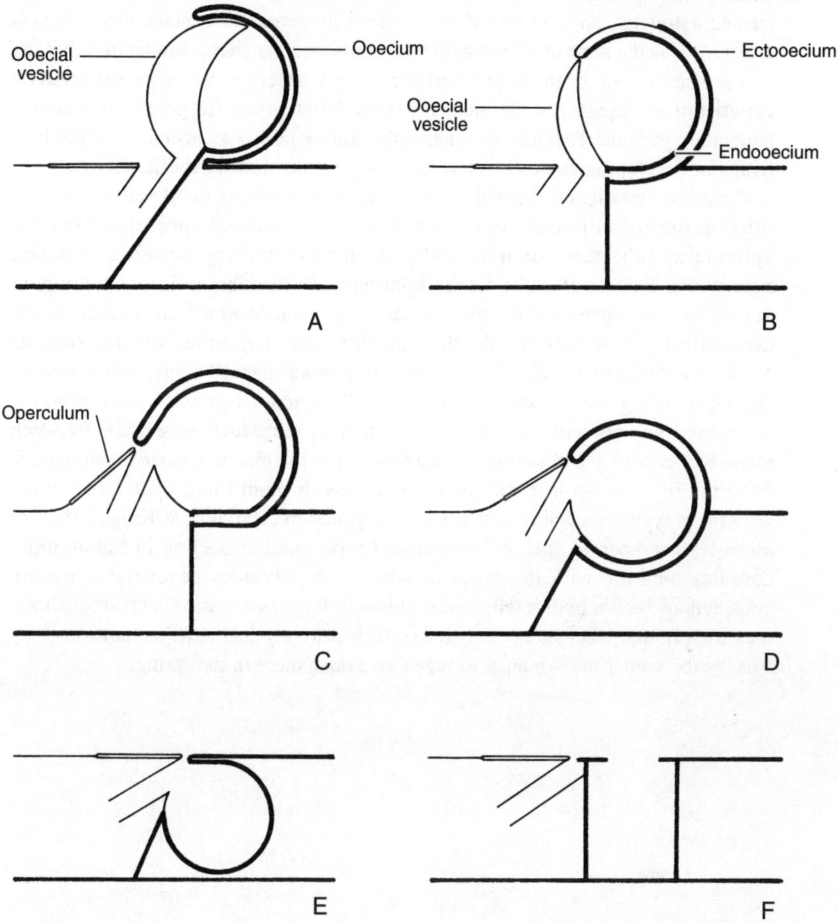

Fig. 13. Different types of ovicell in diagrammatic longitudinal section.
A, Independent hyperstomial, with the ooecium closed by the ooecial vesicle. B, Dependent hyperstomial, with partially calcified ectooecium. C, Dependent, prominent, closed by the autozooid operculum. D, Dependent, subimmersed, closed by the autozooid operculum. E, Immersed (endozooidal), opening below the maternal operculum. F, Immersed (endotoichal), with independent opening.

neanic (juvenile) zooids. The development of a colony from the ancestrula is termed **astogeny**, and the transitional zone from neanic to ephebic autozooids is referred to as the zone of **astogenetic change**. Once the characteristic morphology of the species has been established the colony enters a period of **astogenetic repetition** as essentially identical modules differentiate from each succeeding generation of buds. The growth zone, at the colony periphery, usually marked by a fringe of developing zooids is referred to as the zone of **ontogenetic change**.

Cycles of growth and reproduction vary as widely among the Bryozoa as in any other group of colonial organisms. Weedy, *r*-selected species encrusting ephemeral substrata, such as *Celleporella hyalina* (p. 94) on *Laminaria saccharina* fronds, produce larvae continuously through spring and summer, giving rise to numerous short-lived generations, each of which grows, breeds and dies within a few months. At the other extreme perennial species such as *Pentapora fascialis* (p. 200), which may live more than 10 years, and extensive sheet encrusting species such as *Schizoporella unicornis* (p. 220) are essentially *K*-selected species with perhaps limited, annual reproductive periods. Between these two adaptive end points are numerous intermediates. *Chartella papyracea* colonies live for about three years, and pass through three cycles of colony growth, polypide recycling and sexual reproduction (Dyrynda & King, 1982). In almost all temperate species growth and reproduction decline in the autumn; colonies may die with the onset of winter, or polypides degenerate, leaving conspicuous brown bodies. However, although large portions of encrusting sheets may die off, and erect shoots may die or be lost to physical attrition, portions may survive the winter in dormancy to regenerate once more in the spring.

Ecology

Bryozoans are almost ubiquitous in the marine benthic realm, occurring from the intertidal zone to the greatest abyssal depths. With a very few, very specialised exceptions the only requirement for their presence is a substratum to which an ancestrula can attach. Uniform, intertidal muds are inhospitable to bryozoans, but subtidal muddy sands harbour rare interstitial species, and abyssal oozes support quite large, erect colonies whose founding ancestrulae attach to the smallest pieces of substratum. In continental shelf seas, however, the abundance and taxonomic diversity of benthic bryozoan faunas are directly related to substratum. Coarse sub-littoral sands are usually rather barren, with just a few species, notably *Conopeum reticulum* and *Escharella variolosa* (p. 132), encrusting isolated patches of hard substratum, such as dead oyster valves. Offshore shell gravels, especially those with a diverse range of shell types and sizes, support a rich and varied cheilostomate fauna, and different suites of species may predominate according to the shell type. Rocky grounds are also rich in bryozoans, with dense encrusting communities and numerous erect species as well. In current-swept areas on western coasts *Pentapora fascialis* (p. 200) may be especially common, and its large, rigid colonies provide attachment for numerous other erect and encrusting bryozoans, as well as shelter for many other species of invertebrates. Shaded vertical walls and caves, where sessile organisms do not have to compete with crustose algae, are often densely populated by filter-feeding colonial animals, including many bryozoans.

Sheltered rocky shores support a surprisingly large number of bryozoans, representing all three orders. Some are essentially subtidal, but others are primarily intertidal in distribution, or at least limited to very shallow sublittoral and lower intertidal habitats. *Cauloramphus spiniferum* and *Haplopoma graniferum* (p. 314), for example, seem to be largely limited to intertidal habitats. Both occur higher on the shore than any other cheilostomates, to MTL, and may be very common on the undersides of stones in shallow pools. *Cryptosula pallasiana* (p. 194) and *Umbonula littoralis* (p. 104) seem also to be restricted to inshore and intertidal habitats. They may be extremely common beneath large stones on the lower shore, together with *Schizoporella unicornis* (p. 220), *Schizomavella linearis* (p. 282), *Escharoides coccinea* (p. 114) and *Callopora lineata*. All of these species are also frequently found in large kelp holdfasts, together with another dozen or so erect and encrusting species. Tufted colonies of *Bugula* and *Scrupocellaria*, with attendant chains of *Aetea* and *Scruparia*, are common among hydroids, coralline turfs, or shrubby red algae, and also in shaded areas beneath rock shelves, where they may form a dense turf with erect cyclostomates and ctenostomates, and various algae. Competition for space may be intense among encrusting, filter-feeding animals, and the diversity of particular communities will reflect the competitive abilities, and growth and reproductive strategies, of the species present.

The ecological requirements of hard substratum, intertidal bryozoans are almost entirely unknown. However, rocky coasts support a rich fauna of algal-encrusting species whose abundance and persistence renders them extremely convenient as experimental animals, and which have thus contributed disproportionately to bryozoan research. The sea mat, *Membranipora membranacea*, lives almost exclusively on the smooth-fronded kelps, growing swiftly to develop very extensive unilaminar colonies. Its planktonic cyphonautes larva, at settlement, orientates itself in response to water flow along the kelp frond, and colony growth is directed towards the perennial meristem of the plant, and away from the annually shed frond. Similar directional growth is seen in *Electra pilosa*, which initially develops stellate patches on a wide range of red and brown algae. *Electra pilosa* is found commonly on the wrack, *Fucus serratus*, where it is competitively inferior to three ctenostomate species, *Flustrellidra hispida, Alcyonidium gelatinosum* and *A. hirsutum*. These obligate epiphytes live almost exclusively on *F. serratus*, and their life cycles are to some degree attuned to that of their substratum. Larval release is synchronised within a population, occurring over a relatively short period. The larvae have a short free-swimming existence of just a few hours, appear able to select their host alga, and display phototactic, rugophilic and rheophilic behavioural responses at settlement, which ensure that the population is maintained in its optimum habitat. Surface texture, surface chemistry, and perhaps the proximity of conspecific colonies, may be important in the settlement of hard substratum species. But larval behaviour at settlement and metamorphosis is probably of prime importance only to species occupying limited, ephemeral habitats.

It is still unclear what are the most important food sources for bryozoans. All species obtain their food as fine particles filtered from the sea; *Membranipora membranacea*, and perhaps a minority of other species, is able to absorb dissolved organic substances originating from kelp mucus (De Burgh & Fankboner, 1978), but the majority depend on ingested material. Colonies have been reared in aquaria on diets of cultured dinoflagellates such as *Oxyrrhis marina*, but growth rates and breeding success often do not match those observed in wild populations. Conversely, Hunter & Hughes (1993a), showed that the single-celled alga *Rhodomonas baltica* was a highly successful food for *Celleporella hyalina* (p. 94). Populations of the bryozoan reared in aquaria, through several generations, grew as well as wild colonies, entered the reproductive phase within six weeks of larval settlement, and produced viable larvae. Many phytoplankton species are clearly unsuitable as food for bryozoans. Diatom frustules have often been observed in the gut of actively feeding zooids, but in many cases appear to be voided undamaged. There are many observations of zooids appearing to select particular cells, while rejecting others. What is certain is that the range and volume of food particles available to inshore and intertidal sessile organisms are immense (Best & Thorpe, 1994). Much of this consists of minute organic particles which result from the decay and breakdown of coastal algae. This is resuspended with every tidal movement. The epiphytic ctenostomate *Flustrellidra hispida*

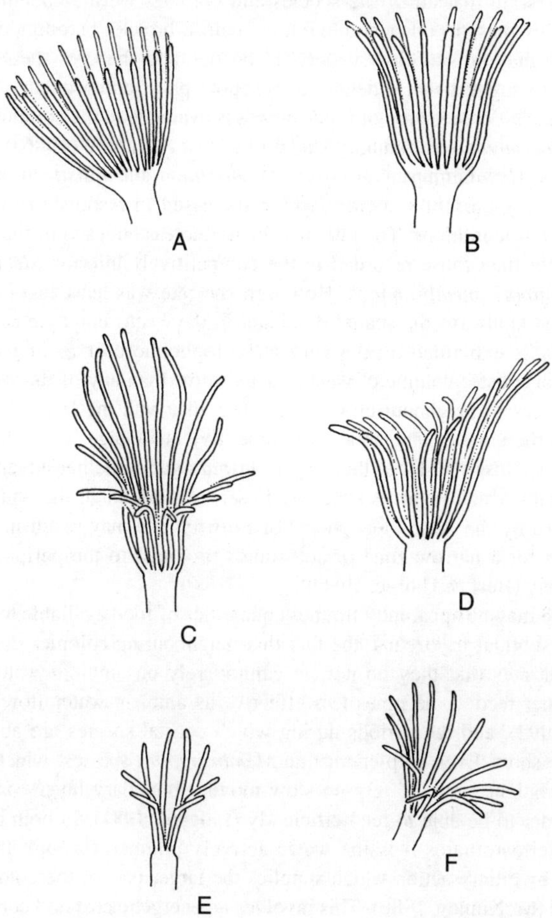

Fig. 14. Lophophore morphology.
A-B, Campanulate. C-D, Obliquely truncate. E-F, Campylonemidan. [After Winston, 1978].

increases its feeding rate as the density of particles in the surrounding water increases (Best & Thorpe, 1983), and it is not unreasonable to infer that intertidal and shallow sublittoral bryozoans feed in response to food availability. While levels of food available to coastal bryozoans do not appear to be limiting in any sense, there is nonetheless good evidence that within multispecies communities there is competition for food resources. In this context, the size and shape of the lophophore (Fig. 14), and its consequent potential filtering capacity are the major determinants of a species, competitive ability. In an encrusting community in which space was dominated by a few species, Ryland (1975) found only a narrow range of variation in size and diameter of the lophophore, between and within species, suggesting that all of the species and colonies were competing for the same food resources. In other communities, with a broader taxonomic range of species, there may be sufficient variety in the size and shape of the constituent lophophores to suggest that a degree of resource partitioning occurs (Winston, 1978), or that a wider selection of food sources is available to the community.

In the *Fucus serratus* community studied by Best & Thorpe (1986a), the three ctenostomates *Alcyonidium gelatinosum, A. hirsutum* and *Flustrellidra hispida* displayed feeding current velocities which increased in response to increasing food particle concentrations. The rate of velocity increase in each of these species was far greater than those recorded in the competitively inferior *Electra pilosa* and *Membranipora membranacea*. However, the rate was least in *Flustrellidra hispida*, which is always the spatial dominant in the *Fucus* epiphyte community. This anomaly is explained by the very large lophophore of *F. hispida* which processes a far greater volume of water per unit time than any of its competitors. Where the edges of neighbouring colonies abut, that with the larger lophophore may overtop the smaller and obtain a competitive advantage by depleting it of food resources (Buss, 1981). In this way *F. hispida* gains another advantage over the species with which it shares space on *F. serratus*, although the wider current flow generated by the large lophophores of *Flustrellidra* may result in enhanced food supplies for a narrow zone of autozooids proximal to the periphery of the adjacent colony (Best & Thorpe, 1986b).

While there may be apparently limitless quantities of food available to intertidal and shallow subtidal bryozoans, the fact that neighbouring colonies compete for food demonstrates that they do not, or cannot, rely on ambient water flow to supply it. Filter-feeding becomes less effective as ambient water flow increases (Okamura, 1992), and the periods during which coastal species are able to feed may be quite short. Even kelp-encrusting *Membranipora* species, which occur in relatively turbulent habitats, rely on slow-moving boundary layers on the kelp surface in order to be able to feed efficiently (Lidgard, 1981). In both encrusting and erect cheilostomates it is the water actively pumped through the colony lophophores by ciliary action which supplies the larger part of the colony's food requirements (McKinney, 1986). This involves an energetic cost and consequently colony forms and patterns of cilia-generated water flow over the colony surface are related to efficient processing of nutrient currents. For example, in erect, branching, fan-shaped colonies seen in *Bugula* and other cellularines, branch width and spacing between branches vary minimally, being at the optimum for efficient spacing of lophophores (McKinney, 1981). These generate a colony-wide feeding current which passes proximally from the colony tips to its base; here, a cone of stagnant flow occurs which is marked by a proximally increasing density of senescent, and thus non-feeding, autozooids. Stagnation becomes a potential problem with sheet-forming encrusting species, and those which develop erect plates, and colony-wide feeding currents can only be maintained by providing excurrent pathways, or chimneys, for the downflow stimulated by filtration currents. Such chimneys may be simply areas of the colony surface marked by groups of inactive autozooids; more often these areas are bordered by autozooids with asymmetrical lophophore bells (Cook & Chimonides, 1980) which serve to deflect surface water flow upwards, away from the colony (Lidgard, 1981). Erect, branching ascophorines may have autozooids with dimorphic lophophores along the branch margins, promoting excurrent flow along the edges of the colony, but the ability to develop excurrent chimneys away from the branch edge allows the growth of broad plates (McKinney, 1989), and has been an important influence on growth forms throughout the evolution of the Bryozoa (McKinney, 1986).

Among encrusting communities competing for food resources obvious advantages accrue to those species which occupy the greatest area in the shortest time, and maintain their dominance in spite of intense spatial competition. Rapid growth, combined with a broad, peripheral multizooidal growing edge and an ability to overgrow neighbouring colonies all contribute to success in competitive interactions, although no single species ever achieves hierarchical dominance. Instead, multispecies encrusting communities tend to consist of competitive networks (Jackson, 1979; Rubin, 1982, 1985), in which the competitive abilities of the constituent species and colonies are modulated by additional factors, such as the angle at which opposing colony edges intersect, the thickness of the colony, the growth pattern – whether radial or directional – and the regenerative capacity of each species. Extrinsic factors, such as interaction with other sessile organisms, predation and physical perturbation, all contribute to the maintenance of the network. However, some bryozoan communities, such as those found on offshore shell deposits studied by Ward & Thorpe (1989) seem exceptionally rich in species, with upwards of 60 occurring in single bottom samples. It is probable that such species diversity is achieved and sustained at least partly through the avoidance of spatial competition. While such communities may include competitively dominant perennial species characterised by large colony size, but low reproductive output, and a catholicity in substratum preference, many others are adapted to narrowly defined microhabitats on temporally or spatially ephemeral substrata. These typically achieve sexual maturity in relatively few astogenetic generations and pass through two or more sexual generations in each breeding season.

Distribution patterns of marine bryozoans are at the best imprecisely drawn, partly because ecological requirements of individual species remain unknown, and it is thus impossible to estimate where the distributional limits of a species might lie, but also and largely because the raw data on which patterns might be based, viz. faunistic records, are so often derived from past misidentifications. Depth *per se* does not seem to be a major factor affecting the distribution of particular species. Distinctive, deep, zoned faunas can be defined beyond the continental shelf break (Hayward, 1981), and coastal species are almost certainly depth-limited according to the distribution of their substratum. A very few hard substratum species seem to be restricted to intertidal habitats, e.g., *Cauloramphus spiniferum* and *Haplopoma graniferum* (p. 314), or to shallow coastal waters with fluctuating salinity, e.g., *Conopeum reticulum* and *Aspidelectra melolontha.* These species ought to prove useful candidates for researching the ecological parameters determining microgeographical distributions in non-epiphytic species. For the most part, however, encrusting sublittoral cheilostomate species appear to be limited only by substratum, and many species are distributed from below the kelps to the edges of the continental shelf. Accepted geographical distributions are constantly being revised as taxonomic precision improves, especially among the ascophorines, although there are distinct differences between the faunas on the east and west coasts of the British Isles, and to a less obvious extent between the north and south. The southeast differs most markedly from any other region of Britain. The generally richer faunas of the west include a preponderance of species whose geographical range extends from the western end of the Mediterranean northwards to Shetland. A very few of these range north to the Faroes. Many other species are found commonly in the Mediterranean, and southwards to northwest Africa, but no further north than western Ireland, and are usually absent from the Irish Sea. A small number of species appears limited to the southwest British Isles, and at present, are known to occur nowhere else, while rather fewer are found sporadically off northwest Scotland and more commonly northwards towards the Arctic circle. Bryozoan species occurring on the east coast of the British Isles, and through the central North Sea are generally distributed around the whole of the islands, but a minority appears to be limited otherwise to the northwest coasts of Scotland, but becomes more common towards the Faroes and Arctic Norway. Factors affecting the geographical and microgeographical distribution patterns of the British bryozoan fauna may include physical environmental parameters, such as temperature and salinity, hydrographical factors, including current patterns, biological and ecological factors, involving substratum, competitive exclusion, reproductive periods, and others which relate to the historical biogeography of individual species. It seems no longer possible to draw clear distinctions between Lusitanian/Mediterranean, temperate/boreal and boreo-arctic components in the fauna, despite the fact that individual geographical ranges seem to indicate one or other of these categories. Taxonomic review continues to reveal areas of confusion where two or more similar species have been consistently misidentified, and even in the comparatively well-studied northeast Atlantic region undescribed species continue to be discovered (e.g., Hayward, 1994). Geographical distribution patterns of benthic bryozoans in this region will not become evident until much more systematic study has been achieved.

Local bryozoan faunas for western Europe were cited in the first edition of this synopsis. None has yet been revised and updated, and the use of many is now hampered by their outdated taxonomy, or descriptions based upon incorrect identifications. However, several good new works are now available which in some cases revise or correct some of the errors of the previous faunal lists. Most important of these are Occhipinti Ambrogi (1981), on the lagoonal and coastal Bryozoa of Italy; Zabala (1986), on the bryozoan fauna of the Catalan coast; and Zabala & Maluquer (1988), who provide illustrated keys for the identification of western Mediterranean species. There has also been a considerable number of new systematic and faunistic research papers published since 1977, and reference is made to these under the appropriate species entries.

Predators

Predator : prey relationships in bryozoan dominated sessile communities are potentially one of the most interesting aspects of bryozoan ecology. Browsers and grazers, including sea urchins, fish, crabs and some prosobranchs, are known to include bryozoans in their diet. They obviously affect sessile community structure, creating patches of substratum for recolonisation and probably thus contributing to species diversity, although neither of these effects has been quantified. However, such predators are non-selective and any measurable response is likely to be at the level of the population, or the community. Bryozoans are also the prey of very many small, selective predators, some of which may be adapted to a very narrow spectrum of prey species, and these relationships promise to prove especially interesting. The opisthobranch predators of bryozoans are well known; the malacological literature details many narrowly-defined predator : prey patterns, and the feeding ecology of numerous intertidal seaslugs is now well documented: *Onchidoris muricata* and *Adalaria proxima* feeding on *Electra pilosa*, and *Polycera quadrilineata* feeding on *Membranipora membranacea*, are well known examples. Field observations on opisthobranchs, very many of which prey on Bryozoa, suggest that many (perhaps a majority) have a very limited range of preferred prey, and that feeding techniques may be specialised in response to the morphology of the prey (Thompson & Brown, 1984; Picton & Morrow, 1994), although Chadwick & Thorpe (1981) are the only authors to have experimentally demonstrated prey selection by sea slugs. However, the bryozoan perspective, in terms of response, and especially defensive reactions, to predators has been rarely investigated. Todd & Havenhand (1989) noted that seaslug predation was important in structuring epifaunal communities, and described the fragmentation and regeneration of *Electra pilosa* colonies caused by the feeding of *Adalaria proxima* and *Onchidoris muricata*. Harvell (1990, 1991) studied the effects of sea slug predation on kelp-encrusting colonies of *Membranipora* on the Washington coast. She showed that *Membranipora* colonies developed surface spines following attack by *Doridella steinbergae*, and that the production of spines could be induced in cultured colonies by the proximity of a tethered slug, suggesting that the bryozoan detected, and recognised, a potential predator through a chemosensory receptor. This work is presently unique. Experimental manipulation of bryozoan colonies is well established as a technique for studying spatial competition, and it is perhaps the way ahead for the study of predator : prey relationships.

Experimental studies, and video-recording, may also finally resolve the role of avicularia in bryozoan defensive strategies. Winston (1986, 1991) observed that avicularia appear to respond to very specific stimuli; they may be triggered to close by the passage of small syllid polychaetes, and tube-building amphipods, and seem also to be sensitive to chemical stimuli. Amphipods, isopods, mites and pycnogonids have all been recorded preying on bryozoan colonies, and the feeding strategies of the latter were described by Wyer & King (1973). Winston (1986) suggested that the immense variety of avicularia seen among the cheilostomates may have evolved in response to very particular predation pressures, perhaps largely imposed by small selective predators.

Finally, marine biochemistry has in recent years isolated numerous bioactive compounds from marine invertebrates, including a large number of sessile animals. Bryozoans feature largely: *Flustra foliacea, Chartella papyracea, Bugula neritina* and *Alcyonidium diaphanum* among the British fauna and many others from other regions of the world (Christopherson, 1985). The substances isolated, in some cases, appear to have potentially important pharmacological properties: bryostatin for example, isolated from *Bugula neritina*, appears to have an effect on cancerous tissue. However, the significance of these compounds for the bryozoan seems unknown. Bioactive substances may have an anti-fouling effect, or they may impart a toxic or noxious taste to the colony as a deterrent to predators, or they are perhaps more important to the internal physiology of the animal. Such questions have yet to be addressed experimentally.

Practical methods

Epiphytic bryozoans are generally conspicuous and easily found, but hard substratum species may take a practised eye to locate and a x10 hand lens is always useful in the field, if only to recognise a bryozoan colony for what it is. Erect, bushy species may be found beneath overhangs, or underneath large stones; samples of small, red algae, coralline turf, hydroid clumps and encrusted kelp holdfasts should be taken back to the laboratory and teased apart in seawater. Encrusted stones should be carefully packed in damp seaweed to avoid crushing their sessile fauna. In all instances care should be taken to minimise damage to intertidal habitats and collecting should be strictly limited. Any large stones moved in the search for specimens should always be returned to their original position.

Many good specimens of subtidal species may be obtained from the trash remaining in dredges or trawls. Well-encrusted shell, gravel or stones may be washed in fresh water, or preserved in 70% alcohol, for later examination. Scuba diving is now well-established, both as a leisure pastime and a scientific technique, and allows selective collecting with minimal damage to the environment. Generally, removal of large and obviously long-lived species such as *Pentapora fascialis* (p. 200) and *Flustra foliacea*, should be discouraged as they are often key components of complex, but fragile habitats.

It is always instructive to observe specimens alive whenever possible, when colour, tentacle number and behaviour can be recorded. Fine paint brushes may be useful for removing detritus. With experience it is possible to identify living specimens of cheilostomates, although only very few may be recognised by gross morphological characters and microscopical examination is necessary to identify the majority. In most cases, however, accurate identification requires that a portion of the specimen be cleaned of its soft tissues, and dried, before it is examined. It is fruitless to try to identify cheilostomates unless zooid morphology can be clearly seen, and its nature understood. In the majority of calloporoidean species, it is only necessary to clear the specimen of detritus, using an ultrasonic cleaner if possible; zooid detail may then appear clear in immersed specimens, especially using substage illumination. Dried calloporoideans should be viewed on a black background with direct illumination. Ascophorines may require cleaning before zooidal details, especially the taxonomically important primary orifice, can be seen. It is best to soak a portion of the specimen in weak domestic bleach for about 3 hours, wash it well in fresh water and clear with an ultrasonic cleaner for about 10 minutes. Dried and mounted on black card in a wooden cavity slide, all details should be readily visible; washable blue ink, or a food colourant, painted onto the surface of the specimen will reduce its reflectivity.

Wet specimens of cheilostomate and cyclostomate bryozoans may be fixed in

4% seawater formalin and preserved in 70% ethyl alcohol. The uncalcified ctenostomates are best preserved in 4% seawater formalin. If voucher specimens are required with expanded lophophores, they should be narcotised before fixation. This is best achieved by placing specimens in bowls of seawater, sprinkling menthol crystals on the surface and leaving them undisturbed in a cool, shady position for about 12 hours. Reference collections of calcified bryozoans, for microscopic examination, are most conveniently kept mounted on black card backgrounds in wooden cavity slides. Specimens should be mounted with a water soluble glue, or double-sided adhesive tape, allowing removal if required. Permanent preparations of lightly calcified and non-calcified species are also useful, and may be prepared using standard microscopical techniques.

In recent years, the most valuable contributions to bryozoan morphology and taxonomy have resulted from scanning electron microscopy, and without doubt SEM examination is indispensable in resolving taxonomic problems, especially among the heavily calcified ascophorines. This technique has been employed in the task of revising the British fauna and needs to be more widely used. In fact it is probable that a clear account of all benthic bryozoan faunas, and a new, stable taxonomy will not be achieved until the type species of each genus has been restudied using SEM.

Classification of the Cheilostomatida

The classification adopted here incorporates most recent taxonomic opinion, especially that of Gordon (1984, 1986, 1989, 1993), but is still to some extent interim. Morphological and taxonomic research on the Cheilostomatida is continuing at an accelerating pace, which will undoubtedly be reflected in further changes to their classification.

Order CHEILOSTOMATIDA Busk, 1852

Suborder INOVICELLINA Jullien, 1888

Superfamily Aeteoidea Smitt, 1867

Family Aeteidae Smitt, 1867

Aetea Lamouroux, 1812
anguina (Linnaeus, 1758)
sica (Couch, 1844)
truncata (Landsborough, 1852)

Suborder SCRUPARIINA Silén, 1941

Superfamily Scruparioidea Gray, 1848

Family Scrupariidae Gray, 1848

Scruparia Oken, 1815
ambigua (d'Orbigny, 1841)
chelata (Linnaeus, 1758)

Family Eucrateidae Johnston, 1847

Eucratea Lamouroux, 1812
loricata (Linnaeus, 1758)

Suborder MALACOSTEGINA Levinsen, 1909

Superfamily Membraniporoidea Busk, 1854

Family Membraniporidae Busk, 1854

Membranipora de Blainville, 1830
membranacea (Linnaeus, 1767)

Conopeum Gray, 1848
reticulum (Linnaeus, 1767)
seurati (Canu, 1928)

Family Electridae Stach, 1937

Electra Lamouroux, 1816
crustulenta (Pallas, 1766)
monostachys (Busk, 1854)
pilosa (Linnaeus, 1767)

Pyripora d'Orbigny, 1852
catenularia (Fleming, 1828)

Aspidelectra Levinsen, 1909
melolontha (Landsborough, 1852)

Suborder FLUSTRINA Smitt, 1868

Superfamily Flustroidea Fleming, 1828

Family Flustridae Fleming, 1828

Flustra Linnaeus, 1761
foliacea (Linnaeus, 1758)

Carbasea Gray, 1848
carbasea (Ellis & Solander, 1786)

Chartella Gray, 1848
barleei (Busk, 1860)
papyracea (Ellis & Solander, 1786)

Securiflustra Silén, 1941
securifrons (Pallas, 1766)

Hincksina Norman, 1903
flustroides (Hincks, 1877)

Sarsiflustra Jullien, 1903
abyssicola (G. O. Sars, 1872)

Superfamily Calloporoidea Norman, 1903

Family Calloporidae Norman, 1903

Callopora Gray, 1848
craticula (Alder, 1856)
discreta (Hincks, 1862)
dumerilii (Audouin, 1826)
lineata (Linnaeus, 1767)
rylandi Bobin & Prenant, 1965

Alderina Norman, 1903
imbellis (Hincks, 1860)

Crassimarginatella Canu, 1900
solidula (Hincks, 1860)

Cauloramphus Norman, 1903
spiniferum (Johnston, 1832)

Tegella Levinsen, 1909
unicornis (Fleming, 1828)

Copidozoum Harmer, 1926
exiguum (Barroso, 1920)
planum (Hincks, 1880)
tenuirostre (Hincks, 1880)

Parellisina Osburn, 1940
curvirostris (Hincks, 1862)

Amphiblestrum Gray, 1848
auritum (Hincks, 1877)
flemingii (Busk, 1854)
solidum (Packard, 1863)

Ellisina Norman, 1903
gautieri Fernandez Pulpeiro & Reverter Gil, 1993

Ramphonotus Norman, 1894
minax (Busk, 1860)

Megapora Hincks, 1877
ringens (Busk, 1856)

Ammatophora Norman, 1903
nodulosa (Hincks, 1877)

Membraniporella Smitt, 1873
nitida (Johnston, 1838)

Family Chaperiidae Jullien, 1888

Larnacicus Norman, 1903
corniger (Busk, 1859)

Superfamily Buguloidea Gray, 1848

Family Bugulidae Gray, 1848

Bugula Oken, 1815
avicularia (Linnaeus, 1758)
calathus Norman, 1864
flabellata (Thompson in Gray, 1848)
fulva Ryland, 1960
neritina (Linnaeus, 1758)
plumosa (Pallas, 1766)
purpurotincta Norman, 1868
simplex Hincks, 1886
stolonifera Ryland, 1960
turbinata Alder, 1857

Dendrobeania Levinsen, 1909
fruticosa (Packard, 1863)
murrayana (Bean in Johnston, 1847)
fessa Kluge, 1955

Bicellariella Levinsen, 1909
ciliata (Linnaeus, 1758)

Bicellarina Levinsen, 1909
alderi (Busk, 1859)

Family Beaniidae Canu & Bassler, 1927

Beania Johnston, 1840
mirabilis Johnston, 1840

Family Epistomiidae Gregory, 1903

Epistomia Fleming, 1828
bursaria (Linnaeus, 1758)

Family Candidae d'Orbigny, 1851

Caberea Lamouroux, 1816
boryi (Audouin, 1826)
ellisii (Fleming, 1814)

Notoplites Harmer, 1923
damicornis Hayward & Ryland, 1978
evocata (Jullien, 1882)
harmeri Ryland, 1963
jeffreysii (Norman, 1868)

Scrupocellaria van Beneden, 1845
inermis Norman, 1867
reptans (Linnaeus, 1767)
scabra (van Beneden, 1848)
scrupea Busk, 1852
scruposa (Linnaeus, 1758)

Tricellaria Fleming, 1828
peachii (Busk, 1851)
ternata (Ellis & Solander, 1786)

Superfamily Microporoidea Gray, 1848

Family Microporidae Gray, 1848

Micropora Gray, 1848
coriacea (Johnston, 1847)
normani Levinsen, 1909

Calpensia Jullien, 1888
nobilis (Esper, 1796)

Rosseliana Jullien, 1888
rosselii (Audouin, 1826)

Steraechmella Lagaaij, 1952
buski Lagaaij, 1952

Family Setosellidae Levinsen, 1909

Setosella Hincks, 1877
vulnerata (Busk, 1860)

Superfamily Cellarioidea Lamouroux, 1821

Family Cellariidae Lamouroux, 1821

Cellaria Ellis & Solander, 1786
fistulosa (Linnaeus, 1758)
salicornioides Lamouroux, 1816
sinuosa (Hassall, 1840)

Suborder ASCOPHORINA Levinsen, 1909

Infraorder ACANTHOSTEGOMORPHA Levinsen, 1902

Superfamily Cribrilinoidea Hincks, 1879

Family Cribrilinidae Hincks, 1879

Cribrilina Gray, 1848
annulata (Fabricius, 1780)
cryptooecium Norman, 1903
punctata (Hassall, 1841)

Gephyrotes Norman, 1903
nitidopunctata (Smitt, 1868)

Collarina Jullien, 1886
balzaci (Audouin, 1826)

Puellina Jullien, 1886
arrecta Bishop & Househam, 1987
bifida (d'Hondt, 1970)
corbula Bishop & Househam, 1987
directa Bishop & Househam, 1987
gattyae (Landsborough, 1852)
innominata (Couch, 1844)
modica Bishop & Househam, 1987
praecox Bishop & Househam, 1987
setosa (Waters, 1899)
venusta Canu & Bassler, 1925

Figularia Jullien, 1886
figularis (Johnston, 1847)

Infraorder HIPPOTHOOMORPHA Gordon, 1989

Superfamily Hippothooidea Busk, 1859

Family Hippothoidae Busk, 1859

Hippothoa Lamouroux, 1821
divaricata Lamouroux, 1821
flagellum Manzoni, 1870

Plesiothoa Gordon & Hastings, 1979
gigerium (Ryland & Gordon, 1977)

Celleporella Gray, 1848
hyalina (Linnaeus, 1767)

Haplota Marcus, 1940
clavata (Hincks, 1857)

Trypostega Levinsen, 1909
venusta (Norman, 1864)

Family Chorizoporidae Vigneaux, 1949

Chorizopora Hincks, 1879
brongniartii (Audouin, 1826)

Infraorder UMBONULOMORPHA Gordon, 1989

Superfamily Lepralielloidea Vigneaux, 1949

Family Umbonulidae Canu, 1904

Umbonula Hincks, 1880
littoralis Hastings, 1944
ovicellata Hastings, 1944

Ragionula Canu & Bassler, 1927
rosacea (Busk, 1856)

Family Romancheinidae Jullien, 1888

Arctonula Gordon & Grischenko, 1994
arctica (M. Sars, 1851)

Family Exochellidae Bassler, 1935

Escharoides Milne Edwards, 1836
coccinea (Abildgaard, 1806)
mamillata (Wood, 1844)
jacksoni (Waters, 1900)

Family Lepraliellidae Vigneaux, 1949

Lepraliella Levinsen, 1916
hippopus (Smitt, 1868)

Family Escharellidae Levinsen, 1909

Escharella Gray, 1848
abyssicola (Norman, 1869)
immersa (Fleming, 1828)
klugei Hayward, 1979
labiosa (Busk, 1856)
laqueata (Norman, 1864)
octodentata (Hincks, 1880)
variolosa (Johnston, 1838)
ventricosa (Hassall, 1842)

Hemicyclopora Norman, 1894
microstoma (Norman, 1864)
multispinata (Busk, 1861)
polita (Norman, 1864)
Neolagenipora Vigneaux, 1949
collaris (Norman, 1867)
eximia (Hincks, 1860)

Family Bryocryptellidae Vigneaux, 1949

Porella Gray, 1848
alba Nordgaard, 1906
compressa (Sowerby, 1805)
concinna (Busk, 1854)
laevis (Fleming, 1828)
minuta (Norman, 1868)
patula (M. Sars, 1851)
struma (Norman, 1868)

Palmiskenea Bishop & Hayward, 1989
lorea (Alder, 1864)
plana (Hincks, 1888)
skenei (Ellis & Solander, 1786)

Family Tessaradomidae Jullien, 1903

Tessaradoma Norman, 1869
boreale (Busk, 1860)

Superfamily Arachnopusioidea Jullien, 1888

Family Exechonellidae Harmer, 1957

Anarthropora Smitt, 1868
monodon (Busk, 1860)

Superfamily Adeonoidea Busk, 1884

Family Adeonidae Jullien, 1903

Reptadeonella Busk, 1884
insidiosa (Jullien, 1903)
violacea (Johnston, 1847)

Infraorder LEPRALIOMORPHA Gordon, 1989

Superfamily Schizoporelloidea Jullien, 1883

Family Watersiporidae Vigneaux, 1949

Watersipora Neviani, 1895
complanata (Norman, 1864)

Family Cryptosulidae Vigneaux, 1949

Cryptosula Canu & Bassler, 1925
pallasiana (Moll, 1803)

Family Hippoporinidae Bassler, 1935

Hippoporina Neviani, 1895
pertusa (Esper, 1796)

Codonellina Bassler, 1934
lacunata Hayward & Hansen, 1999

Pentapora Fischer, 1807
foliacea (Ellis & Solander, 1786)

Family Teuchoporidae Neviani, 1895

Phylactella Hincks, 1879
labrosa (Busk, 1854)

Family Schizoporellidae Jullien, 1903

Schizoporella Hincks, 1877
cornualis Hayward & Ryland, 1995
dunkeri (Reuss, 1848)
errata (Waters, 1878)
hesperia Hayward & Ryland 1995
magnifica Hincks, 1886
patula Hayward & Ryland 1995
unicornis (Johnston in Wood 1844)

Schizobrachiella Canu & Bassler, 1920
sanguinea (Norman, 1868)

Arthropoma Levinsen, 1909
cecilii (Audouin, 1826)

Escharina Milne Edwards, 1836
alderi (Busk, 1856)
dutertrei (Audouin, 1826)
hyndmanni (Johnston, 1847)
johnstoni (Quelch, 1884)
vulgaris (Moll, 1803)

Phaeostachys Hayward, 1979
spinifera (Johnston, 1847)

Family Stomachetosellidae Canu & Bassler, 1917

Stomachetosella Canu & Bassler, 1917
cruenta (Busk, 1854)
normani Hayward 1994
sinuosa (Busk, 1860)

Family Gigantoporidae Bassler, 1935

Cylindroporella Hincks, 1877
tubulosa (Norman, 1868)

Superfamily Smittinoidea Levinsen, 1909

Family Smittinidae Levinsen, 1909

Smittina Norman, 1903
affinis (Hincks, 1862)
bella (Busk, 1860)
cheilostoma (Manzoni, 1870)
crystallina (Norman, 1867)
landsborovii (Johnston, 1847)

Smittoidea Osburn, 1952
amplissima Hayward, 1979
marmorea (Hincks, 1877)
reticulata (MacGillivray, 1842)

Parasmittina Osburn, 1952
trispinosa (Johnston, 1838)

Family Bitectiporidae MacGillivray, 1895

Schizomavella Canu & Bassler, 1917
auriculata (Hassall, 1842)
cristata (Hincks, 1879)
cuspidata (Hincks, 1880)
discoidea (Busk, 1859)
hastata (Hincks, 1862)
hondti Reverter Gil & Fernandez Pulpeiro, 1996
linearis (Hassall, 1841)
ochracea (Hincks, 1880)
sarniensis Hayward & Thorpe 1995
teresae Reverter Gil & Fernandez Pulpeiro, 1996

Family Microporellidae Hincks, 1880

Microporella Hincks, 1877
ciliata (Pallas, 1766)
appendiculata (Heller 1867)

Fenestrulina Jullien, 1888
malusii (Audouin, 1826)

Diporula Hincks, 1879
verrucosa (Peach, 1868)

Haplopoma Levinsen, 1909
bimucronatum (Moll, 1803)

graniferum (Johnston, 1847)
impressum (Audouin, 1826)
planum Ryland, 1963
sciaphilum Silén & Harmelin, 1976

Superfamily Celleporoidea Johnston, 1838

Family Celleporidae Johnston, 1838

Cellepora Linnaeus, 1767
pumicosa (Pallas, 1766)

Celleporina Gray, 1848
decipiens Hayward, 1976
hassallii (Johnston, 1847)
pygmaea (Norman, 1868)
tubulosa (Hincks, 1880)

Lagenipora Hincks, 1877
lepralioides (Norman, 1868)

Turbicellepora Ryland, 1963
avicularis (Hincks, 1860)
magnicostata (Barroso, 1919)
boreale Hayward & Hansen 1999

Omalosecosa Canu & Bassler, 1925
ramulosa (Linnaeus, 1767)

Buskea Heller, 1867
dichotoma (Hincks, 1862)
quincuncialis (Norman, 1867)

Palmicellaria Alder, 1864
elegans Alder, 1864

Buffonellaria Canu & Bassler, 1927
divergens (Smitt, 1873)

Family Hippoporidridae Vigneaux, 1949

Hippoporidra Canu & Bassler, 1927
lusitania Taylor & Cook, 1981

Hagiosynodos Bishop & Hayward, 1989
latus (Busk, 1856)

Family Phidoloporidae Gabb & Horn, 1862

Reteporella Busk, 1884
beaniana (King, 1846)
couchii (Hincks, 1878)
incognita Hayward & Ryland, 1996
septentrionalis (Harmer, 1933)
watersi (Nordgaard, 1907)

Schizotheca Hincks, 1877
divisa (Norman, 1864)
fissa (Busk, 1856)
serratimargo Hincks, 1886

Rhynchozoon Hincks, 1895
bispinosum (Johnston, 1847)

Stephanollona Duvergier, 1921
armata (Hincks, 1862)

Key to Genera of Cheilostomatida

It is important to be familiar with the general structure of bryozoans before attempting to identify them. Cheilostomate structures important in their identification have been described and illustrated in the preceding pages of this synopsis, and a glossary of terms is provided at the back of the volume (p. 389). Marginal figures within the key refer to major diagnostic characteristics and are not intended as realistic depictions of particular species. This key allows the identification of all British cheilostomate genera, described within this volume or in the second edition of its companion (Synopsis 10). Once you have obtained a generic name, check the diagnosis before proceeding to identification of the species. This is an important step as certain diagnostic characters may not have been utilised in the key. Finally, the specific identification should be checked with the species description and figures, and care should be taken to ensure that the provenance of the specimen accords with the known geographical range and habitat occurrence recorded for the species. Precision is essential if accurate distribution maps are to be developed for bryozoan species. The zoogeography, as well as the systematics, of cheilostomate bryozoans has too often been confused by misidentification.

Many of the figures presented in the Systematic Account of species have been reprinted from the first edition of this *Synopsis*, but a considerable proportion has been redrawn, and some species are now also illustrated by SEM micrographs. The range of astogenetic and ontogenetic variation seen in each species has been described and figured whenever possible, but it is essential to remember that structures such as spines, avicularia and ovicells may be missing, or damaged, or not developed in some individual colonies. The size ranges for zooids of each species have been measured from actual specimens but, in samples from the extreme north or extreme south of the British sea area, zooid sizes may vary slightly beyond the ranges cited here. However, as a general rule, and particularly among the ascophorines, numbers of oral spines, and the shape and proportions of the primary orifice do not vary significantly between populations. Thus, more or fewer spines, a broader or narrower orifice, or differently shaped condyles from those described for a particular species generally indicate a different species.

1. **Colony consisting of uniserial chains of autozooids**, either repent and entirely encrusting; **or** with slender, tubular units rising vertically from encrusting uniserial chains; **or** with erect uniserial chains rising from encrusting chains or from slender stolons
.... **2**

Colony not as described
.... **9**

2.(1) Colony consists of branching, uniserial chains of fusiform autozooids, each with an erect, tubular distal portion, which may be straight or curved and spoon-shaped at its tip
.... *Aetea* (*Synopsis* 10, p. 99)

Colony not as described
.... **3**

3.(2) **Frontal shield of autozooid partly membranous**, with the operculum set at its distal end (anascan)
.... **4**

Frontal shield of autozooid entirely calcified, with the operculum closing a defined orifice (ascophoran)
.... **6**

4.(3) Colony consists of erect, branching chains of horn-shaped autozooids, rising from an encrusting autozooid chain, or from a slender, encrusting stolon
.... *Scruparia* (*Synopsis* 10, p. 107)

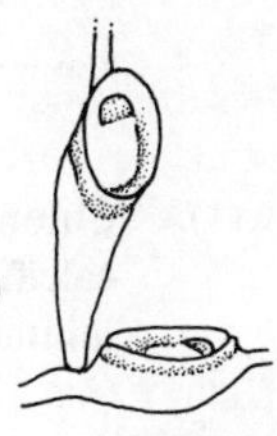

Colony entirely encrusting
.... **5**

5.(4) Colony consists of branching, uniserial chains of thickly calcified, pyriform autozooids. Oval frontal membrane constitutes almost half total autozooid length. No spines
.... *Pyripora* (*Synopsis* 10, p. 136)

Colony consists of branching, uniserial chains of lightly calcified fusiform autozooids. Elongate frontal membrane constitutes more than half autozooid length, bordered by thin spines
.... *Beania mirabilis* (*Synopsis* 10, p. 244)

6.(3) Colony consists of erect chains of lightly calcified autozooids, arising from encrusting autozooid chains
.... ***Haplota*** (p. 96)

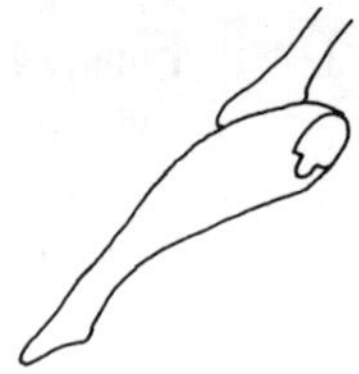

Colony entirely encrusting. Autozooids generally well calcified
.... **7**

7.(6) Autozooids oval to hexagonal; frontal shield granular, with distinct marginal pores, and scattered frontal pores. Small adventitious avicularia present
.... ***Escharina alderi*** (p. 228)

Autozooids elongate oval to fusiform; frontal shield smooth or corrugate, but without pores. No avicularia
.... **8**

8.(7) Autozooids fusiform, with a filiform proximal portion of varying length. Autozooid chains branch in a cruciform pattern
.... ***Hippothoa*** (p. 84)

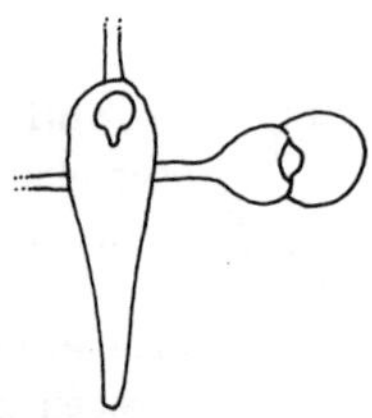

Autozooids elongate oval, closely spaced; branching apparently irregular, with adjacent autozooids often linked by short tubular processes
.... ***Plesiothoa*** (p. 90)

9.(1) Colony erect, branching; **bushy, flabellate or twiggy. Lightly calcified and flexible, or with more substantial calcification but jointed**
.... **10**

Colony **erect**: branching, reticulate or plate-like, but rigid, **not flexible; or nodular; or encrusting**, sheet-like
.... **29**

10.(9) **Colony flexible; branching dichotomously but unjointed. Unilaminar**, with distinct frontal and abfrontal surfaces
.... **11**

Colony flexible and unjointed, but bilaminar, with autozooid orifices on both faces of frond; or with distinct chitinous joints at each branching point
.... **18**

11.(10) **Small adventitious avicularia present**, sessile or shortly stalked
.... **12**

Avicularia absent
.... **17**

12.(11) Frontal surface of colony with small sessile avicularia; basal surface with large vibracula, bearing long, toothed, setae
.... ***Caberea*** (*Synopsis* **10**, p. 251)

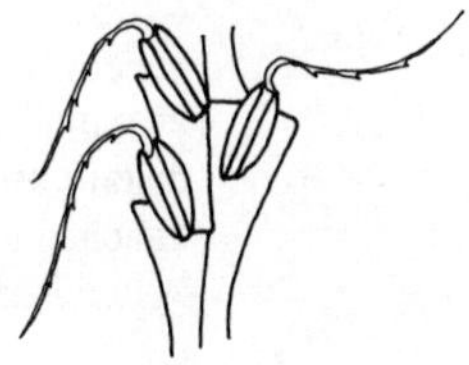

Avicularia all pedunculate, i.e., on short stalks
.... **13**

13.(12) Avicularia shaped like a tobacco pipe. Autozooids arranged in pairs, without spines
.... ***Epistomia bursaria*** (*Synopsis* **10**, p. 248)

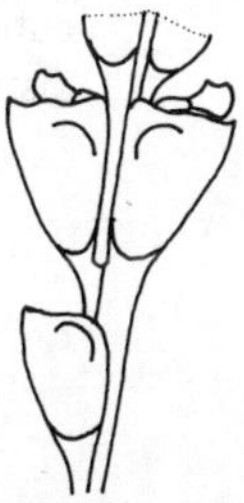

Avicularia of the bird's-head type. Autozooids alternating along the length of the branch, not in pairs, biserial or multiserial
.... **14**

14.(13) Colony branches strictly biserial, the autozooids long and slender, with oval or rounded triangular opesiae, oblique to long axis of branch and alternating left and right along it
.... **15**

Colony branches biserial to multiserial, the autozooids more or less rectangular, with elongate opesia overlapping laterally or in adjacent rows
.... **16**

15.(14) Autozooids with four to nine long, curved spines, up to twice autozooid length, at distal end of opesia. Ovicell globular, attached by a stalk halfway along inner edge of opesia
.... ***Bicellariella*** (*Synopsis* **10**, p. 240)

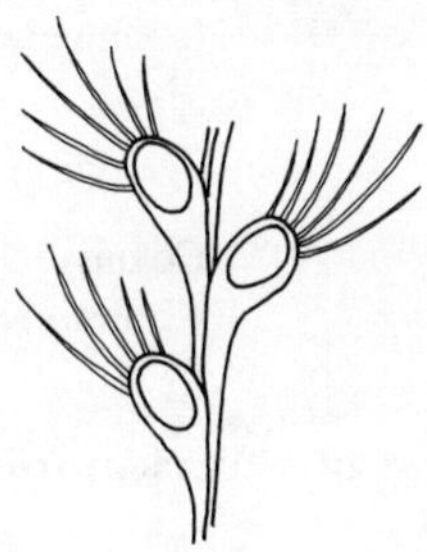

One or two long spines on the outer distal corner of the autozooid. Ovicell hemispherical, at distal end of opesia
.... ***Bicellarina*** (*Synopsis* **10**, p. 242)

16.(14) Proximal ends of autozooids, in basal view, distinctly forked. Spines at distal corners of autozooids only. Avicularia attached to lateral walls of autozooids
.... ***Bugula*** (*Synopsis* **10**, p. 207)

Proximal ends of autozooids, in basal view, quite straight. Spines at distal corners of autozooids, or also along lateral walls. Avicularia attached to proximal portion of frontal wall
.... ***Dendrobeania*** (*Synopsis* **10**, p. 233)

17.(11) Colony frondose, consisting of alternating rows of linguiform autozooids, with largely membranous frontal walls. Operculum distinct; no ovicells
.... ***Carbasea*** (*Synopsis* **10**, p. 144)

Colony branches biserial, autozooids elongate, with short proximal gymnocyst. No operculum; small globular ovicells present
.... ***Bugula neritina*** (*Synopsis* **10**, p. 220)

18.(10) Colony stiff, white; the branches consisting of slender cylinders linked by conspicuous, yellowish brown chitinous joints, branching dichotomously. Autozooids well calcified beneath a glistening frontal membrane, arranged in whorls, opening around the whole axis of the cylinder
.... ***Cellaria*** (*Synopsis* **10**, p. 302)

Colony branches not consisting of stiff, jointed cylinders
.... **19**

19.(18) Colony branches slender, consisting of elongate autozooids in back-to-back pairs, with an oval opesia comprising just less than half autozooid length. Branching profusely to give a dense bushy form
.... ***Eucratea*** (*Synopsis* **10**, p. 112)

Colony branches bilaminar and frondose, or unilaminar and biserial
.... **20**

20.(19) **Colony frondose, bilaminar**
.... **21**

Colony branches biserial, unilaminar
.... **25**

21.(20) Avicularia absent. Autozooids rectangular, each with a pair of short distal spines
.... ***Chartella papyracea*** (*Synopsis* **10**, p. 146)

Interzooidal or vicarious avicularia present
.... **22**

22.(21) Avicularia smaller than autozooids, with short, rounded or subtriangular mandibles
.... **23**

Avicularia as large as autozooids, with elongate, spatulate mandibles
.... ***Sarsiflustra abyssicola*** (*Synopsis* **10**, p. 154)

23.(22) Autozooids with stout, distal spines. Avicularia with broad, semi-oval mandibles, and subtriangular, proximal frontal membrane. Colony generally thick, horny
.... ***Flustra foliacea*** (*Synopsis* **10**, p. 141)

Autozooids without spines. Colony lightly calcified, rather flimsy
.... **24**

24.(23) Autozooids broadly rectangular. Avicularia with obliquely orientated, semi-oval mandibles. Colony with broad, lobate, fronds
.... ***Chartella barleei*** (*Synopsis* **10**, p. 148)

Autozooids elongate rectangular. Avicularia with distally-orientated subtriangular mandibles. Colony with narrow, wedge-shaped fronds
.... ***Securiflustra securifrons*** (*Synopsis* **10**, p. 150)

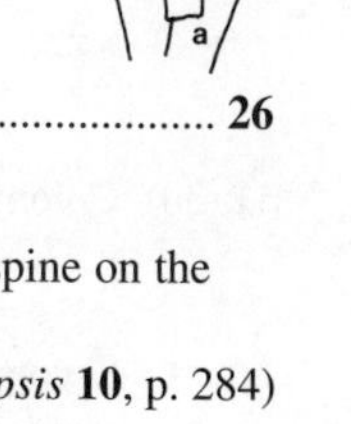

25.(20) Vibracula present on basal surfaces of autozooids; avicularia present frontally, and often also along lateral margins of autozooids
.... ***Scrupocellaria*** (*Synopsis* **10**, p. 268)

Avicularia present or absent, but no vibracula
.... **26**

26.(25) Avicularia absent. Autozooids slender, each with a short spine on the outer distal corner
.... ***Tricellaria peachii*** (*Synopsis* **10**, p. 284)

Avicularia present
.... **27**

27.(26) Joints at bifurcations inconspicuous, in basal view seen to traverse autozooids GK and FJ
.... ***Notoplites*** (*Synopsis* **10**, p. 256)

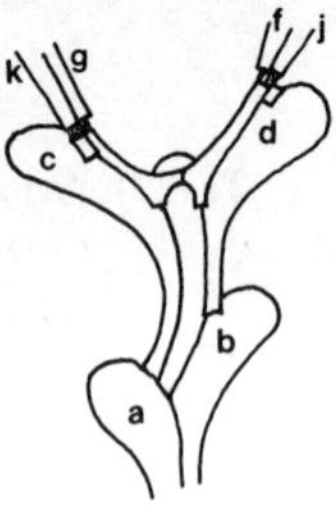

Joints at bifurcations clear, in basal view seen to traverse autozooids CG and DF
.... **28**

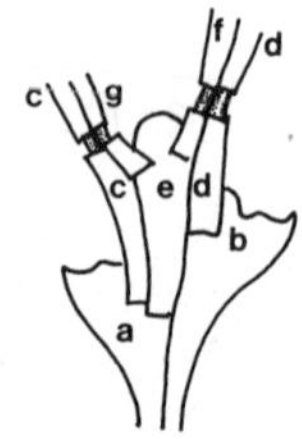

28.(27) Autozooids with large, prominent, lateral avicularia, each as wide as the autozooid
.... ***Tricellaria*** (*Synopsis* **10**, p. 281)

Autozooids with small lateral avicularia, much narrower than the autozooid
.... ***Scrupocellaria scabra*** (*Synopsis* **10**, p. 272)

29.(9) **Colony erect, rigid; branching, reticulate or plate-like**
.... **30**

Colony encrusting, sheet-like; or forming nodules
.... **42**

30.(29) Colony reticulate, forming a shallow cup, an irregularly curved fan, or irregularly folded and anastomosed to form a three-dimensional bun-shape
.... ***Reteporella*** (p. 366)

Colony with slender or stout branches, of anastomosed, bilaminar plates. Not reticulate
.... **31**

31.(30) Colony slender, twiggy, branching dichotomously
.... **32**

Colony with stout, cylindrical, lobate or knobbly branches, dividing regularly or irregularly; or of broad, anastomosed plates, developing a honeycomb structure
.... **34**

32.(31) Each autozooid with a conspicuous spiramen just proximal to the tubular peristome. Frontal calcification appears striated. Avicularia distributed along the margins of the autozooids

.... ***Tessaradoma*** (p. 180)

Autozooids lacking spiramina. Calcification smooth. Avicularia associated with the autozooid orifice

.... **33**

33.(32) Autozooids strictly ordered in alternating, back-to-back pairs, each with a small suboral avicularium on a prominent umbonate process

.... ***Palmicellaria*** (p. 352)

Autozooids in whorls of three or more, each with a suboral avicularium incorporated in a deep peristome. Additional frontal avicularia present, small, or large and spatulate

.... ***Buskea*** (p. 346)

34.(31) **Colony with stout, cylindrical, lobate or knobbly branches**, dividing regularly or irregularly

.... **35**

Colony consists of flat or folded bilaminar plates, fused to form a three-dimensional honeycomb-like structure

.... **41**

35.(34) Each autozooid with a conspicuous, crescentic ascopore proximal to the orifice and a single, lateral avicularium. Colony branches thick, smooth, and cylindrical

.... ***Diporula*** (p. 302)

Autozooids without ascopores

.... **36**

36.(35) Avicularia polymorphic, numerous; one proximo-lateral to each autozooid orifice; others frontal, small or large, the mandibles short triangular, rounded, or enlarged and spatulate. Colony nodular or irregularly branched

.... ***Turbicellepora*** (p. 334)

Avicularia monomorphic, one adjacent to each autozooid orifice; similar avicularia sometimes along autozooid margins (generally few), rarely an enlarged type with spatulate mandible present. Colony always erect and branching

.... **37**

37.(36) A single median suboral avicularium present, proximally directed, becoming enclosed in a deep peristome. Primary orifice with a distinct lyrula, ovicell imperforate
.... ***Porella*** (p. 154)

Avicularium median and/or lateral suboral in position. Primary orifice sinuate, bell-shaped or oval, without a lyrula. Ovicell perforate or imperforate
.... **38**

38.(37) Primary orifice sinuate (Fig. 15. 26-51). Avicularium lateral-suboral, incorporated in a medially notched peristome. Additional small, or enlarged, spatulate avicularia on margins of autozooids in older parts of colony. Ovicell perforate
.... ***Buskea*** (p. 346)

Primary orifice bell-shaped (Fig. 15.1) or oval
.... **39**

39.(38) Primary orifice transversely oval. Each autozooid with a single lateral suboral avicularium, on a prominent spiked umbo, incorporated in a flared peristome. Ovicell imperforate
.... ***Omalosecosa*** (p. 342)

Primary orifice bell-shaped, with distinct condyles
.... **40**

40.(39) Avicularium median suboral, sometimes on a short, spiked umbone; one or more lateral oral avicularia may be present, umbonate or not. Ovicell with irregular frontal perforations. Colony with slender, bifurcating branches
.... ***Palmiskenea*** (p. 172)

Avicularium lateral suboral, single, not umbonate. Primary orifice later hidden by sinuate secondary orifice. Ovicell imperforate. Colony with broad, lobate branches
.... ***Ragionula*** (p. 110)

FIG. 15. Orifice size and shape in some ascophorine autozooids.
1, *Cryptosula pallasiana*. 2, *Pentapora fascialis*. 3, *Hippoporina pertusa*. 4, *Watersipora complanata*. 5, *Hemicyclopora polita*. 6, *Lepraliella hippopus*. 7, *Hagiosynodos latus*. 8,

Hippoporidra lusitanica. 9, *Trypostega venusta*. 10, *Palmiskenea skenei*. 11, *Palmiskenea lorea*. 12, *Neolagenipora eximia*. 13, *Porella alba*. 14, *Porella minuta*. 15, *Porella concinna*. 16, *Smittina cheilostoma*. 17, *Smittina landsborovii*. 18, *Smittina crystallina*. 19, *Smittoidea reticulata*. 20, *Smittoidea marmorea*. 21, *Parasmittina trispinosa*. 22, *Escharella labiosa*. 23, *Escharella immersa*. 24, *Escharella variolosa*. 25, *Escharella abyssicola*. 26, *Schizobrachiella sanguinea*. 27, *Schizoporella unicornis*. 28, *Schizoporella hesperia*. 29, *Schizoporella dunkeri*. 30, *Schizomavella cuspidata*. 31, *Schizomavella linearis*. 32, *Schizomavella hastata*. 33, *Schizomavella discoidea*. 34, *Escharina vulgaris*. 35, *Escharina johnstoni*. 36, *Escharina dutertrei haywardi*. 37, *Phaeostachys spinifera*. 38, *Arthropoma cecilii*. 39, *Stomachetosella sinuosa*. 40, *Rhynchozoon bispinosum*. 41, *Schizotheca fissa*. 42, *Stephanollona armata*. 43, *Reteporella couchii*. 44, *Reteporella beaniana*. 45, *Celleporina hassallii*. 46, *Turbicellepora avicularis*. 47, *Turbicellepora magnicostata*. 48, *Buskea dichotoma*. 49, *Buskea quincuncialis*. 50, *Omalosecosa ramulosa*. 51, *Celleporella hyalina* - autozooid. 52, *C. hyalina* - ovicellate female zooid. 53, *C. hyalina* - male zooid. 54, *Chorizopora brongniartii*. 55, *Microporella ciliata*. 56, *Fenestrulina malusii*.

41.(34) Primary orifice large, elongate bell-shaped, with smooth distal rim (Fig. 15.2). Frontal shield uniformly porous. Avicularia small, adventitious, median suboral, sporadic. Ovicell perforate
.... ***Pentapora*** (p. 202)

Primary orifice small, subcircular, with denticulate distal rim. Frontal shield with inconspicuous marginal pores only. Avicularia vicarious. Ovicell imperforate or with a small frontal fissure
.... ***Schizotheca*** (p. 382)

42.(29) Autozooids very lightly calcified, the frontal surfaces entirely membranous, except for a pair of short rounded tubercles proximally; entire polypide visible through the transparent frontal wall. Forming broad lacy colonies on kelps, and occasionally on other large brown algae
.... ***Membranipora*** (Synopsis 10, p. 115)

At least part of the autozooid frontal wall calcified; membrane present or absent but not constituting whole of frontal surface
.... **43**

43.(42) **Frontal surface of autozooid with a distinct frontal membrane**, within which is set the operculum; bordered by gymnocystal calcification and/or erect spines; partly or wholly underlain by cryptocystal calcification, or partly hidden by an overarching basket of fused spines, but always present
.... **44**

Frontal surface of autozooid entirely calcified, except for the orifice, which is closed by the operculum
.... **72**

44.(43) Frontal membrane entirely underlain by a concave, granular cryptocyst, the small opesia exactly the same size as the operculum
.... **45**

Cryptocyst not underlying whole of membrane, the opesia larger than the operculum
.... **46**

45.(44) Autozooids hexagonal to oval, the lateral walls raised adjacent to the orifice to form prominent knobs. Opesiules small. Prominent umbonate ovicells present

.... ***Micropora*** (*Synopsis* **10**, p. 287)

Autozooids elongate rectangular, lateral walls not raised lateral to orifice. Opesiules large. Ovicells absent

.... ***Calpensia*** (*Synopsis* **10**, p. 292)

46.(44) **Avicularia absent**

.... **47**

Adventitious and/or vicarious avicularia present

.... **57**

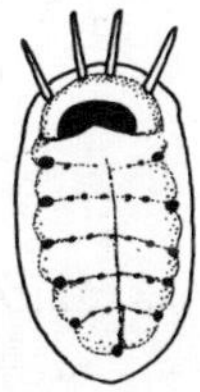

47.(46) Frontal membrane partly hidden by an overarched shield of fused, flattened spines

.... **48**

Frontal membrane with or without a border of spines, but not overarched by them

.... **50**

48.(47) A single stout spine on each side of the autozooid orifice. Ovicells absent

.... ***Aspidelectra*** (*Synopsis* **10**, p. 138)

Four or five slender oral spines present. Prominent ovicells occur

.... **49**

49.(48) Ovicell hemispherical, with a raised transverse ridge, often borne by frontally budded dwarf autozooids

.... ***Cribrilina*** (*Synopsis* **10**, p. 310)

Ovicell globular, with a narrow longitudinal ridge. No dwarf autozooids. A slender seta on each side of the orifice, and a median, proximal lacuna below

.... ***Collarina*** (*Synopsis* **10**, p. 318)

50.(47) At least half length of frontal membrane underlain by cryptocystal calcification; operculum set in the distal half, the opesia. Gymnocyst present or absent

.... **51**

Frontal membrane may be bordered by an underlying rim of cryptocyst, but not substantially underlain by it. Membrane and opesia equivalent to more than half total autozooid length

.... **54**

51.(50) Autozooid with semicircular opesia constituting about half total length, gymnocyst not apparent. Prominent interzooidal nodules diagnostic
.... ***Ammatophora*** (*Synopsis* **10**, p. 200)

Opesia semielliptical or trifoliate. No interzooidal nodules
.... **52**

52.(51) Opesia semielliptical, with straight proximal edge. Gymnocyst developed laterally and proximo-laterally. Ovicells cap-like, inconspicuous. Small basal pore chambers present
.... ***Rosseliana*** (*Synopsis* **10**, p. 294)

Opesia trifoliate, with straight or convex proximal edge. Gymnocyst developed proximally. Ovicell globular, prominent
.... **53**

53.(52) Autozooid with distal oral spines. Ovicell with a thickened, longitudinal median ridge
.... ***Megapora*** (*Synopsis* **10**, p.196)

No oral spines. Ovicell short, cap-like, with a single pore close to its disto-basal edge
.... ***Steraechmella*** (*Synopsis* **10**, p. 296)

54.(50) Autozooids with narrow, but thickly calcified gymnocyst and cryptocyst; with distal oral spines only. Prominent globular ovicells present
.... **55**

Gymnocyst and cryptocyst variably developed, but both lightly calcified; spines, if present, developed around the whole border of the opesia, or just at the proximal end. No ovicells
.... **56**

55.(54) Opesia elliptical, with a narrow cryptocystal rim. Gymnocyst with one or more low, rounded knobs. Ovicell smooth, with a rectangular area of exposed entooecium frontally, bordered by a low ridge
.... ***Crassimarginatella*** (*Synopsis* **10**, p. 170)

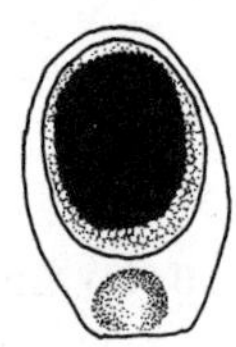

Opesia broadest proximally, pyriform or rounded triangular; cryptocyst especially broad proximally. Gymnocyst smooth, without knobs. Ovicell granular, with exposed entooecium occupying most of frontal surface
.... ***Alderina*** (*Synopsis* **10**, p. 168)

56.(54) Gymnocyst minimally developed. Spines absent; **or** two to 12 thin, incurved spinules evenly distributed around the opesia. Small, round or triangular kenozooids may be interspersed among the autozooids. On hard substrata in coastal or estuarine habitats, or in brackish water .... ***Conopeum*** (*Synopsis* **10**, p. 119)

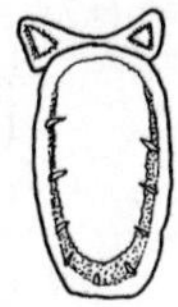

Gymnocyst comprising one-sixth or more of autozooid length. One median spine at proximal end of opesia, or several spines, of which the medio-proximal is the longest and/or stoutest. Often epiphytic; in marine or brackish water .... ***Electra*** (*Synopsis* 10, p. 127)

57.(46) Frontal membrane overarched by a basket-like shield of fused, flattened spines, with lines of pores between; the shield comprising the whole of the frontal surface, or bordered proximally by smooth gymnocyst **58**

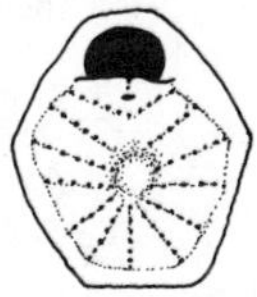

Spines present or absent, but not fused to form a shield **61**

58.(57) Avicularia small, adventitious **59**

Avicularia larger, vicarious **60**

59.(58) Spines forming frontal shield only lightly fused, with long, narrow fissures between. Avicularia on lateral or proximal borders of the autozooid .... ***Membraniporella*** (*Synopsis* **10**, p. 202)

Spines forming frontal shield closely fused at numerous points, with round foramina between. Avicularia latero-proximal to orifice .... ***Gephyrotes*** (*Synopsis* **10**, p. 316)

60.(58) Cribrate shield constituting larger part of frontal surface. Avicularia with pointed, triangular mandibles. Ovicells with longitudinal ridge, or an umbo; imperforate .... ***Puellina*** (*Synopsis* **10**, p. 320)

Cribrate shield bordered proximally by a smooth gymnocyst. Avicularia with broad, spatulate mandibles. Ovicells with two large frontal fenestrae .... ***Figularia*** (*Synopsis* **10**, p. 338)

61.(57) Avicularia small, interzooidal, with long, setiform mandibles. Cryptocyst extending three-quarters length of frontal membrane, with a pair of slit-like opesiules
.... ***Setosella*** (*Synopsis* **10**, p. 298)

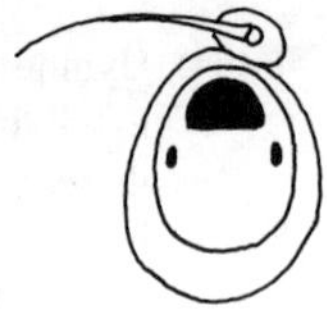

Avicularia variable, but not with setiform mandibles
.... **62**

62.(61) Avicularia pedunculate, club-shaped, with terminal mandibles; sparse. Cryptocyst narrow; opesia bordered by numerous thin, pointed spines. Ovicell cap-like, inconspicuous. Intertidal, on hard substrata
.... ***Cauloramphus*** (*Synopsis* **10**, p. 172)

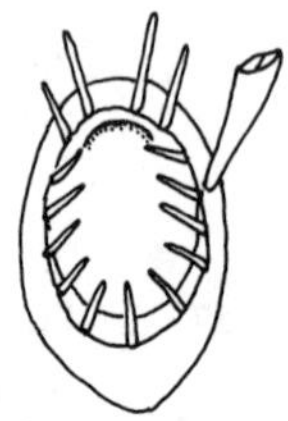

Avicularia sessile
.... **63**

63.(62) **Avicularia adventitious**, on the autozooid gymnocyst; often associated with ovicells
.... **64**

Avicularia interzooidal or vicarious
.... **67**

64.(63) Cryptocyst well developed, especially broad proximally, defining a trifoliate or pyriform opesia
.... **65**

Cryptocyst a narrow granular band bordering an oval opesia
.... **66**

65.(64) Ovicell with an arched, medially peaked frontal ridge edging an exposed area of entooecium. Avicularia developed along the edges of the autozooids
.... ***Amphiblestrum*** (*Synopsis* **10**, p. 185)

Ovicell uniformly granular. Avicularia transversely orientated on the gymnocyst of the autozooid, capping the ovicell when present
.... ***Ramphonotus*** (*Synopsis* **10**, p. 194)

66.(64) Basal pore chambers present, especially distinct at growing edges of colony. One to four spines at the distal end of the autozooid, or up to 20 evenly spaced around the opesia. Several common species
.... ***Callopora*** (*Synopsis* **10**, p. 156)

No pore chambers; vertical walls of autozooids with large, multiporous septula. Two to four spines distally. One species occurs in northern Britain
.... ***Tegella*** (*Synopsis* **10**, p. 174)

67.(63) Avicularium with semi-elliptical mandible
.... **68**

Avicularium with curved or triangular, pointed mandible
.... **70**

68.(67) Avicularia as large as autozooids, with knob-like hinge condyles. Autozooids without spines, each with one or two rounded umbones on the gymnocyst
.... ***Crassimarginatella*** (*Synopsis* **10**, p. 170)

Avicularia much smaller than autozooids, frequent. Autozooids bearing spines
.... **69**

69.(68) Frontal membrane without an underlying cryptocyst, bordered by short, stout spines. Ovicell immersed
.... ***Hincksina*** (*Synopsis* **10**, p. 152)

Frontal membrane underlain by a broad cryptocystal rim. Cavity of cleaned autozooid crossed distally by calcified struts (occlusor laminae). Two pairs of oral spines, the proximal pair bifurcate. Ovicell prominent, spherical
.... ***Larnacicus*** (*Synopsis* **10**, p. 204)

70.(67) Avicularium with strongly curved rostrum, and a distal kenozooid. Autozooids mostly without spines
.... ***Parellisina*** (*Synopsis* **10**, p. 182)

Avicularium with straight rostrum
.... **71**

71.(70) Avicularium situated distal to the autozooid, with short triangular rostrum directed obliquely distally
.... ***Ellisina*** (*Synopsis* **10**, p. 192)

Avicularium lateral to autozooid, with sharply pointed, abruptly tapered rostrum directed distally
.... ***Copidozoum*** (*Synopsis* **10**, p. 177)

72.(43) **Frontal shield of autozooid evenly perforated**; the pores densely or widely spaced, but not confined to the autozooid margins
.... **73**

Frontal shield of autozooid with marginal pores only; occasionally indistinct, **or apparently quite imperforate**
.... **93**

73.(72) **Avicularia absent**
.... **74**

Avicularia present
.... **86**

74.(73) Primary orifice hidden by a tall, smooth, cylindrical peristome, projecting well above the surface of the autozooid; a slender, tubular spiramen situated at the base of the peristome. Ovicells small, spherical, imperforate. Colonies generally small, rounded, unilaminar
....***Cylindroporella*** (p. 248)

Peristome, if present, not as described. No spiramen
.... **75**

75.(74) Frontal shield a concave cryptocyst overlain by a membranous outer frontal wall (glistening in dry material), with paired opesiules proximo-lateral to orifice
.... **76**

Frontal shield not a cryptocyst, convex, without an outer frontal membrane
.... **77**

76.(75) Autozooids hexagonal to oval, the lateral walls raised adjacent to the orifice to form prominent knobs. Opesiules small. Prominent umbonate ovicells present
.... ***Micropora*** (*Synopsis* **10**, page 287)

Autozooids elongate rectangular, lateral walls not raised lateral to orifice. Opesiules large. Ovicells absent
.... ***Calpensia*** (*Synopsis* **10**, page 292)

77.(75) Primary orifice more or less D-shaped, with straight proximal edge; a small ascopore immediately proximal to orifice
.... ***Haplopoma*** (p. 304)

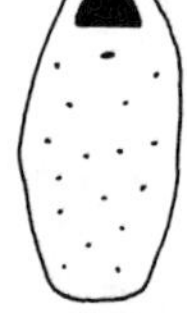

Primary orifice not D-shaped, no ascopore
.... **78**

78.(77) Primary orifice with a distinct lyrula on proximal edge (Fig. 15, 14-25) **79**

Primary orifice without a lyrula **80**

79.(78) Primary orifice with a broad, rectangular lyrula, enveloped by a low, thickened peristome with a deep medio-proximal notch .... ***Smittina*** (p. 250)

Primary orifice with a narrow, pointed lyrula, enveloped by a broad, deep, flared peristome, especially prominent proximally .... ***Phylactella*** (p. 204)

80.(78) Primary orifice with a sinus, narrow and slit-like, or broad and shallow (Fig. 15, 26-42) **81**

Primary orifice bell-shaped or cleithridiate (Fig. 15, 1-9) **82**

81.(80) Primary orifice D-shaped, with a narrow, slit-like, medio-proximal sinus. Frontal shield thinly calcified and densely perforate. Ovicell prominent, smooth, white, imperforate .... ***Arthropoma*** (p. 224)

Primary orifice with a U-shaped sinus, broad and shallow, or narrower, almost parallel-sided. Frontal shield thick, sparsely perforated. Ovicell rugose, with a small median pore .... ***Stomachetosella*** (p. 240)

82.(80) Primary orifice with prominent, thin, lateral condyles; appearing cleithridiate (Fig. 15, 9). Dwarf zooids with small round orifices scattered among the autozooids .... ***Trypostega*** (p. 98)

Primary orifice broadly bell-shaped **83**

83.(82) Primary orifice with a broad border of smooth, thickened calcification, occasionally developing rounded knobs proximo-laterally. Frontal shield flat, densely perforate. No ovicells .... ***Watersipora*** (p. 192)

Primary orifice without such a border, although proximo-lateral knobs may occur. Frontal shield convex, often nodular, or with a median-suboral umbo. Large hyperstomial ovicells present **84**

84.(83) Primary orifice with conspicuous lateral condyles. Ovicell thick, umbonate or ridged, with few, scattered peripheral pores
.... ***Hagiosynodos*** (p. 362)

Primary orifice with conspicuous proximo-lateral condyles. Ovicell globular, smooth, evenly perforated
.... **85**

85.(84) Primary orifice as wide as long, bordered by a flaring peristome. Ovicell short, equivalent to length of orifice
.... ***Codonellina*** (p. 198)

Primary orifice wider than long, peristome scarcely developed. Ovicell large, at least twice orifice length
.... ***Hippoporina*** (p. 196)

86.(73) Primary orifice enveloped by a low, thickened peristome, with a median distal avicularium and a median proximal avicularium on its rim
.... ***Anarthropora*** (p. 190)

Peristome and avicularia not as described
.... **87**

87.(86) Primary orifice with a distinct proximal lyrula (Fig. 15, 16-18), and a single median suboral avicularium
.... ***Smittina*** (p. 250)

Primary orifice sinuate or bell-shaped
.... **88**

88.(87) Primary orifice bell-shaped
.... **89**

Primary orifice sinuate (Fig. 15, 26-42)
.... **90**

89.(88) Primary orifice with convex proximal border, with a low, thickened peristome. Large hyperstomial ovicells present. Sublittoral, on rocky grounds, developing an erect, honeycomb-shape colony
.... ***Pentapora*** (p. 200)

Primary orifice with concave proximal border, with a deep, flared peristome. No ovicells. Intertidal and shallow sublittoral, developing extensive flat sheets
.... ***Cryptosula*** (p. 194)

90.(88) Primary orifice with V-shaped sinus and bluntly pointed condyles. Avicularia tiny, developed along the distal borders of the autozooids (sutural)

.... ***Schizobrachiella*** (p. 222)

Sinus shallowly concave, or a narrower U-shape, or a parallel-sided slit; condyles rounded, pointed or angular, not especially prominent. Avicularia larger, on swollen cystids, not sutural

.... **91**

91.(90) Primary orifice bordered by thick, basally jointed spines. Avicularium proximo-lateral in position, distant from orifice, directed obliquely proximally. Often with a short frontal umbo bearing a stout, chitinous spine. Ovicell imperforate

.... ***Phaeostachys*** (p. 238)

Oral spines few or absent. Avicularia at distal end of autozooid, associated with orifice. Ovicell with frontal or marginal perforations

.... **92**

92.(91) Avicularium median suboral in position. Ovicell with frontal perforations

.... ***Schizomavella*** (p. 272)

Avicularium lateral to orifice, single or paired. Ovicell with marginal pores, or pronounced, radiating fluting

.... ***Schizoporella*** (p. 206)

93.(72) **Avicularia absent**

.... **94**

Avicularia present

.... **99**

94.(93) Each autozooid with a median ascopore proximal to the orifice

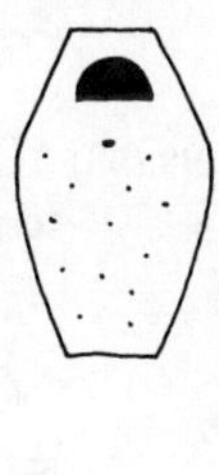

.... **95**

No ascopore

.... **96**

95.(94) Ascopore small, circular; ovicell with scattered pores
.... ***Haplopoma*** (p. 304)

Ascopore large, crescentic, with denticulate rim; ovicell with marginal pores and fluting, imperforate centrally
.... ***Fenestrulina*** (p. 298)

96.(94) Primary orifice with a well developed lyrula (Fig. 15, 22-25), and bordered by four to eight spines. Ovicell imperforate
.... ***Escharella*** (p. 120)

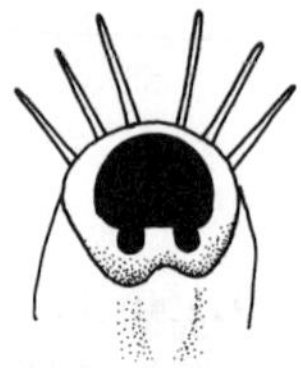

Primary orifice without a lyrula; occasionally with an indistinct median convexity. Ovicell perforate or imperforate
.... **97**

97.(96) Colonies lightly calcified, in small round patches. Autozooids contiguous, but with small gaps between. Ovicells borne on short, tapered female zooids, with broad, dimorphic orifices
.... ***Celleporella*** (p. 92)

Colonies thickly calcified, no gaps between autozooids. Ovicells borne on normal autozooids
.... **98**

98.(97) Primary orifice bordered by a deep, flared peristome; spines few, small and inconstant. Ovicell with scattered pores
.... ***Neolagenipora*** (p. 147)

Primary orifice with minimal development of peristome, but with well developed spines; **or** with short, tubular peristome and no spines. Ovicell always smooth, thick, imperforate
.... ***Hemicyclopora*** (p. 140)

99.(93) Each autozooid with a conspicuous ascopore proximal to the primary orifice
.... **100**

No ascopore
.... **101**

100.(99) Colony encrusting; autozooids with basal pore chambers, visible at growing edges. Ovicell imperforate. Avicularium with a slender, setiform mandible

.... ***Microporella*** (p. 292)

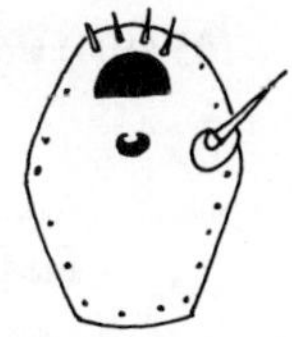

Colony erect, branching; autozooids with large multiporous septula in vertical walls. Ovicell perforate. Avicularium with short triangular mandible

.... ***Diporula*** (p. 302)

101.(99) **Avicularia small, adventitious, monomorphic**; close to the orifice or along the margins of the autozooid

.... **102**

Avicularia variable: adventitious and/or interzooidal or vicarious; often polymorphic, sometimes very frequent

.... **115**

102.(101) Autozooid with D-shaped orifice; proximal to it a single, distally-directed, triangular avicularium, above a single or twinned frontal spiramen. Enlarged brooding gonozooids present

.... ***Reptadeonella*** (p. 184)

No frontal spiramen. No gonozooids. Medio-proximal avicularium, if present, not distally directed

.... **103**

103.(102) Orifice with a distinct lyrula (Fig. 15, 13-21)

.... **104**

Orifice without a lyrula

.... **106**

104.(103) Avicularium lateral-suboral in position, the rostrum directed medio-proximally onto the autozooid peristome. Ovicell with scattered pores

.... ***Parasmittina*** (p. 270)

Avicularium median-suboral in position, the rostrum directed proximally

.... **105**

105.(104) Colony encrusting. Ovicells with conspicuous frontal pores
.... ***Smittoidea*** (p. 262)

Colony encrusting or erect. Ovicells imperforate, or with a single median pore
.... ***Porella*** (p. 154)

106.(103) Orifice large, shielded proximally by a prominent lip, frequently with small teeth on its inner face. Avicularia paired, lateral to orifice
.... **107**

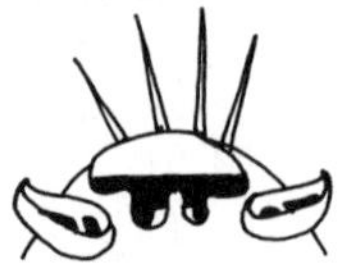

Orifice bell-shaped or sinuate, with or without a variably developed peristome, but not developed as a prominent proximal lip
.... **108**

107.(106) Avicularia with hooked, triangular rostra, directed laterally or disto-laterally. A conspicuous oral shelf in the distal half of the orifice, below the spines
.... ***Escharoides*** (p. 112)

Avicularia with short, rounded rostra, distally directed. No oral shelf
.... ***Arctonula*** (p. 108)

108.(106) Primary orifice D-shaped. Autozooids contiguous, linked by short tubular processes with small spaces between; each with a small avicularium immediately distal to the orifice. Forms glistening white sheets
.... ***Chorizopora*** (p. 100)

Primary orifice bell-shaped, orbicular or sinuate
.... **109**

109.(108) Primary orifice bell-shaped, with or without associated avicularia
.... **110**

Primary orifice orbicular or sinuate
.... **113**

110.(109) Primary orifice becomes partly obscured by a deep peristome with a medio-proximal notch (pseudosinus); avicularium lateral suboral in position, single. Develops erect, lobate branches
.... ***Ragionula*** (p. 110)

Primary orifice not obscured; **or** enveloped proximally and laterally by a flared, umbonate peristome without a pseudosinus
.... **111**

111.(110) Primary orifice large, with pronounced lateral condyles; with a thin, flared peristome, not hiding the orifice. Avicularium, when present, single, median suboral
.... ***Umbonula*** (p. 103)

Avicularium, if single, median suboral, mounted on an umbonate process incorporated in a deep, proximal peristome
.... **112**

112.(111) Avicularia lateral to orifice, but not upon a peristome, or along autozooid margins. Ovicell nodular, imperforate. Colony encrusting
.... ***Lepraliella*** (p. 152)

Avicularia median-suboral, with one or more also lateral to the orifice, all on the rim of a flared, often umbonate peristome. Develops an erect branching colony
.... ***Palmiskenea*** (p. 172)

113.(109) Primary orifice with a well marked sinus; peristome thin and flared, or absent. Avicularia single or paired, distal to orifice; or proximally, close to the margins of the autozooid. Ovicell imperforate. Colony a flat sheet
.... ***Escharina*** (p. 226)

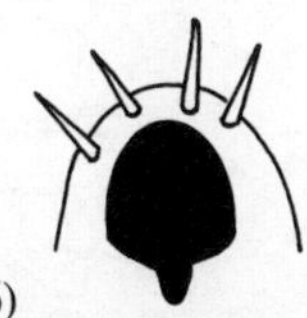

Primary orifice orbicular or sinuate, bordered by a deep peristome often bearing cylindrical avicularia. Ovicell imperforate. Colony nodular or pisiform
.... **114**

114.(113) Orifice sinuate. Peristome bearing two (rarely one) tall, cylindrical avicularia with small rounded mandibles. Enlarged vicarious avicularia may also occur. Ovicell with a perforated frontal tabula
.... ***Celleporina*** (p. 322)

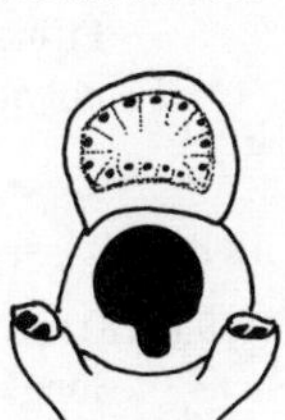

Orifice orbicular. Peristome with one tiny avicularium (may be lacking). No vicarious avicularia. Ovicell with an irregular frontal foramen
.... ***Lagenipora*** (p. 332)

115.(101) Forms nodular colonies on gastropod shells inhabited by hermit crabs. Autozooids dimorphic, with cleithridiate primary orifices of different sizes. Adventitious and vicarious avicularia present
.... ***Hippoporidra*** (p. 358)

Colony and autozooids not as described
.... **116**

116.(115) Colony a thin, glistening white sheet. Autozooids linked by short tubes, separated by small spaces, each with a small interzooidal avicularium distal to the D-shaped orifice
.... ***Chorizopora*** (p. 100)

Colony and autozooids not as described
.... **117**

117.(116) Distal rim of primary orifice finely denticulate (Fig. 15, 41)
.... **118**

Distal rim of primary orifice smooth
.... **120**

118.(117) Primary orifice orbicular. Ovicell imperforate with a frontal fissure. Autozooids in radiating lines, with groups of interzooidal avicularia between
.... ***Schizotheca*** (p. 382)

Primary orifice sinuate. Ovicell not fissured. Avicularia adventitious, polymorphic
.... **119**

119.(118) Orifice with a low peristomial rim, not obscuring it. Oral spines long, basally jointed, spatulate. Avicularia lateral oral, single or paired; small, with semi-elliptical mandible, or enlarged, with slender, spatulate mandible
.... ***Stephanollona*** (p. 386)

Orifice with a deep, notched or umbonate peristome, incorporating a lateral suboral avicularium, and with a stout tooth (uncinate process) on its inner surface. Frontal adventitious avicularia sporadic, with narrow, triangular mandible
.... ***Rhynchozoon*** (p. 378)

120.(117) Primary orifice transversely oval, bordered by a deep peristome incorporating a single, medio-proximal avicularium, often with a spiked umbonate process. Ovicell smooth, with a few irregular perforations close to its aperture. No enlarged avicularia

.... ***Cellepora*** (p. 318)

Primary orifice distinctly sinuate. Avicularia usually polymorphic. Ovicell with radial fluting, a perforated frontal tabula, or numerous pores

.... **121**

121.(120) No peristome. Avicularia lateral oral, sessile, with subtriangular mandible; or overlapping lateral margins of autozooids, with swollen cystids supporting slender, tapered rostra. Ovicell with radial fluting

.... ***Buffonellaria*** (p. 356)

Peristome well developed, incorporating columnar lateral or suboral avicularia. Ovicell perforate or with a frontal tabula

.... **122**

122.(121) Ovicell with a perforated frontal tabula. Oral avicularia paired, lateral to orifice (rarely single)

.... ***Celleporina*** (p. 322)

Ovicell with numerous scattered pores. Avicularium single, lateral suboral

.... ***Turbicellepora*** (p. 334)

Systematic Part

Superfamily HIPPOTHOOIDEA Busk, 1859

Family HIPPOTHOIDAE Busk, 1859

Colony encrusting or erect. Frontal wall gymnocystal, thin, imperforate. Primary orifice schizoporelloid, without spines. Vicarious avicularia present or absent. Pore-chambers small, within the wall, or tubular. Ovicells prominent, developed in relation to special female zooids; the combined orifice closed by the zooidal operculum.

Genus HIPPOTHOA Lamouroux, 1821

Colony adnate, with branching, uniserial chains of autozooids. Polymorphs include feeding autozooids, female zooids bearing ovicells, and minute zooeciules of unknown function. Autozooids club-shaped, comprising a distal dilatation and a thread-like proximal portion (known as the cauda). Frontal wall smooth, non-porous. Orifice with a sinus and paired condyles; that of female zooids not greatly different from that of autozooids. Pore-chambers accommodated within the lateral walls, not tubular. Polypide without gizzard.

Type species: *Hippothoa divaricata* Lamouroux, 1821

Key to the British species of *Hippothoa*

1. Branching from autozooids in the form of an ‘x’ (up to two buds each side, spreading diagonally). Female zooids caudate, the dilated portion as long as in an autozooid. Sinus of orifice small, V-shaped
.... ***Hippothoa divaricata*** (p. 86)

Branching from autozooids in the form of a ‘+’ (not more than one bud each side). Female zooids almost ecaudate, the entire autozooid short and inversely triangular. Sinus of orifice wide, almost semicircular
.... ***Hippothoa flagellum*** (p. 88)

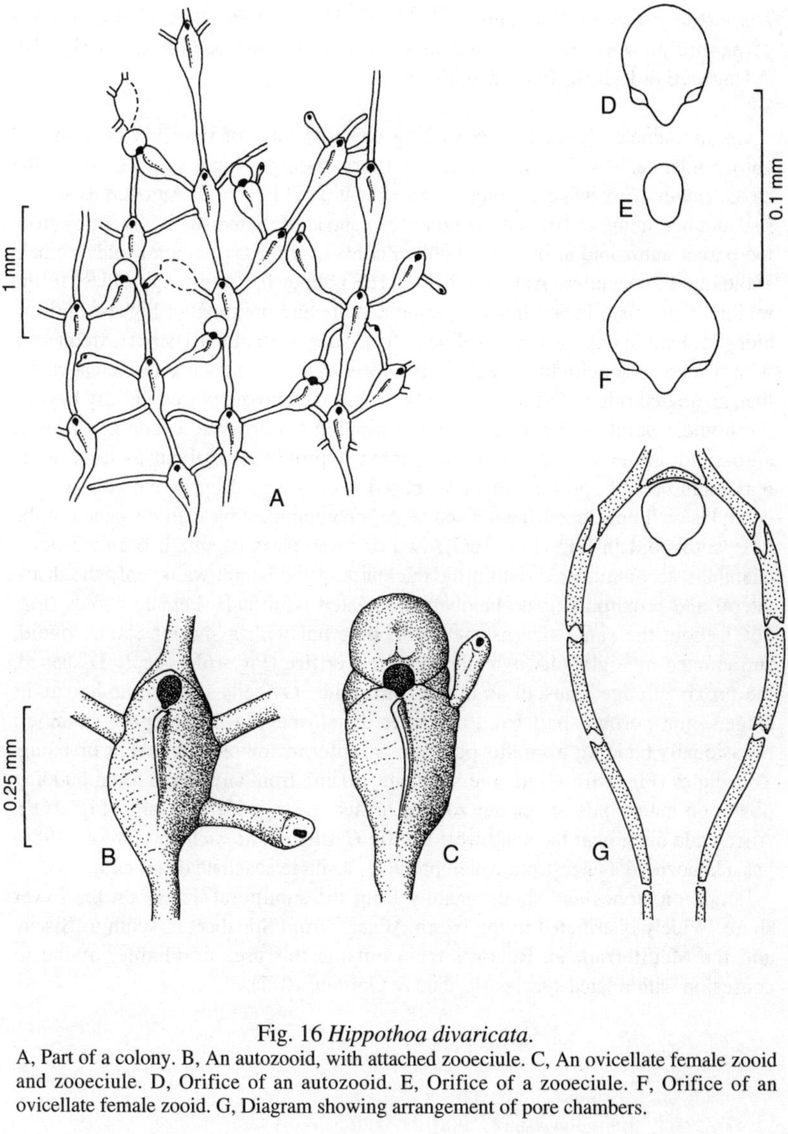

Fig. 16 *Hippothoa divaricata.*
A, Part of a colony. B, An autozooid, with attached zooeciule. C, An ovicellate female zooid and zooeciule. D, Orifice of an autozooid. E, Orifice of a zooeciule. F, Orifice of an ovicellate female zooid. G, Diagram showing arrangement of pore chambers.

Hippothoa divaricata Lamouroux
(Figs 16; 17A, B)

Hippothoa divaricata Lamouroux 1821: 82; Hincks 1880: 288, pl. 44, figs 1-3 (excluding vars. *conferta* and *patagonica*); Ryland & Gordon, 1977: 19; Hayward & Ryland, 1979: 246, fig. 106

Colony adnate, delicate, of branching uniserial lines of slenderly club-shaped autozooids, each with a distal dilatation terminating a long, slender cauda; the cauda one-half to twice as long as the dilatation (Fig. 16A). Autozooids giving rise to a maximum of four lateral buds, the caudae of these usually diverging from the parent autozooid at oblique angles. Zooids of three types: autozooids, female zooids and zooeciules. Autozooid (Fig. 16B) about 0.3 mm long, widest in the region of the disto-lateral buds, tapering distally and proximally; highest midway along its length; calcification hyaline at first, later white, faintly striate, frequently with a median longitudinal keel (in the form of an embossed linear crest rather than an angled ridge). Orifice (Fig. 16D) longer than broad, placed distally beyond the highest point of the autozooid and dipping towards the cauda of the next autozooid. Anter rounded distally, narrower proximally, about as long as its maximum breadth; poster (sinus) V-shaped, narrow and deep, proximally obtuse; condyles well-developed. Two distal septula communicating with the cauda of the next autozooid in line (Fig. 16G); two or three pairs of small, rounded pore-chambers accommodated within the thickness of the lateral walls, only the disto-lateral and proximo-lateral chambers associated with buds. Female zooids (Fig. 16C) about the same size as the autozooids but with a shorter cauda, ovoid, surmounted by a globular ovicell. Combined orifice (Fig. 16F) deeply D-shaped, the proximal edge sinusoid as in the autozooids. Ovicells smooth, umbonate or ridged, non-porous apart from a minute distal opening from a pore-chamber: occasionally budding from this position. No information on embryos or breeding. Zooeciules (Fig. 16B, C) minute, elongate, arising from any of the usual budding points on autozooids or female zooids; orifice ovate, widest distally (Fig. 16E). Ancestrula unknown; the southwest Pacific *H. divaricata pacifica* Gordon (1984) has a kenozooidal ancestrula which produces a single ecaudate distal bud.

Found on stones and shells, mainly from the sublittoral, rarely on the lower shore. Widely distributed in the North Atlantic from Spitzbergen south to Biscay and the Mediterranean. Records from outside this area unreliable, owing to confusion with related species (Ryland & Gordon, 1977).

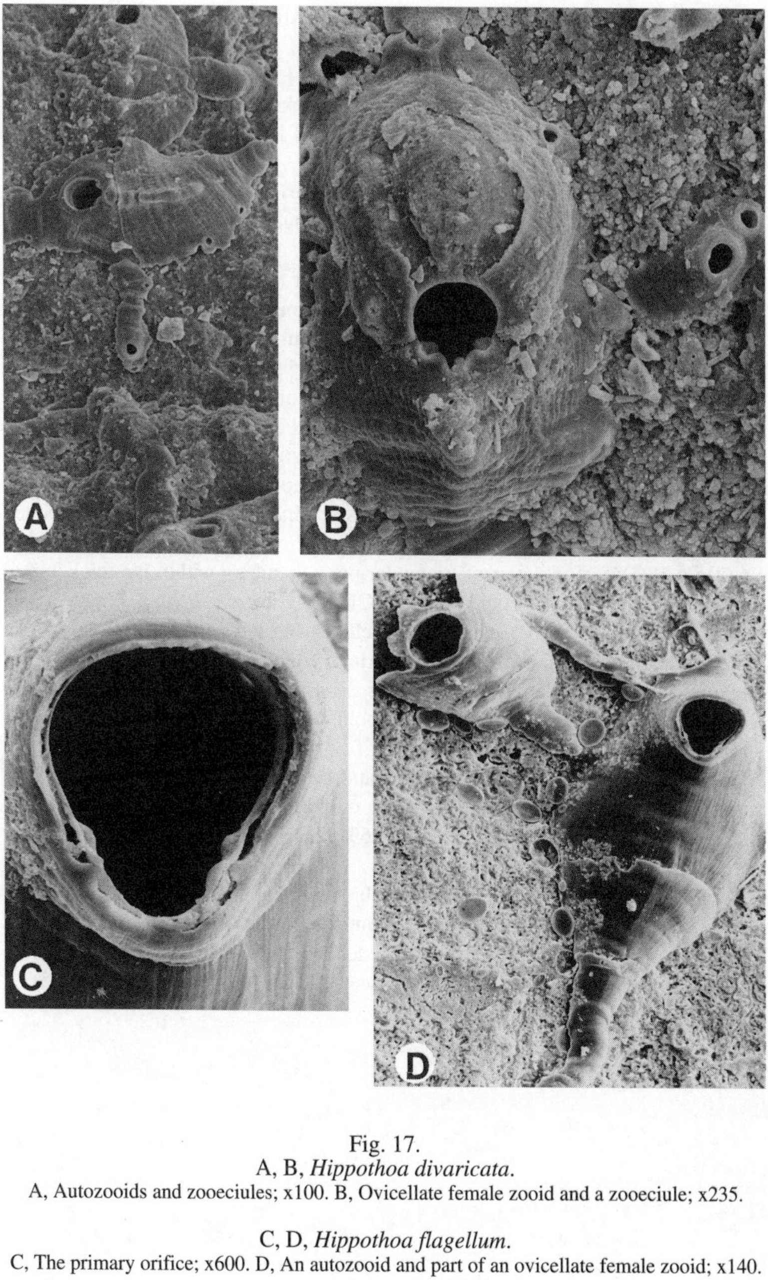

Fig. 17.
A, B, *Hippothoa divaricata*.
A, Autozooids and zooeciules; x100. B, Ovicellate female zooid and a zooeciule; x235.

C, D, *Hippothoa flagellum*.
C, The primary orifice; x600. D, An autozooid and part of an ovicellate female zooid; x140.

Hippothoa flagellum Manzoni
(Figs 17C, D; 18)

Hippothoa flagellum Manzoni 1870b: 328; Hincks 1880: 293, pl. 44, figs 5-6; Ryland & Gordon 1977: 22; Hayward & Ryland 1979: 248, fig. 107

Colony adnate, delicate, of branching uniserial lines of autozooids, each consisting of a distal dilatation terminating a long, thread-like cauda; the cauda one to four times as long as the dilatation (Fig. 18A, B). Branching basically cruciform, each autozooid giving rise to a pair of midlateral buds, the caudae of which diverge from the parent autozooid approximately at right angles. Zooids of three types: autozooids, female zooids and zooeciules. Dilatation of the autozooid (Fig. 18C) about 0.3 mm long, widest in the region of the midlateral branches, tapering distally and proximally; highest at the orifice; calcification hyaline at first, later white, neither keeled frontally nor markedly striate transversely. Orifice (Fig. 18D) longer than broad, rimmed, level with the substratum or dipping distally. Anter subcircular; poster (sinus) U-shaped, wide and deep; condyles small (and scarcely evident at magnifications obtainable with a stereomicroscope). A single distal septulum communicating with the cauda of the next autozooid in line (Fig. 18G); a single pair of triangular pore-chambers within the thickness of the lateral walls. Female zooids (Fig. 18C) with short cauda and wedge-shaped dilatation, surmounted by a globular ovicell; arising laterally from an autozooid. Combined orifice D-shaped, with the proximal edge concave (Fig. 18F). Ovicells smooth, non-porous; sometimes budding distally; embryos yellow. Zooeciules (Fig. 18C) arising laterally from an autozooid, minute, elongate, distally upturned so that the orifice opens horizontally; orifice almost perfectly oval (Fig. 18E). Ancestrula kenozooidal, small, circular and devoid of an orifice (Fig. 18H); a single mid-distal bud is produced beyond a transverse partition.

Around the Isle of Man, Eggleston (1969) recorded embryos most abundantly during the autumn.

Found particularly on shells and other biogenic carbonates, sublittorally, from 10-275 m (at least). Distributed throughout the North Atlantic from Norway to West Africa, and widely reported from other regions. This species appears to be practically cosmopolitan in shallow shelf seas (Moyano, 1986; Hayward, 1995).

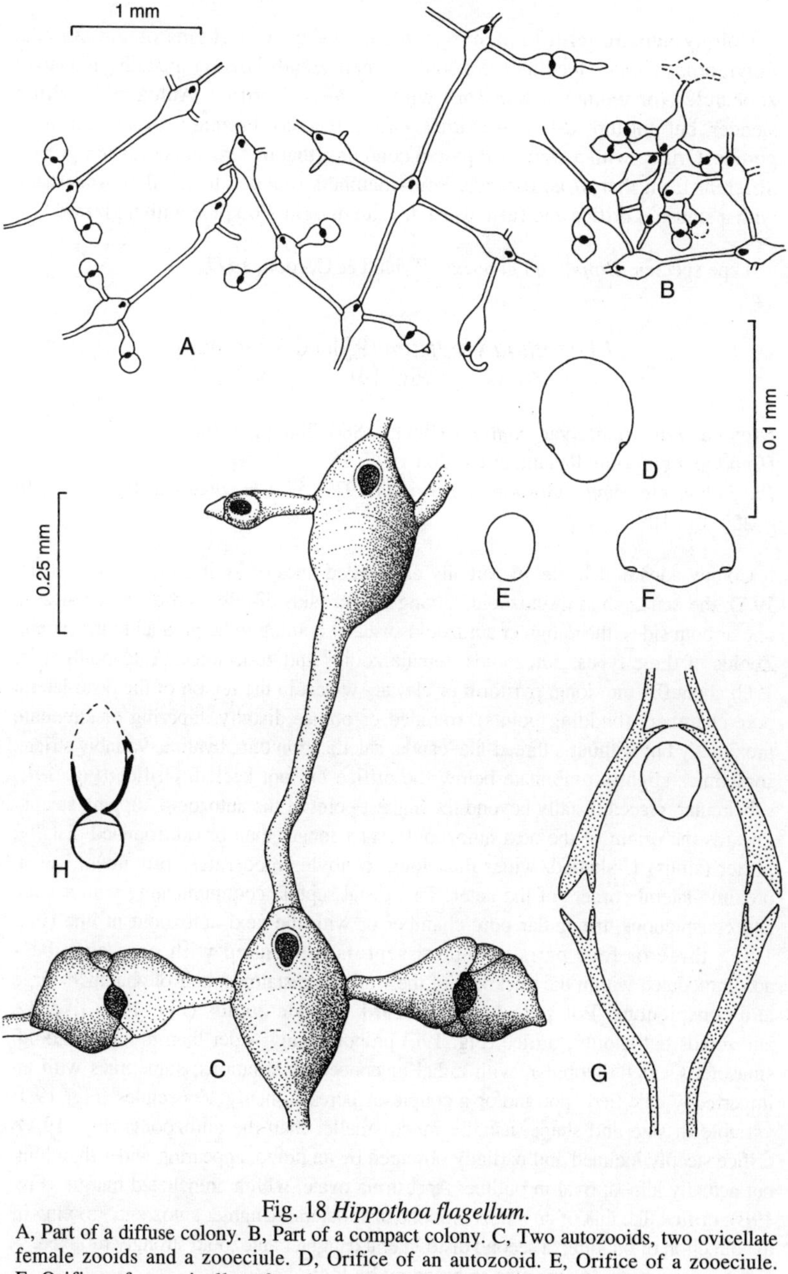

Fig. 18 *Hippothoa flagellum.*
A, Part of a diffuse colony. B, Part of a compact colony. C, Two autozooids, two ovicellate female zooids and a zooeciule. D, Orifice of an autozooid. E, Orifice of a zooeciule. F, Orifice of an ovicellate female zooid. G, Diagram showing arrangement of pore chambers. H, Ancestrula and part of first zooid.

Genus PLESIOTHOA Gordon & Hastings, 1979

Colony adnate, with branching, uni- or multiserial chains of autozooids. Polymorphs include feeding autozooids, female zooids bearing ovicells, and small zooeciules (of unknown function) with a reversed orifice. Autozooids with a slender, but not thread-like, proximal portion (cauda). Frontal wall smooth, non-porous. Orifice with a sinus and paired condyles; that of female zooids not greatly different from that of autozooids. Pore-chambers rounded to tubular. Ancestrula with a sinusoid orifice and (usually) a dentate margin. Polypide with a gizzard.

Type species: *Hippothoa gigerium* Ryland & Gordon, 1977

Plesiothoa gigerium (Ryland & Gordon)
(Fig. 19)

Hippothoa divaricata var. *conferta* Hincks 1880: 288, pl. 1, fig. 2, pl. 44, fig. 4
Hippothoa gigerium Ryland & Gordon 1977: 28
Plesiothoa gigerium : Gordon & Hastings 1979: 577; Hayward & Ryland 1979: 250, fig. 108

Colony adnate, delicate, of partially aggregated lines of pyriform autozooids (Fig. 19A); the cauda short. Autozooids giving rise to buds distally and disto-laterally on one or both sides, the daughter autozooids usually coming to lie parallel to the parent. Zooids of three types: autozooids, female zooids and zooeciules. Autozooids (Fig. 19C) about 0.4 mm long, pyriform or clavate, widest in the region of the disto-lateral pore-chambers (budding points), rounded or obtuse distally, tapering or attenuate proximally but without a thread-like cauda; calcification thin, hyaline, variably striate, sometimes slightly umbonate below the orifice but not keeled. Orifice (Fig. 19E) subcircular, placed distally beyond the highest point of the autozooid, dipping steeply towards the origin of the next autozooid. Anter longer than broad, rounded distally; poster (sinus) U-shaped, wider than long; condyles moderately prominent, set in proximo-lateral corners of the anter. Two distal septula communicating with a wide and conspicuous, triangular pore-chamber or with the next autozooid in line (Fig. 19G); three or four pairs of lateral septula associated with pore-chambers accommodated within the thickness of the wall, the distalmost pair of chambers large and conspicuous. Polypide with a gizzard. Female zooids (Fig. 19C) like the autozooids but shorter; orifice (Fig. 19F) proportionally wider than in an autozooid, sinusoid. Ovicells globular, with radial or concentric striations; sometimes with an imperfectly calcified spot and/or a couple of pores frontally. Zooeciules (Fig. 19D) variable in size and shape, usually much smaller than the autozooids (Fig. 19A); orifice steeply inclined and partially obscured by an umbo, appearing wider than long but actually almost oval in outline. Ancestrula ovate, with a crenellated margin (Fig. 19B); orifice like that of an autozooid; one disto-lateral daughter autozooid growing in the direction of budding, a second disto-lateral daughter autozooid strongly reflexed.

No information on embryos or breeding biology.

Found on red algae such as *Palmaria palmata* epiphytic on *Laminaria hyperborea* stipes, associated with some rather similar species (e.g., *Celleporella*

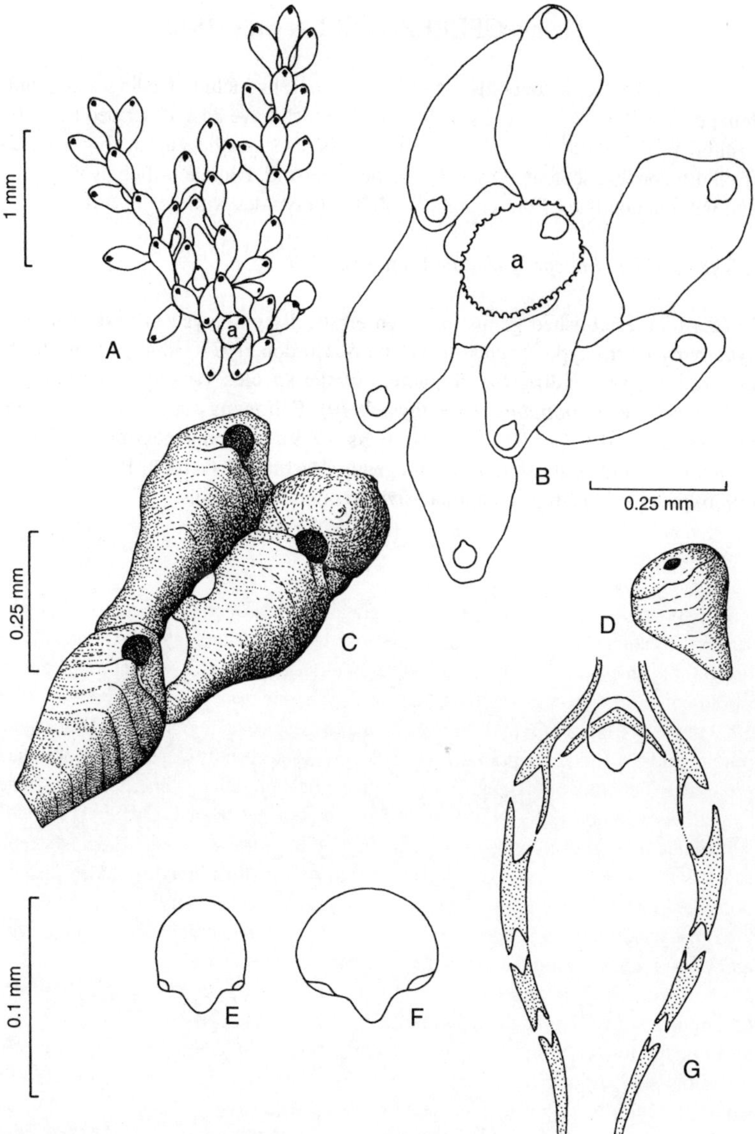

Fig. 19 *Plesiothoa gigerium.*
A, Part of a colony, ancestrula marked "a". B, Ancestrula and first zooids from same colony. C, Two autozooids and an ovicellate female zooid. D, Zooeciule. E, Orifice of an autozooid. F, Orifice of an ovicellate female zooid. G, Diagram showing arrangement of pore chambers.

hyalina, Haplopoma impressum, H. bimucronatum; also *Electra pilosa*). In the shallow sub-littoral, south-west England and the Channel Islands; extending to South Africa. The genus is otherwise known from Chile, New Zealand (Moyano & Gordon, 1980), the South Shetland Islands and Antarctica (Hayward, 1995).

Genus CELLEPORELLA Gray, 1848

Colony crustose, autozooids coherent. Polymorphs include feeding autozooids, female ovicellate zooids, and smaller zooids which have been described as males. Frontal wall smooth, non-porous. Autozooids sausage-shaped, lacking a cauda, frontally convex, separated by sulci; orifice sinusoid. Female orifice usually wide and without condyles. Pore-chambers tubular. Polypides without gizzard.

Type species: *Cellepora hyalina* Linnaeus, 1767

This long-established genus has been erratically employed in past years (see synonymy of the type species in Ryland & Gordon, 1977: 38-40) owing to the uncertainty surrounding the characters of the species variously attributed to *Celleporella* and *Hippothoa* (Hastings, 1979). Following substantial systematic revision (Ryland & Gordon, 1977; Hastings, 1979), it became possible to redefine *Celleporella* and establish it as a valid genus (Hastings & Gordon, 1979). Moyano (1986) provided the most comprehensive account of the genus.

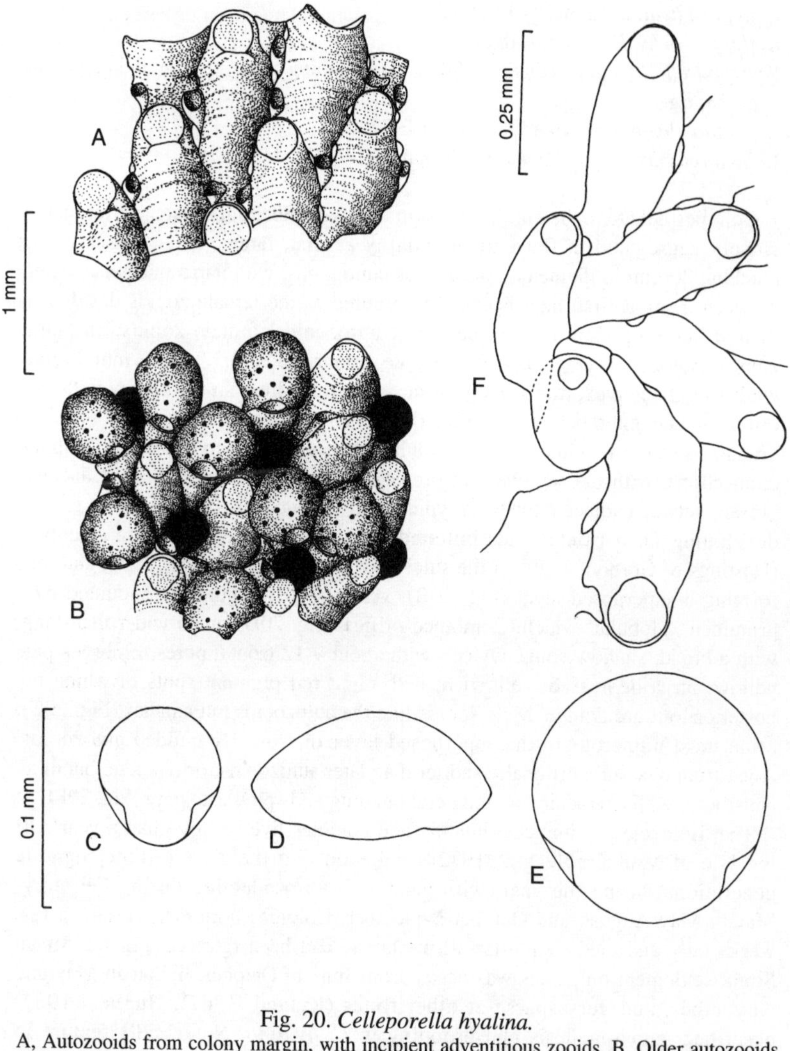

Fig. 20. *Celleporella hyalina.*
A, Autozooids from colony margin, with incipient adventitious zooids. B, Older autozooids (blacked orifices) with adventitious, male and ovicellate female zooids. C, Orifice of a male zooid. D, Orifice of an ovicellate female zooid. E, Orifice of an autozooid. F, Ancestrula and periancestrular zooids.

Celleporella hyalina (Linnaeus)
(Figs 20; 21)

Cellepora hyalina Linnaeus 1767: 1286
Celleporella hyalina : Gray 1848: 128
Schizoporella hyalina : Hincks 1880: 271 (excluding vars. *cornuta* and *depressa*), pl. 18, fig. 1
Hippothoa hyalina : Ryland & Gordon 1977: 38
Celleporella hyalina : Hayward & Ryland 1979: 252, fig. 109

Colonies adnate, coherent, in the form of small subcircular patches; frequently closely aggregated in flat clumps. Glassy at first, later opaque, ivory-white; reaching 20 mm in diameter. Autozooids contiguous, with narrow lacunae visible between them at first, later becoming obscured as the female zooids develop by frontal budding. Zooids of three types: autozooids, female zooids, and male zooids. Autozooids (Fig. 20A, B) elongate-oblong, about 0.37 x 0.17 mm. Frontal wall strongly convex, transversely striate at first, the lines strengthening with age; often a lunate umbo below the orifice (Fig. 20E). Orifice almost circular, the sinus broad and shallow; without inset condyles. Pore-chambers in the form of tubular connections, with the septular end projecting into the parent autozooid; lacunae present between adjacent tubes. Polypide with 12 tentacles. Female zooids usually developing from frontal buds initiated from the roof holes of pore-chambers (Hastings & Gordon, 1979) in the sulci between autozooids (Fig. 20A), and thus forming a superposed layer (Fig. 20B); very short, triangular; surmounted by a prominent, globular ovicell. Combined orifice (Fig. 20D) much wider than long, with a broad, shallow sinus. Ovicell with about 9-12 frontal pores. Embryos pale yellow; larva depressed, yellowish, with eight red pigment spots of which the posterior four are double. Male zooids like the autozooids but smaller (Fig. 20C); often most numerous in the superposed layer of frontally budded autozooids. Ancestrula oval, proportionally shorter than later autozooids; orifice like that of an autozooid; with characteristic unilateral budding and spiral astogeny (Fig. 20F).

The life cycle is completed quickly and colonies live for less than a year. Off the Isle of Man, Eggleston (1972a) believed that there were three separate generations during the year, with peak embryo production during February-March, May-August, and October-November. Colonies bore embryos for a few weeks only and died soon after all the larvae had been released. For the Menai Strait, settlement on panels was heavy from June to October, lighter in May and November, and very sparse at other times (Ryland, 1967). Hughes (1987) described gametogenesis and brooding in *C. hyalina*. Nurse cells feature in oogenesis, and fertilisation occurs internally during late vitellogenesis. Self-fertilisation can occur within isolated colonies although offspring appear to have reduced fitness (Hunter & Hughes, 1993a). The developing embryo receives placental nutrition from the brooding female zooid and increases in volume 15-fold during embryogenesis. Each female zooid is capable of brooding a maximum of four successive embryos, and reproductive output under optimum conditions must be impressive. Under suboptimal conditions reproduction declines and

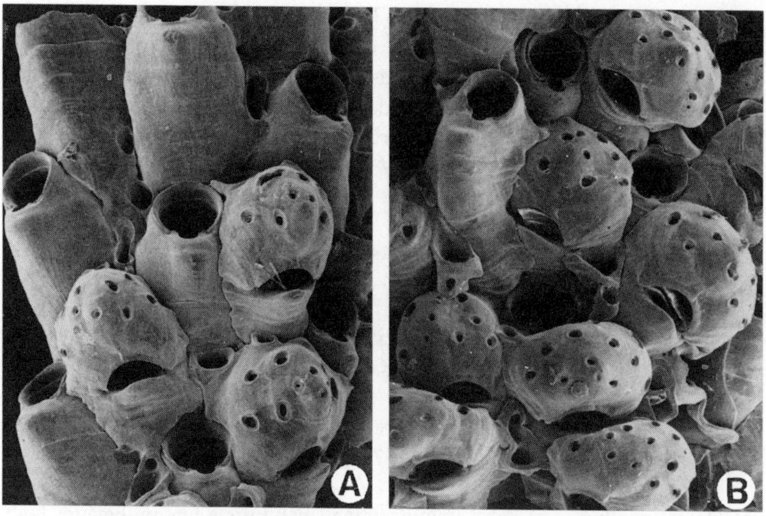

Fig. 21. *Celleporella hyalina.*
A, Growing edge of a colony; x120. B, Colony in later ontogeny, with numerous female zooids; x120. (Specimen from Lundy)

colony growth pattern, ratio of zooid types and zooid size deviate from normal (Hunter & Hughes, 1993b; 1994). Peak larval release occurs shortly after sunrise, allowing larvae maximum opportunity to find a suitable settlement site within their 12 hour free-swimming period (Cancino *et al.*, 1994).

Essentially on species of algae in shallow water, often in shelter reaching great abundance on *Laminaria*, especially *L. saccharina*, on which the colonies clump in the dimples on the frond; occasionally on stones and shells (some British records of *C. hyalina* are almost certainly derived from misidentified *Haplopoma graniferum*). Intertidal and sublittoral: to 55 m off the Isle of Man (Eggleston, 1969).

Some doubt surrounds the geographical distribution of *C. hyalina*, but it certainly extends from the Arctic down European coasts as far as the Bay of Biscay and down the Pacific coast of America to California, parts of Mexico affected by cold water upwelling and Chile. Present on the eastern seaboard of Canada, Maine and Massachusetts; records from south of Cape Cod are largely, perhaps entirely, erroneous.

Genus HAPLOTA Marcus, 1940

Colony erect, with chains of autozooids arising from repent lines of autozooids. Polymorphs include feeding autozooids, female ovicellate zooids and small-orificed zooids similar to those described as males in *Celleporella*. Frontal wall smooth, non-porous. Autozooids usually with a short cauda; the orifice sinusoid. Female orifice wide, curved proximally.

Type species: *Scruparia clavata* Hincks, 1857

Haplota clavata (Hincks)
(Fig. 22)

Scruparia clavata Hincks 1857: 175; 1880: 24, pl. 3, figs 5-8
Haplota clavata : Hayward & Ryland 1979: 256, fig. 110

Colony initially repent, later erect, forming small tufts a few millimetres high. Repent autozooids forming linear series, arising in a "fish-tail" under the proximal autozooid, each budding the next autozooid in line distomedially; elongate-oval, rounded distally, not or scarcely caudate (i.e., without a slender proximal portion); without lateral buds or producing one or an opposite pair of repent or upright caudate autozooids from wide, triangular bases. Erect chains of autozooids basically uniserial, though sometimes almost biserial when associated with ovicellate zooids and short branches, the autozooids then lying back to back. These autozooids with a cauda, club-shaped (hence *clavata*), varying considerably in length according to the extent of the cauda; about 0.37 (0.29-0.50) x 0.13 mm in specimens from the Firth of Clyde. Successive autozooids in series arise from a baso-distal "fish-tail" (Fig. 22A, B); branches emanating not laterally but from a subcircular origin in the concavity of a "fish-tail". Ovicellate zooids arise similarly from within the fork of a "fish-tail" or are budded baso-distally from another female zooid, wedge-shaped, adnate to the branch (Fig. 22C) or standing free (Fig. 22D); overall length including ovicell 0.33-0.40 mm. Calcification smooth, with occasional transverse growth marks, non-porous. Orifice of autozooid subcircular, with a shallow proximal sinus (Fig. 22E), without interior condyles; about 88 x 73 µm. Tentacle number unknown. Occasional, slightly smaller zooids branch from the main series; they have an orifice of the same shape but only about 45 µm across: from analogy with *Celleporella* these are presumed to be males. Female orifice D-shaped, about 77 µm across, with a broad, scarcely indented sinus (Fig. 22F). Ovicells subglobular, often with a faint, sutured crest; perforated by a few large, irregularly-shaped pores. Embryo colour and seasonal occurrence unknown. Ancestrula oval, with sinusoid orifice, budding a single distal autozooid.

Found on algae, various hydroids (particularly, in the Clyde, *Sertularia polyzonias*) and bryozoans, such as *Crisidia cornuta*. From low water mark to 73 m (Marcus, 1940).

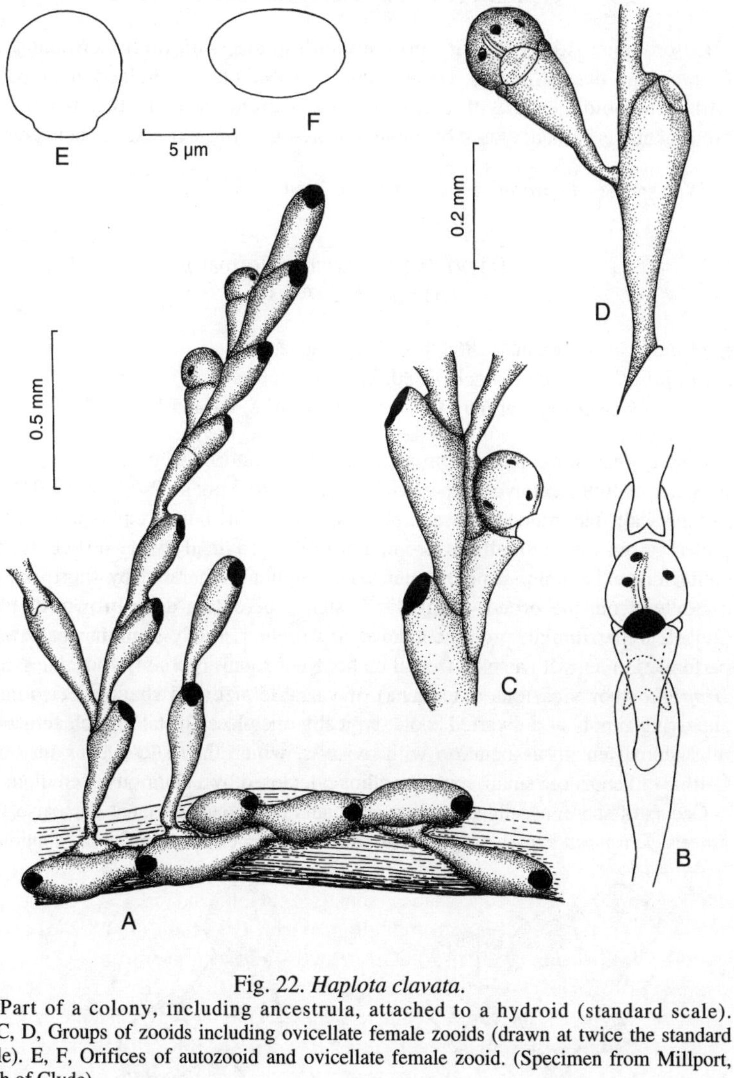

Fig. 22. *Haplota clavata*.
A, Part of a colony, including ancestrula, attached to a hydroid (standard scale). B, C, D, Groups of zooids including ovicellate female zooids (drawn at twice the standard scale). E, F, Orifices of autozooid and ovicellate female zooid. (Specimen from Millport, Firth of Clyde).

Recorded from northern Britain, from the Clyde and localities on the north-east coast (Hincks, 1880; no recent records). The species is inconspicuous and easily overlooked, but we have found it in reasonable quantity off Little Cumbrae in the Firth of Clyde, just where Hincks obtained his type specimen over a century ago. Marcus (1940) reported it from 46 m in the Skagerrak; it is also known from Iceland and North America between the Gulf of St. Lawrence and Cape Cod.

Genus TRYPOSTEGA Levinsen, 1909

Colony encrusting. Primary orifice cleithridiate, with prominent condyles. Frontal wall evenly porous. Heterozooids ("zooeciules") regularly interspersed with autozooids, each with an orifice and operculum. Ovicell hyperstomial, prominent, perforated; closed by zooidal operculum. Basal pore-chambers present.

Type species: *Lepralia venusta* Norman, 1864

Trypostega venusta (Norman)
(Figs 23; 24A, B)

Lepralia venusta Norman 1864: 84, pl. 10, figs 2, 3
Schizoporella venusta : Hincks 1880: 276, pl. 30, figs 6, 7
Trypostega venusta : Harmer 1957: 953; Hayward & Ryland 1979: 258, fig. 111

Colony encrusting, forming broad, thin sheets. Autozooids oval to hexagonal or irregular in outline, convex, separated by shallow grooves; 0.42-0.55 x 0.2-0.3 mm. Frontal wall fine grained, smooth, perforated by numerous, evenly-spaced, round pores; raised into a small but prominent umbo proximal to the orifice. Orifice cleithridiate (keyhole-shaped), anter semi-elliptical, separated by short, pointed condyles from the broad, triangular poster. Operculum deep brown, distinct. Ovicell hyperstomial, prominent, oval or irregular, finely granular and evenly perforated, often with a small frontal umbo. Kenozooids (possibly zooeciules, as in *Hippothoa*, or vicarious avicularia) of variable size and shape, occasionally appearing simply as a dwarfed zooid; typically one placed distal to each autozooid, most consistent in association with ovicells, which they succeed or surround. Orifice of kenozooid small, round or elliptical, closed by a chitinous operculum.

Occurs offshore, usually encrusting shell, and often quite frequent on coarse shell gravels. *T. venusta* is a warm temperate species with a circum-tropical distribution. In

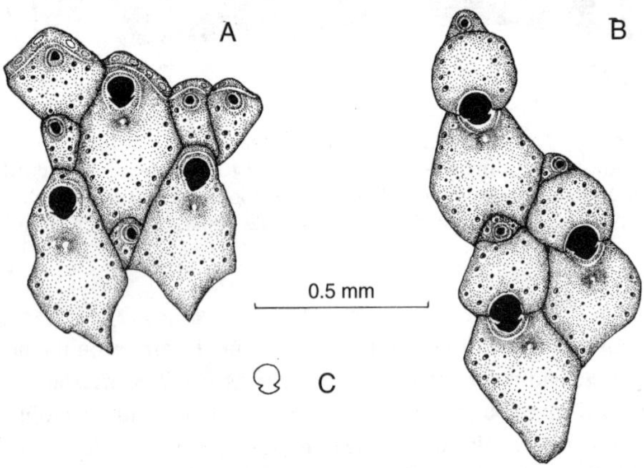

Fig. 23. *Trypostega venusta*.
A, Autozooids from the colony margin, with zooeciules of various sizes. B, Ovicellate zooids. C, Outline diagram of the primary orifice.

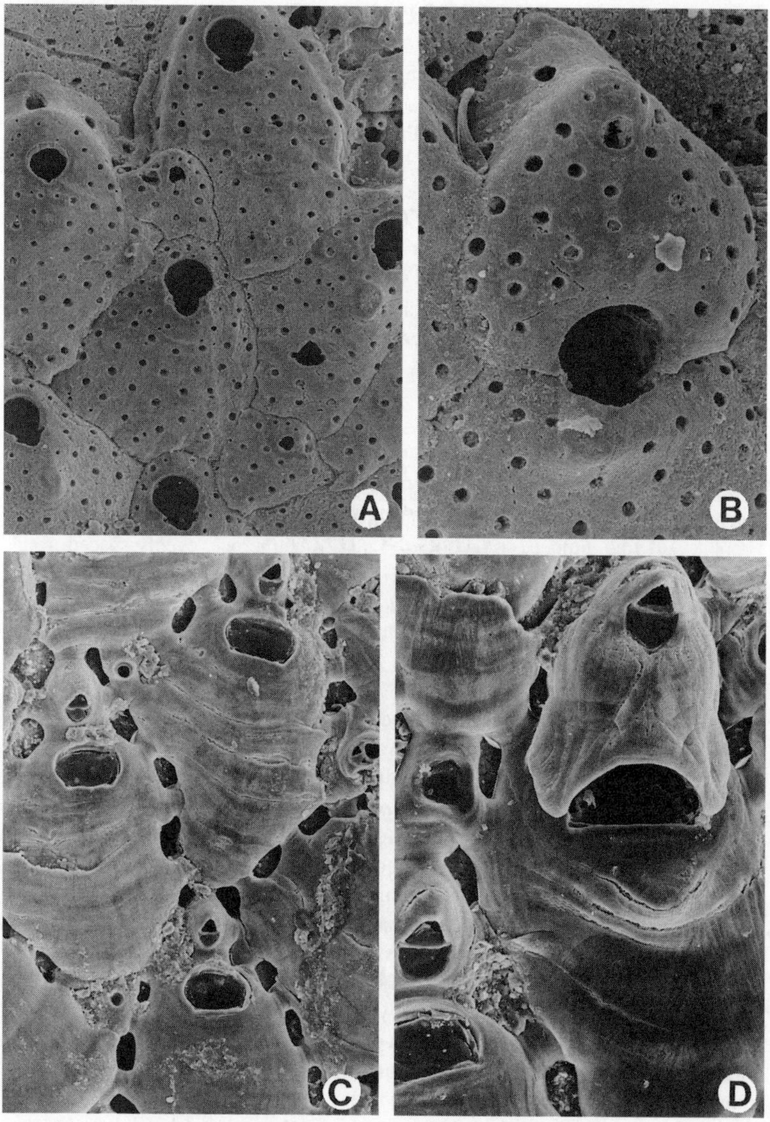

Fig. 24. A, B. *Trypostega venusta*.
A, Autozooids and kenozooids at the colony margin; x90. B, Detail of an ovicellate zooid; x180. (Specimen from Lundy). C, D, *Chorizopora brongniartii*. C, Group of autozooids; x90. D, An ovicellate zooid; x150. (Specimen from Martins Haven)

the eastern Atlantic it ranges from West Africa to the south-west British Isles; it has recently been recorded from Majorca but elsewhere in the Mediterranean it is replaced by the mainly Indo-Pacific *T. claviculata* (Hincks) (Zabala & Maluquer, 1988). In British waters, it is found in the western end of the English Channel, the Gulf of St. Malo and as far north as Lundy in the Bristol Channel; not recorded from the Irish Sea.

Family CHORIZOPORIDAE Vigneaux, 1949

Colony encrusting. Frontal wall gymnocystal, thin, imperforate. Primary orifice D-shaped, without spines. Avicularia vicarious. Pore-chambers tubular. Ovicell imperforate, hyperstomial, closed by the zooidal operculum.

Genus CHORIZOPORA Hincks, 1879

Colony encrusting. Primary orifice semicircular. Frontal wall thin, imperforate. Avicularia vicarious, small. Ovicell hyperstomial, prominent, closed by zooidal operculum. Interzooidal communication via tubular extensions of the pore-chambers. Small kenozooids present.

Type species: *Flustra brongniartii* Audouin, 1826

Chorizopora brongniartii (Audouin)
(Figs 24C, D; 25)

Flustra brongniartii Audouin 1826: 240, pl. 10, fig. 6
Chorizopora brongniartii : Hincks 1880: 224, pl. 32, figs 1-4; Hayward & Ryland 1979: 240, fig. 103

Colony forming broad, spreading sheets, thin and translucent, and often clearly visible only when dried. Autozooids oval, occasionally pear-shaped or irregular, broadest at the distal end, flat or slightly convex, separated by shallow grooves; 0.45-0.6 x 0.25-0.3 mm. Adjacent autozooids linked by short tubules, extensions of the pore chambers (cf. *Hippothoa*). Frontal wall fine grained, vitreous and glistening in young autozooids, becoming more granular as calcification continues; imperforate, marked with transverse striations or ridges and often with a prominent, conical frontal umbo. Primary orifice semicircular, subterminal, with a thin rim but no peristome. Avicularia vicarious; cystid small, elliptical, linked to adjacent autozooids by communication tubes, mandible at an acute angle to the frontal plane, triangular, directed distally. Characteristically, a single avicularium situated distal to each autozooid, with a few additional ones occurring elsewhere. Ovicell prominent, elongate, hyperstomial; fine grained, occasionally with a longitudinal frontal ridge; the distal avicularium intimately associated with the ovicell. Small, irregularly shaped kenozooids may be present, linked by tubules to surrounding autozooids; generally small, with a circular or oval orifice but no operculum. Embryos red. Ancestrula oval, similar to later autozooids but with about four spines frontally, and four orally.

Eggleston (1969) reported breeding throughout the year, with a peak in June-September, in Manx waters. Gautier (1962) found ancestrulae in April-August, in the western Mediterranean.

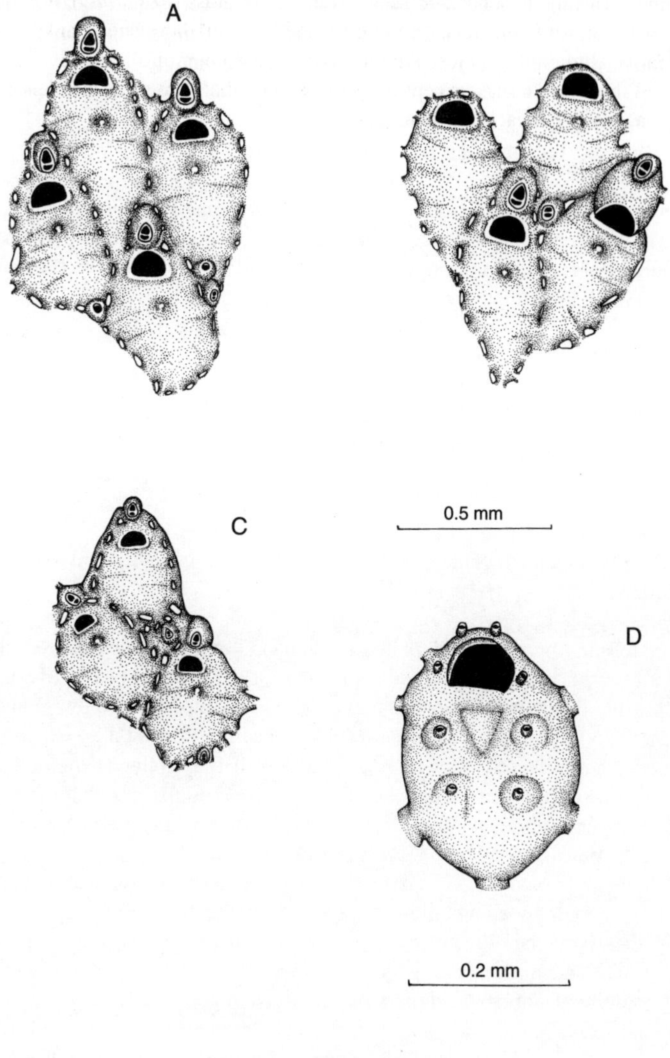

Fig. 25. *Chorizopora brongniartii.*
A, Four typical autozooids. B, A group of autozooids, including one with an ovicell. C, Three smaller autozooids from a Mediterranean specimen. D, The ancestrula.

Widespread and abundant off all British coasts, *C. brongniartii* is essentially a sublittoral species and ranges to at least 100 m, and perhaps deeper. Although occurring on calcareous algae and Mediterranean eel-grass (*Posidonia*), it is most characteristic of hard substrata, particularly shell. On offshore shell banks it can be the most abundant species. Apparently almost cosmopolitan, distributed through all the world's seas, except for polar waters; there are great differences in size over its geographical range (Fig. 25C).

Superfamily LEPRALIELLOIDEA Vigneaux, 1949

Classification of this superfamily is still under debate. D. P. Gordon has proposed that the species here assigned to the families Exochellidae and Escharellidae should, instead, all be included within the Exochellidae. (D. P. Gordon, *in litt.,* Feb. 1998).

Family UMBONULIDAE Canu, 1904

Colony encrusting; or erect, bilaminar. Frontal shield with conspicuous marginal areolar pores, bordered by interareolar ribs. Secondary orifice simple, flared, or with one or more denticles on its inner face. Oral spines present or absent. Avicularia present, adventitious. Ovicell hyperstomial, perforate; or absent. Vertical walls with multiporous septula.

Genus UMBONULA Hincks, 1880

Colony encrusting, unilaminar; or partly erect, lobed or foliaceous, folded or bilaminar. Oral spines absent. Avicularia median suboral. Ovicells present or absent.

Type species: *Cellepora verrucosa* Esper, 1790 = *Umbonula ovicellata* Hastings, 1944

Lepralia arctica M. Sars, 1851, formerly included in *Umbonula*, was referred to a new genus, *Arctonula*, by Gordon & Grischenko (1994).

Key to the British Species of *Umbonula*

1. Frontal shield with few (6-12) marginal areolae, with very pronounced interareolar ridges converging on the suboral umbo. Ovicell present
.... ***Umbonula ovicellata*** (p.106)

Frontal shield with many (14-20) marginal areolae; interareolar ridges feebly developed. No ovicells
.... ***Umbonula littoralis*** (p.104)

Umbonula littoralis Hastings
(Figs 26A; 27)

Umbonella (*Umbonula* in the plate) *verrucosa* : Hincks 1880: 317 (in part), pl. 39, fig. 1
Umbonula littoralis Hastings 1944: 280, fig. 1B; Hayward & Ryland 1979: 74, fig. 22

Colonies developing broad, spreading, unilaminar sheets, dull orange. Autozooids oval to hexagonal, strongly convex, separated by distinct sutures. Secondary orifice longer than broad, elliptical to rounded quadrangular, thin rimmed, with a slight mid-proximal convexity; marginal sclerite of operculum and extensive frontal membrane clearly visible in frontal view. Frontal shield finely granular, bordered by 14-20 large, closely spaced areolae; interareolar ridges short, not extending far medially, and rarely reaching the umbo. Avicularia sporadically present, situated immediately proximal to orifice, cystid low and tumid, the rostrum oval, perpendicular to plane of orifice, orientated disto-proximally; sometimes absent, when a low conical umbo may be developed in its place.

Embryos red, brooded internally. Recorded October-November on the Northumberland coast (Eggleston, 1975) and September to February around the Isle of Man (Eggleston, 1969).

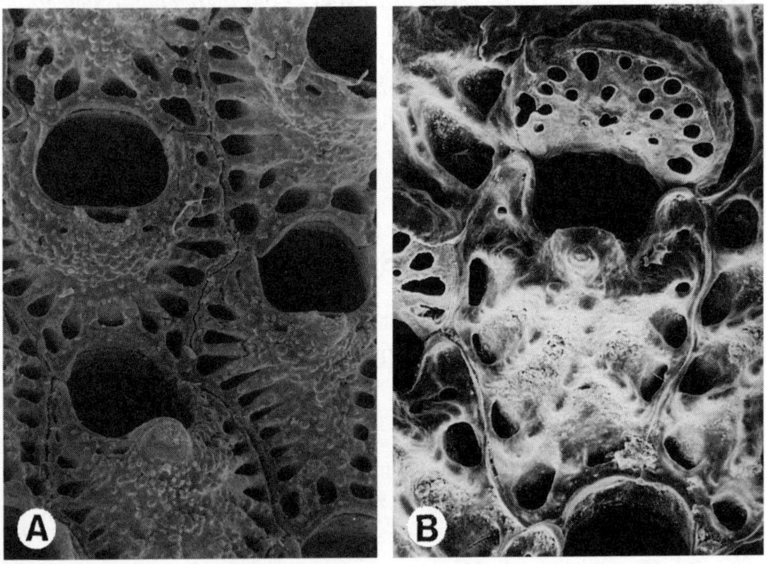

Fig. 26. A, *Umbonula littoralis*; x50. B, *Umbonula ovicellata*; x80.

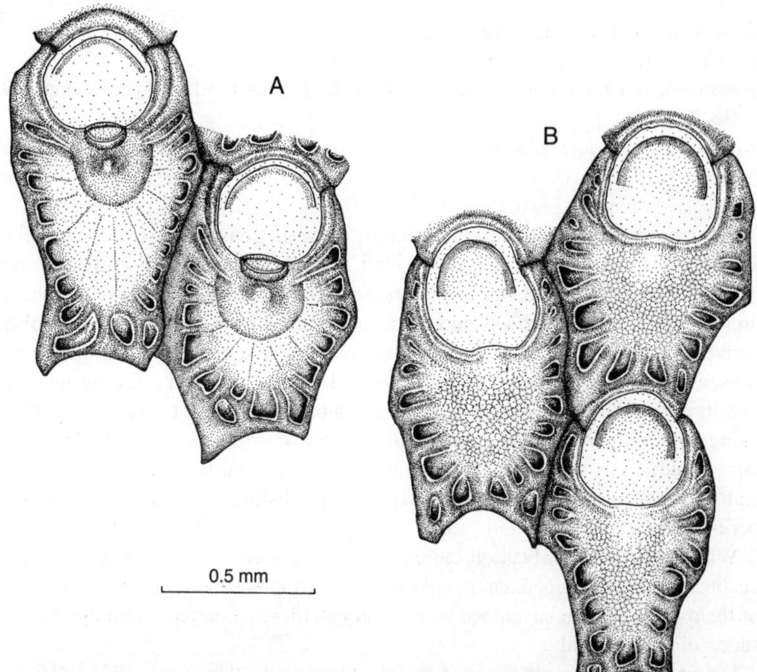

Fig. 27. *Umbonula littoralis*.
A, Autozooids with suboral avicularia and umbones. B, Non-umbonate autozooids.

Umbonula littoralis is a common intertidal species, distributed from MTL into the *Laminaria* zone, but probably no deeper. It encrusts stones and shells, forms extensive sheets on the undersides of boulders or on shaded overhangs, occurs on clumps of larger hydroids and the underside of *Himanthalia* buttons, and is a conspicuous component of the *Laminaria* holdfast fauna. Common on all British coasts, ranging north to Bergen and south to the Channel Islands and west Brittany. It does not occur in the Mediterranean.

Umbonula ovicellata Hastings
(Figs 26B; 28)

Cellepora verrucosa Esper 1790: 239
not *Cellepora verrucosa* Linnaeus 1767: 1286
Umbonella (*Umbonula* in the plate) *verrucosa* : Hincks 1880: 317 (in part), pl. 39, fig. 2
Umbonula ovicellata Hastings 1944: 277, figs 1A, 2A, B; Hayward & Ryland 1979: 72, fig. 21

Autozooids oval to hexagonal, strongly convex, separated by deep grooves; 0.56-0.78 x 0.36-0.5 mm. Frontal wall initially fine-grained and vitreous, becoming coarse and nodular; imperforate centrally with six to 12 large peripheral areolae. Thickened ridges extend from the vertical walls, between the areolae, converging on the proximal border of the aperture. A suboral avicularium usually present; cystid swollen, umbonate; mandible situated distally, facing into the aperture, semicircular, perpendicular to frontal surface of autozooid. Aperture elongated, oval, with a thin, raised rim. Vertical walls containing a single line of large multiporous septula close to the base. Ovicell prominent, broader than long and flattened frontally; perforated by pores in radiating linear series, the largest peripherally.

With continued calcification the umbonate avicularium and interareolar ridges become greatly thickened and may bear numerous knobs or ridges. The distal half of the ovicell may be enveloped by nodular calcification derived from the distally succeeding autozooid.

Orange embryos were reported in July from Scilly (Hayward, 1971). Gautier (1962) stated that in the western Mediterranean embryos may be observed from April to November.

A southern species, widespread in the Mediterranean (Gautier, 1962; Hayward, 1974), reaching the northern limit of its range in the south-west British Isles. It has been reported from the Channel Islands and from the coasts of Cornwall, Devon and Dorset, as far east as Swanage (Hastings, 1944), but does not seem to reach Lundy or south-west Wales. Distributed in the Mediterranean from the shore to 50 m, most frequently on red or brown algae and *Posidonia* leaves. In British waters, it is strictly sublittoral, occurring from the lower levels of the kelp beds downwards, on *Laminaria* holdfasts, *Crisia, Pentapora* and other hard substrata. *U. ovicellata* was formerly confused with *U. littoralis*, being regarded by Hincks (1880) as a deep water variety of *"U. verrucosa"*.

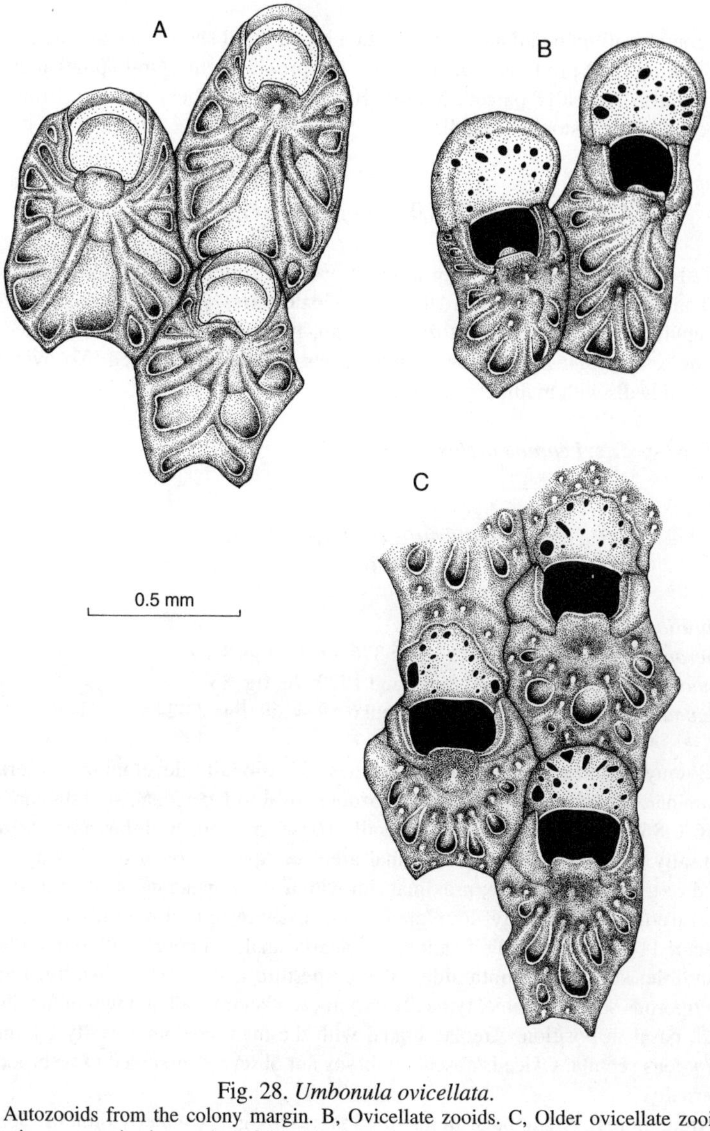

Fig. 28. *Umbonula ovicellata*.
A, Autozooids from the colony margin. B, Ovicellate zooids. C, Older ovicellate zooids, showing progressive immersion of the ovicells.

Family ROMANCHEINIDAE Jullien, 1888

Frontal wall umbonuloid. Orifice with a broad distal shelf extending below the operculum, and prominent denticles on the proximal rim. Oral spines present. Avicularia typically paired, lateral. Basal pore-chambers present. Embryos brooded in hyperstomial ovicells.

Genus ARCTONULA Gordon & Grischenko, 1994

Colony encrusting, unilaminar; or erect, foliaceous. Autozooids with umbonuloid frontal shield bordered by a single series of large areolae; aperture occupying about one-third autozooid length, its proximal rim bearing a median denticle. No spines. Avicularia adventitious, lateral-oral, paired. No ovicells. Vertical walls with multiporous septula.

Type species: *Lepralia arctica* M. Sars, 1851

Arctonula arctica (M. Sars)
(Fig. 29)

Lepralia arctica M. Sars 1851: 149
Mucronella pavonella : Hincks 1880: 376, pl. 39, figs 8-10
Umbonula arctica : Hayward & Ryland 1979: 76, fig. 23
Arctonula arctica : Gordon & Grischenko 1994: 68, figs 23-24

Colony forming broad rounded patches, occasionally developing into erect, bilaminar, foliaceous expansions. Autozooids oval to hexagonal, slightly convex, 0.56-0.86 x 0.4-0.5 mm. Frontal wall coarse grained, nodular, imperforate centrally with 10-16 distinct marginal areolae. Aperture oval, comprising one-third of the frontal surface; proximal rim with a short, quadrate, median denticle. Interareolar ribs weakly developed. Sutures between autozooids becoming obscured by secondary calcification. Avicularia small, elliptical, with semicircular mandible; situated on both sides of the aperture and directed distally. Large multiporous septula visible, typically two in each lateral wall and two in the distal wall, basal in position. Areolae linked with the main coelomic cavity by small uniporous septula. Ovicells absent; embryos not observed, assumed to be brooded internally.

Arctic-boreal, circumpolar in distribution, reaching its southerly limit off north-east England. Formerly reported offshore from Northumberland and Yorkshire, and from the Dogger Bank (Hincks, 1880). There are no recent reports and it may be that its geographical range has contracted in the present century.

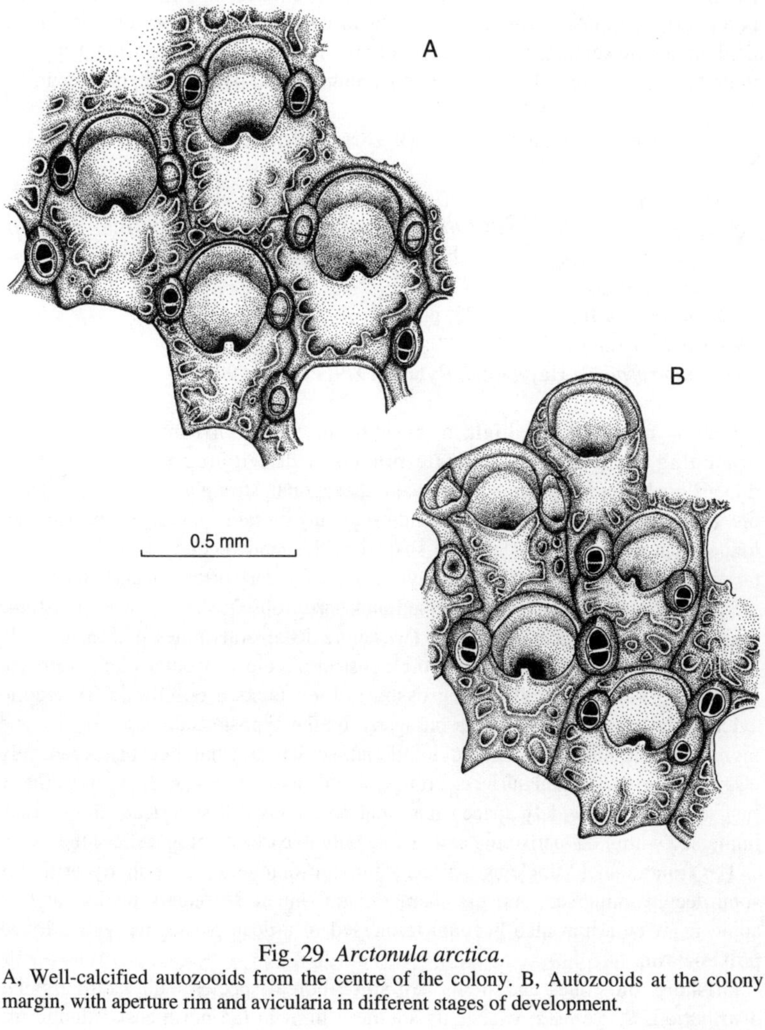

Fig. 29. *Arctonula arctica*.
A, Well-calcified autozooids from the centre of the colony. B, Autozooids at the colony margin, with aperture rim and avicularia in different stages of development.

Genus RAGIONULA Canu & Bassler, 1927

Colony erect, bilaminate, branching, attached by encrusting sheets of autozooids. Frontal shield of autozooids umbonuloid, with indistinct marginal pores. Primary orifice lepralioid, with small condyles, secondary orifice sinuate. Oral spines present. Avicularia adventitious, lateral suboral and frontal. Ovicell imperforate, not closed by autozooid operculum. Basal pore chambers present.

Type species: *Eschara rosacea* Busk, 1856

Ragionula rosacea (Busk)
(Figs 30; 31C, D)

Eschara rosacea Busk 1856c: 33, pl. 1, fig. 4
Escharoides rosacea : Hincks 1880: 336, pl. 47, figs 5-9
Ragionula rosacea : Hayward & Ryland 1979: 212, fig. 90

Colony up to 20 mm high, often with an equal horizontal spread; rigid, branching, with broadly palmate branches, developing a compact three-dimensional structure. Autozooids oval to hexagonal, strongly convex; boundaries distinct at growing edge, elsewhere indicated only by faint sutures. Primary orifice longer than wide, more or less bell-shaped, with rounded condyles and a concave proximal edge; developing a narrowly sinuate secondary orifice in early ontogeny, which is then progressively immersed and partly obscured by a deep peristome with a pronounced proximal lobe. Two short distal oral spines present in early ontogeny. Avicularium lateral suboral; rostrum acute to frontal plane, directed laterally, oval, with a slender crossbar which lacks a columella. A second avicularium may develop in later ontogeny, on the opposite side of the orifice and more proximal than the first, and additional, similar avicularia occur sporadically elsewhere on the frontal surface. Frontal shield finely granular, thick, indistinctly rugose, with few, widely spaced marginal pores. Ovicell spherical, imperforate, immersed within the peristome and visible only in earliest ontogenetic stages.

The frontal shield thickens markedly through ontogeny; the primary orifice is soon deeply immersed, and the characteristic sinuate secondary orifice, and its adjacent avicularium also become immersed in a deep peristome with a lobed proximal rim.

Offshore, on stones and shells. An arctic-boreal species with a circumpolar distribution, *R. rosacea* reaches its southern limit in the north-east Atlantic off Shetland and northern Scotland.

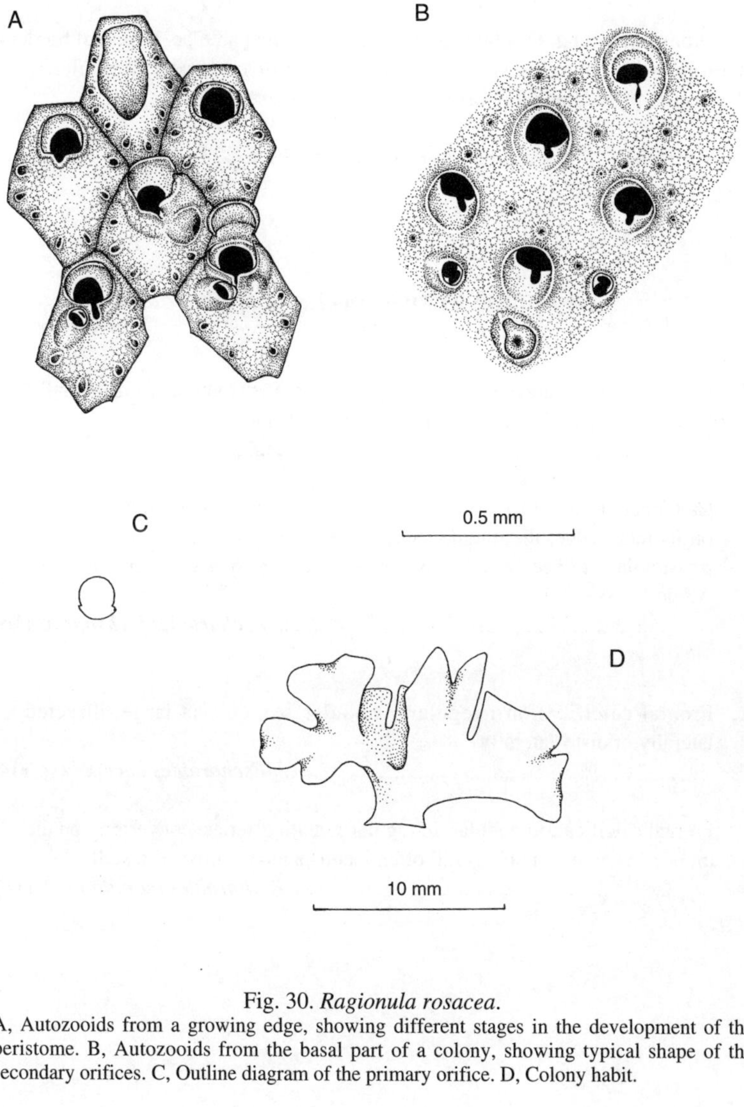

Fig. 30. *Ragionula rosacea*.
A, Autozooids from a growing edge, showing different stages in the development of the peristome. B, Autozooids from the basal part of a colony, showing typical shape of the secondary orifices. C, Outline diagram of the primary orifice. D, Colony habit.

Genus ESCHAROIDES Milne-Edwards, 1836

Colony encrusting. Frontal wall with large marginal areolae. Proximal border of aperture with denticles; oral spines present on distal border. Avicularia well developed, paired, lateral to orifice. Ovicell prominent, hyperstomial.

Type species: *Cellepora coccinea* Abildgaard, 1806

Key to the British species of *Escharoides*

1. Calcification of autozooids coarsely nodular. Peristome peaked medially, with three similarly-sized denticles on its inner face
.... **2**

Calcification of autozooids granular. Peristome with three blunt denticles on its inner edge, the middle usually the largest; above this the proximal peristomial rim becomes deeply notched. A large species; deep water and Arctic
.... ***Escharoides jacksoni*** (p.116)

2. Frontal calcification irregularly nodular. Avicularia large, directed laterally or disto-laterally
.... ***Escharoides coccinea*** (p.114)

Frontal calcification nodular, in regular radiating series, converging on the umbo. Avicularia small, cystid often inconspicuous, directed distally
.... ***Escharoides mamillata*** (p.118)

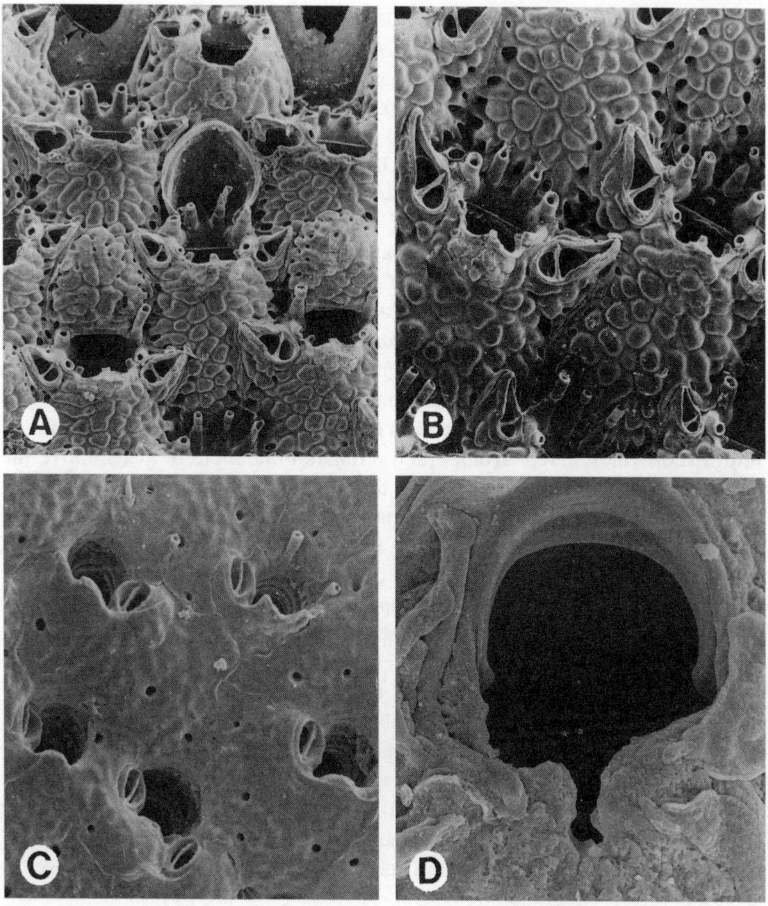

Fig. 31.
A, B, *Escharoides coccinea.*
A, Autozooids at the colony margin; x50. B, Detail; x70.

C, D, *Ragionula rosacea.*
C, Autozooids in late ontogeny; x90. D, The secondary orifice; x300.

Escharoides coccinea (Abildgaard)
(Figs 31A, B; 32)

Cellepora coccinea Abildgaard 1806: 30, pl. 146, figs 1, 2
Mucronella coccinea : Hincks 1880: 371, pl. 34, figs 1-3
Peristomella coccinea : Marcus 1940: 235
Escharoides coccinea : Hayward & Ryland 1979: 78, fig. 24

Colony forming extensive orange patches. Autozooids broad, quadrangular, convex, separated by deep grooves; 0.6-0.7 x 0.3-0.4 mm. Frontal wall initially smooth, imperforate, with large and distinct marginal areolae; rapidly becoming coarsely granular and nodular as secondary calcification proceeds, obscuring the areolae. Aperture elliptical; six thick tubular spines in a disto-lateral arc, at first long and prominent but usually broken in older autozooids. A broad shelf extends below the operculum from the distal rim of the aperture; the proximal rim is raised into a short thick mucro from which a quadrate denticle projects above the operculum. Avicularia paired, lateral to the aperture, directed disto-laterally; cystid extensive, swollen, with large marginal pores and nodular calcification; mandible elongate-triangular. In fertile areas of the colony one of the avicularia may be reduced or completely suppressed (Fig. 30B). Ovicell prominent, flattened frontally; at first thin, hyaline and imperforate, rapidly covered by secondary calcification to give the same granular aspect as the autozooid frontal wall; becoming rather immersed. A double series of small round pores present around the periphery of the ovicell; indistinct, usually visible only in newly developed individuals or in mounted material. Embryos red. Ancestrula tatiform, oval, with narrow cryptocyst and 13-14 marginal spines.

Embryos may be present in all months of the year but are abundant, in Manx waters, only in July (Eggleston, 1972a). In the western Mediterranean, Gautier (1962) reported their presence in May and November, with settlement in July and November. *E. coccinea* is a perennial species and most colonies probably live for at least two years (Eggleston, 1972a).

A common and characteristic intertidal species of rocky shores all around the British Isles, found on a wide range of substrata. Particularly associated with *Laminaria* holdfasts, there forming a conspicuous, constant, and often significant, proportion of the bryozoan epifauna; but also below the kelp beds, on stones and shells, to an undetermined depth.

A boreal species, widespread and often common around the British Isles and Ireland; to the south it extends into the Mediterranean, as far as the Aegean (Hayward, 1974) and Madeira. To the north it reaches Shetland and the west coast of Norway but is apparently absent from the Faeroes and Arctic waters.

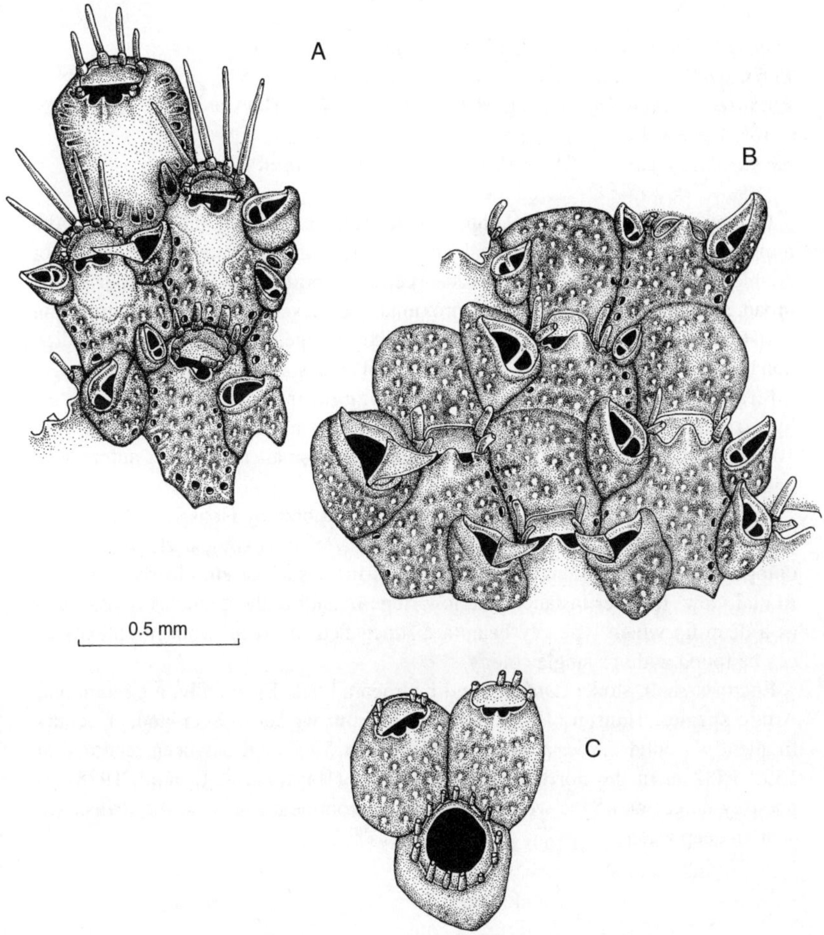

Fig. 32. *Escharoides coccinea.*
A, Autozooids at the colony margin. B, Well-calcified, ovicellate zooids. C, The ancestrula, with the first autozooids budded from it.

Escharoides jacksoni (Waters)
(Fig. 33)

Smittia jacksoni Waters 1900: 87, pl. 12, fig. 18
Peristomella bidenkapi Kluge 1946: 200, pl. 2, fig. 6
Escharoides bidenkapi : Kluge 1962: 566, fig. 400; Hayward & Ryland 1978: 150, fig. 5A, B
Escharoides jacksoni : Kluge 1962: 568, fig. 401; Hansen 1962: 33

Colony forming a thick, unilaminar crust. Autozooids large, 0.8-1.0 x about 0.8 mm, visible to unaided eye. Frontal shield finely granular, bordered by a single or double series of small, round areolae. Aperture suborbicular; distal border with a broad, straight edged suboral shelf; proximal edge developed as a thick, projecting peristome with thickened denticles on its inner edge, and often with a median notch in the rim. Three to five short distal oral spines. Avicularia single or paired, laterally directed, rostrum narrowly triangular, hooked at tip. Ovicell hyperstomial, globular, finely granular. Ancestrula tatiform, 0.5 mm long, with 10-12 peripheral spines; periancestrular autozooids smaller than later autozooids, with six to eight oral spines.

The peristome shows considerable variation, as noted by Hansen (1962). In the form described by Kluge (1946) as *Peristomella bidenkapi*, perhaps an early ontogenetic phase, the three denticles on its inner surface are clearly visible in frontal view. In other instances the low ridge bounding the denticles is produced as a deep lip which typically bears a distinct median notch. Both morphologies can be found within a single colony.

Encrusts shell, stone, *Lophelia* and hydrocorallines. Essentially, a circumpolar Arctic species. Hansen (1962) recorded it from western Greenland; it occurs frequently around Faroe, at depths greater than 50 m and has been recorded at 1390-1582 m in the northern Bay of Biscay (Hayward & Ryland, 1978). It possibly ranges along the western edge of the continental shelf in the British sea area, in deep water.

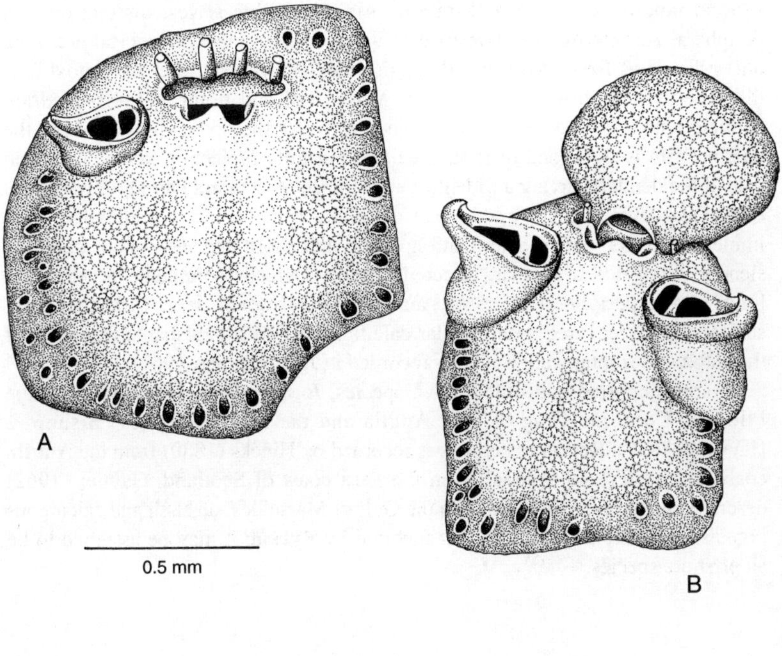

Fig. 33. *Escharoides jacksoni*.
A, Typical autozooid. B, An ovicellate zooid.

Escharoides mamillata (Wood)
(Fig. 34)

Lepralia mamillata Wood 1844: 19
Mucronella coccinea var. *a (mamillata)* : Hincks 1880: 373, pl. 34, figs 4-6
Escharoides mamillata : Lagaaij 1952: 86; Hayward & Ryland 1979: 80, fig. 25

Autozooids broad, oval to quadrangular, 0.5-0.8 x 0.4-0.6 mm; strongly convex and separated by deep grooves. Frontal wall in peripheral autozooids smooth, vitreous and fine-grained, rapidly obscured by nodular secondary calcification, forming regular radiating series of ridges and knobs converging on a prominent suboral mucro. Areolae small, round, often in double series; distinct only in peripheral autozooids. Aperture oval to quadrate; six spines in a distal arc, long and pointed at first, short and thick in older autozooids, the distalmost pair displaced when an ovicell develops. Mucro tall and blunt, a short quadrate denticle with a concave free edge projecting above the operculum; laterally, the aperture rim is thickened to form a pair of inwardly-projecting knobs, one each side of the denticle, giving a tridentate appearance. Avicularia paired, lateral to the orifice and level with the middle pair of spines; cystid small, not globular, often immersed in the frontal calcification of the autozooid and indistinct; mandible slender, elongated triangular, directed distally. Ovicell prominent, broader than long, with a peripheral series of small, indistinct pores, seen with difficulty; surface rapidly covered with nodular calcification, indistinguishable from that of the autozooid frontal wall. Embryos recorded in July (Gautier, 1962).

A rarely reported and little-known species, *E. mamillata,* is known from the Pliocene Coralline Crag of East Anglia and the Low Countries (Bishop & Hayward, 1989). Recent material was recorded by Hincks (1880) from the Antrim coast and by Ryland (1963a) from the west coast of Scotland. Gautier (1962) described material from 40-80 m in the Gulf of Marseilles, on shell and calcareous organisms, a habitat similar to that reported by Ryland; it may be assumed to be an offshore species.

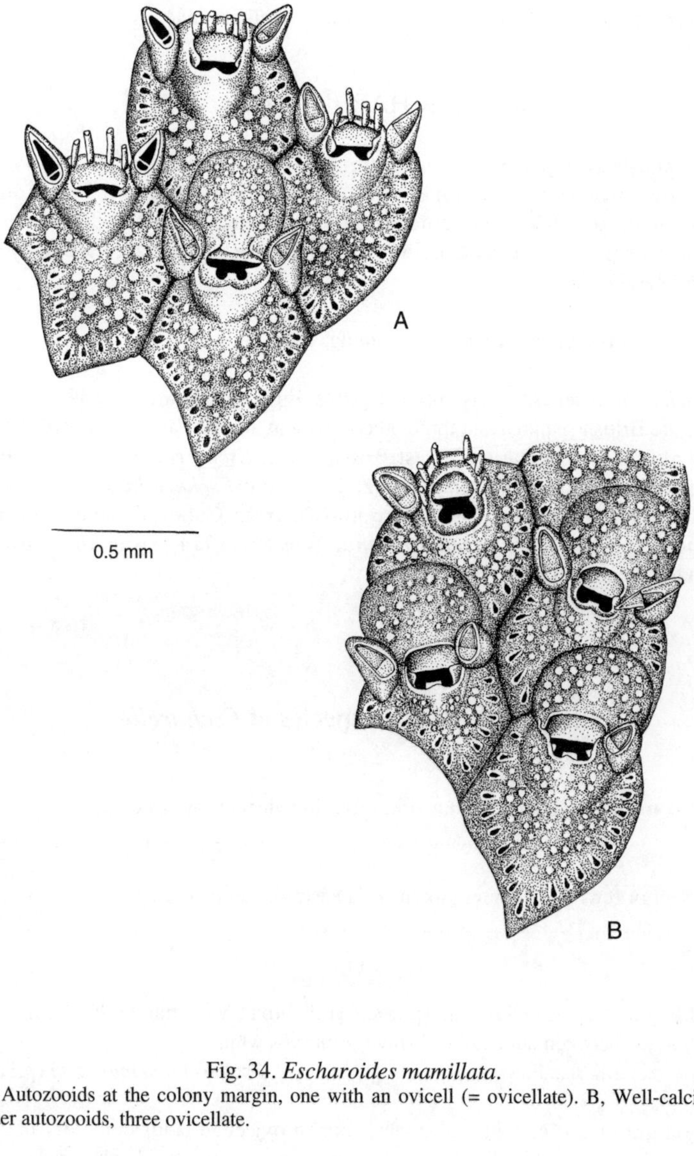

Fig. 34. *Escharoides mamillata*.
A, Autozooids at the colony margin, one with an ovicell (= ovicellate). B, Well-calcified, older autozooids, three ovicellate.

Family ESCHARELLIDAE Levinsen, 1909

Frontal wall with marginal pores only. Pore chambers present. Primary orifice typically with lyrula and condyles; enclosed by well developed peristome and with conspicuous oral spines. No avicularia. Ovicell prominent.

Genus ESCHARELLA Gray, 1848

Frontal wall imperforate, with marginal areolae. Orifice with large lyrula; condyles, where present, small. Oral spines present. Ovicell imperforate, prominent, not closed by zooidal operculum. Numerous small basal pore-chambers present. No avicularia. Ancestrula produces one distal followed by two disto-lateral buds.

Type species: *Lepralia immersa* Fleming, 1828

Escharella is an especially speciose genus. Eight species are presently recorded from the British sea area and three others, present around Faroe (Hayward, 1994), may also occur off the northwest British Isles. *Escharella praealta* (Calvet) (López de Cuadra & Garcia-Gómez, 1993) and *Escharella acuta* (Zabala, Maluquer & Harmelin, 1993) are known from Catalonia and Gibraltar respectively and *Escharella rylandi* (Geraci 1974) from the western Mediterranean.

Key to the British species of *Escharella*

1. Six to eight oral spines (sometimes four in ovicellate autozooids)
.... **2**

Four or fewer oral spines (six in periancestrular autozooids)
.... **3**

2. Margin of orifice with eight spines. Lyrula broad with straight distal edge. Sutures between autozooids distinct. Embryos white
.... ***Escharella octodentata*** (p.126)

Margin of orifice with six spines, four in ovicellate autozooids. Lyrula broad with concave distal edge. Sutures between autozooids obscured by secondary calcification. Embryos pink
.... ***Escharella immersa*** (p.122)

3. Four oral spines present
.... **4**

Three or fewer spines
.... **6**

4. Autozooids small, up to 0.5 mm long. Lyrula longer than wide. Peristome produced into a slender, tapered mucro
.... ***Escharella labiosa*** (p. 124)

Autozooids larger than 0.5 mm. Lyrula broader than long
.... **5**

5. Lyrula truncate, corners not upturned. Peristome extending laterally as far as the spine bases, produced proximally as a slender, tapered mucro. Restricted to northern sites
....***Escharella klugei*** (p.130)

Lyrula quadrate, with upturned, cuspidate corners. Peristome incomplete centrally, often forming an abrupt, quadrate mucro. Widespread and common
....***Escharella ventricosa*** (p.128)

6. Autozooids up to 0.8 mm long. Lyrula slender, anvil-shaped. Mucro often bifid. Autozooids flat with conspicuous aerolae. Widely distributed and common
.... ***Escharella variolosa*** (p.132)

Autozooids up to 1.3 mm long. Lyrula low and broad. Mucro truncate. Areolae conspicuous or hidden. Northern or deep water only
.... **7**

7. Autozooids strongly convex. Areolae inconspicuous, usually seen only in cleaned material
.... ***Escharella abyssicola*** (p.134)

Autozooids flat. Areolae distinct, in all stages of growth, often developing strong interareolar ridges
.... ***Escharella laqueata*** (p. 138)

Escharella immersa (Fleming)
(Figs 35; 37A)

Lepralia immersa Fleming 1828: 533
Mucronella peachii : Hincks 1880: 360, pl. 50, figs 1-5
Escharella immersa : Hayward & Ryland 1979: 138, fig. 53

Colony an extensive white, unilaminar sheet. Autozooids oval to hexagonal, becoming irregular in outline, convex, separated by distinct grooves only when newly formed; short and broad, 0.3-0.4 x 0.2-0.3 mm. Frontal wall initially finely granular, becoming progressively coarser and more rugose with secondary calcification; areolae small, indistinct in newly developed autozooids, accentuated by increasing calcification and becoming very marked, with stout interareolar ridges, the grooves between autozooids being obliterated in later ontogeny. Primary orifice orbicular, with a broad, anvil-shaped lyrula, often with a concave distal edge; blunt, lateral condyles present. Six oral spines, four of which persist in ovicellate autozooids. Peristome well developed, thickened, with a stout, pointed mucro proximally. Ovicell broader than long, granular, recumbent on succeeding autozooid and rapidly immersed in secondary calcification. Tentacles 11-12. Embryos pink or red. Ancestrula oval, 0.3 mm long, tatiform, with an extensive cryptocyst and 10-11 tall, pointed spines.

Embryos present throughout the year in Manx waters, with peak abundance in February to July (Eggleston, 1969). On hard substrata and occasionally *Laminaria* holdfasts; widespread and abundant off all British coasts from the lower shore downwards. Lower bathymetric limit unknown, but uncommon below 100 m. Frequent on offshore shell banks, often in association with *E. ventricosa* and *E. variolosa*. A cold water, boreal-arctic species, *E. immersa* has been reported from numerous localities in the North Atlantic (Powell, 1968) but does not seem to extend far south of the British Isles and is absent from the Mediterranean.

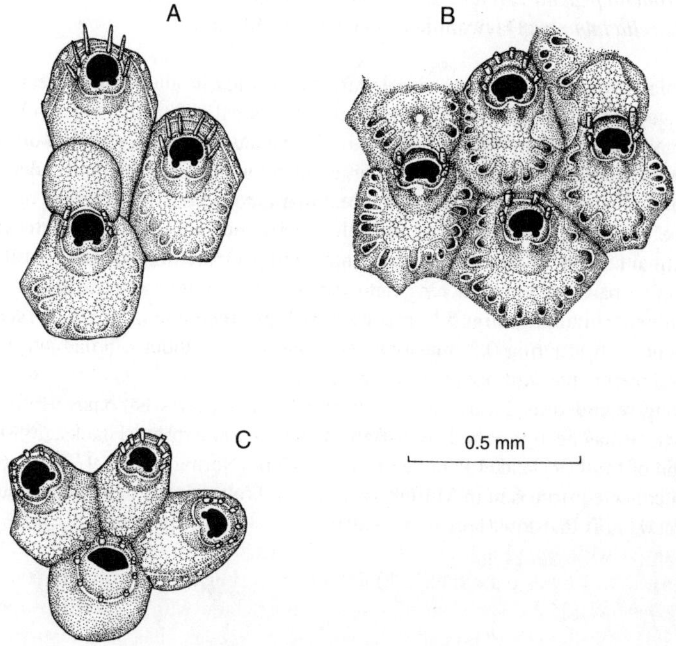

Fig. 35. *Escharella immersa*.
A, Autozooids from the growing edge, one with an ovicell. B, Later autozooids, with the characteristic secondary calcification. C, The ancestrula and periancestrular zooids.

Escharella labiosa (Busk)
(Figs 36; 37B)

Lepralia peachii var. *labiosa* Busk 1856b: 309, pl. 10, figs 5, 6
Mucronella peachii var. *labiosa* : Hincks 1880: 361, pl. 51, fig. 1
Escharella labiosa : Hayward & Ryland 1979: 140, fig. 54

Colony forming small round patches. Autozooids in alternating linear series, oval, convex and separated by deep grooves; small, 0.38-0.5 x 0.24-0.3 mm. Frontal wall granular, areolae small and inconspicuous. Primary orifice orbicular, slightly wider than long; lyrula slender, quadrate with a concave distal edge. Four distal oral spines (five or six in periancestrular autozooids), all of which persist in ovicelled autozooids. Peristome well developed, enclosing the lateral and proximal borders of the orifice, rising into a tall peak proximally; later enveloping the spine bases and encroaching onto the ovicell as a low frontal ridge. Ovicell prominent, granular, slightly broader than long, recumbent on the succeeding autozooid. Ancestrula 0.3 mm long, oval, tatiform, without a broad cryptocyst, frontal membrane surrounded by 11 or 12 spines.

Long regarded as a variety of *E. immersa*, this species is consequently little known. It has been reported from Belfast Bay and Guernsey (Hincks, 1880), the Sound of Mull (Ryland, 1963a), and County Cork (Norton *et al.*, 1971). It has also been collected from 6 m in Milford Haven (Pembrokeshire), encrusting stone, and from 20 m off Padstow (north Cornwall).

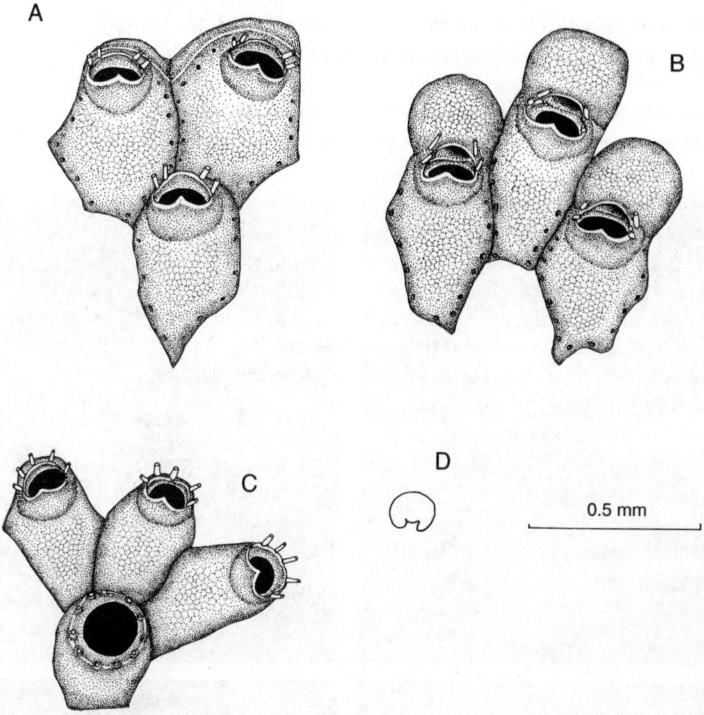

Fig. 36. *Escharella labiosa.*
A, Autozooids at the colony margin. B, Ovicellate zooids. C, The ancestrula and periancestrular zooids. D, Outline diagram of primary orifice.

Escharella octodentata (Hincks)
(Figs 37C; 38)

Mucronella peachii var. (*octodentata*) Hincks 1880: 361, pl. 51, fig. 2
Escharella octodentata : Hayward & Ryland 1979: 142, fig. 55

Colonies forming small, round, silvery patches. Autozooids oval to hexagonal, strongly convex, separated by deep grooves; 0.4-0.65 x 0.3-0.45. Primary orifice

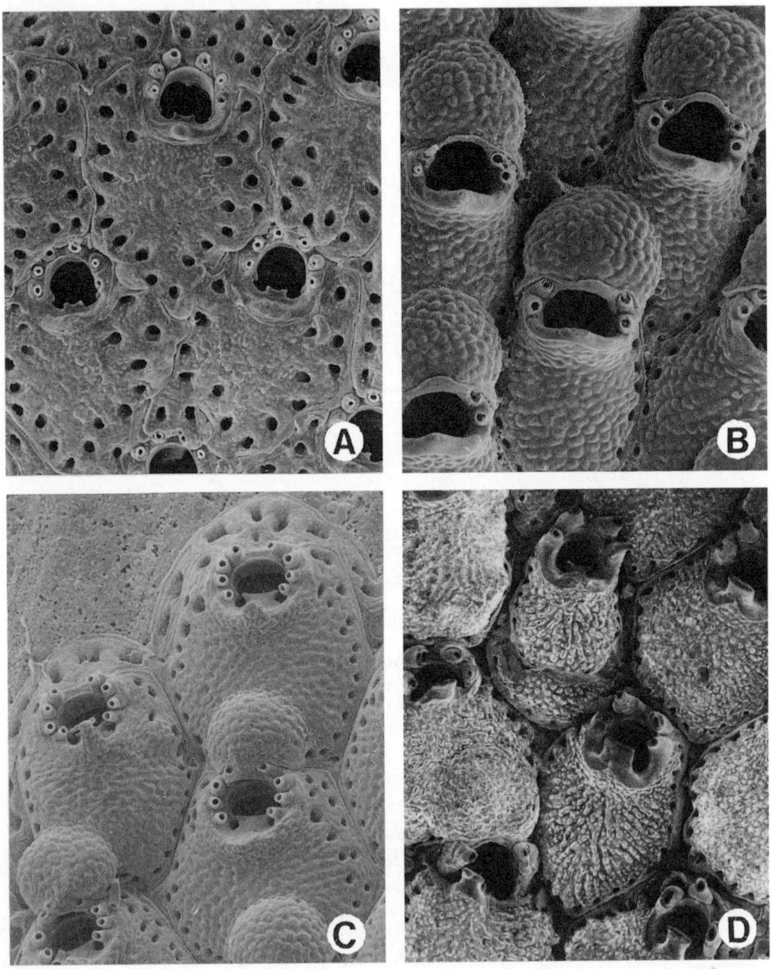

Fig. 37.
A, *Escharella immersa*; x80. B, *Escharella labiosa*; x75.
C, *Escharella octodentata*; x60. D, *Escharella ventricosa*; x50.

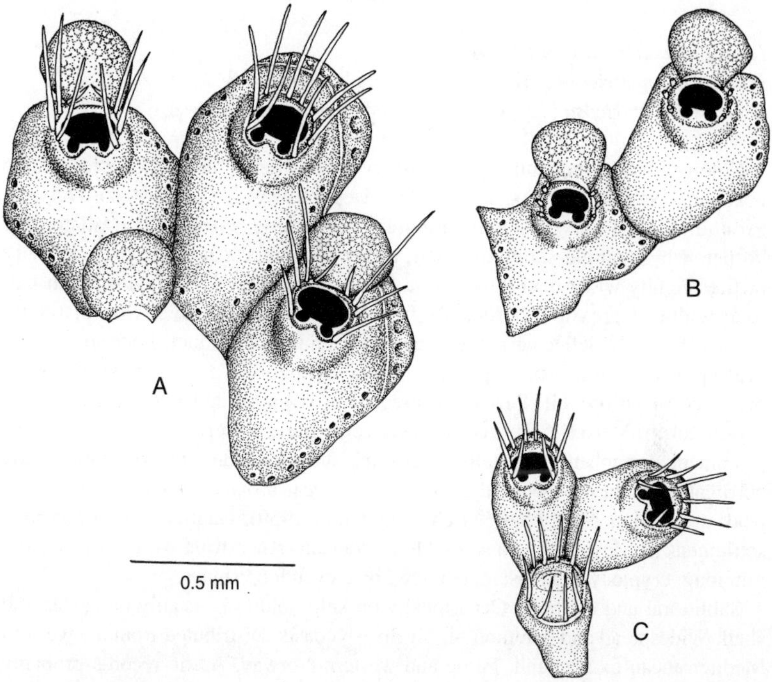

Fig. 38. *Escharella octodentata*.
A, Autozooids at the colony margin, two with ovicells. B, Later autozooids, with oral spines missing. C, The ancestrula, and first two autozooids.

about as wide as long; proximal edge entirely occupied by a large, distinctive, anvil-shaped lyrula with slightly concave distal edge; condyles small, pointed, downcurved. Eight long, slender oral spines equally spaced around distal and lateral borders of orifice; these are long and distally curved when undamaged, often exceeding 0.6 mm, all persist in ovicelled autozooids. Peristome low and thickened, produced mid-proximally as a short, pointed umbo. Frontal shield finely granular, with a single series of small marginal pores, occasionally doubled adjacent to the orifice. Ovicell small, spherical, prominent, surface uniformly granular. Embryos white. Ancestrula 0.3 mm long, tatiform, with an extensive cryptocyst and a transversely oval opesia; bordered by nine long spines.

Offshore, on stones and shell, from 50 m to 200 m depth. Reported from West Norway, Faroe, Shetland and along the western coasts of Britain to the Bay of Biscay. Recorded at depths of 80-200 m in the western Mediterranean (Gautier, 1962).

Escharella ventricosa (Hassall)
(Figs 37D; 39)

Lepralia ventricosa Hassall 1842: 412
Mucronella ventricosa : Hincks 1880: 363, pl. 50, figs 6-8
Escharella ventricosa : Hayward & Ryland 1979: 144, fig. 56

Colony a broad, unilaminar sheet, silvery white when living. Autozooids oval to hexagonal, strongly convex, separated by deep grooves; 0.55-0.7 x 0.45-0.55 mm. Frontal shield finely granular, minutely corrugated, vitreous and glistening, bordered by a single series of small, rather indistinct marginal pores. Primary orifice slightly wider than long; lyrula broad, anvil-shaped, occupying about half total width of proximal border, its free edge slightly concave, with distinctly cusped, frontally deflected corners; condyles minute, indistinct. Four stout distal oral spines present (six in periancestrular autozooids), all persisting in the presence of an ovicell. Peristome present proximally and laterally, typically produced mid-proximally as an erect rectilinear mucro. Ovicell globular, prominent, recumbent on distally succeeding autozooid; calcification granular and corrugated. Embryos recorded throughout the year around the Isle of Man, with peak production November-March (Eggleston, 1969). Gautier (1962) recorded settlement in May in the western Mediterranean. Ancestrula oval, tatiform, 0.3 mm long; cryptocyst extensive, bordered by eleven long spines.

Sublittoral and offshore. Occasionally on kelp holdfasts, usually on stones and shell. Widespread and common off all British coasts. Distributed from the western Mediterranean to Shetland, Faroe and western Norway. Arctic records probably refer to *Escharella klugei* (q.v.).

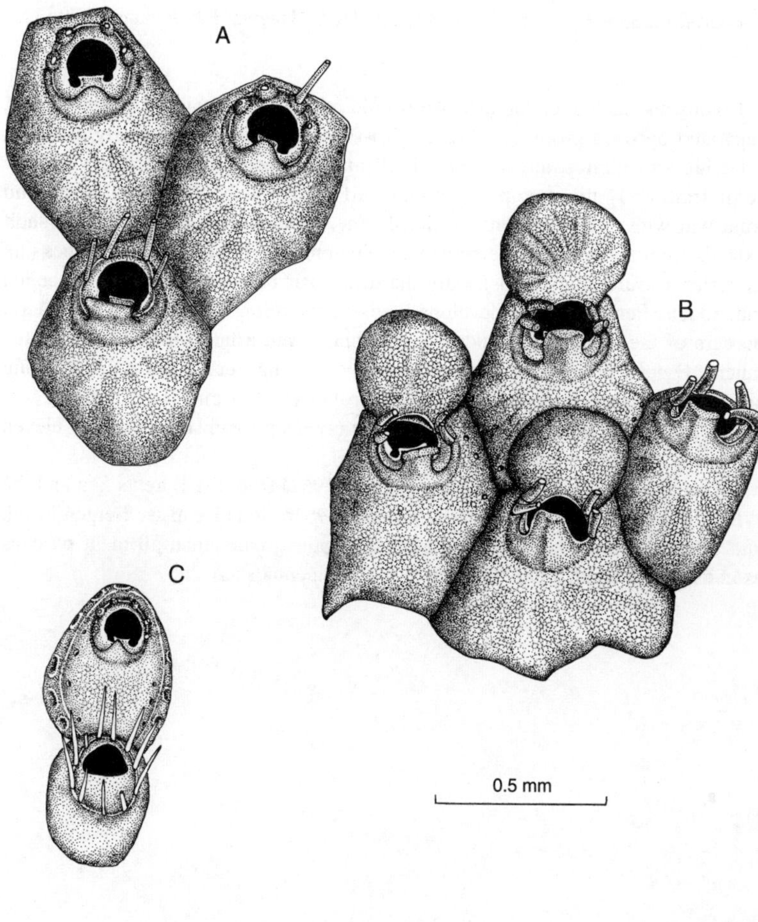

Fig. 39. *Escharella ventricosa.*
A, Autozooids at the colony margin. B, Ovicellate zooids. C, The ancestrula, with the first autozooid budded from it.

Escharella klugei Hayward
(Figs 40; 42A)

Escharella ventricosa var. *peristomata* Kluge 1962: 405, fig. 256; 1975: 488; Ryland 1963b: 26, fig. 8

Escharella klugei Hayward 1979: 82, fig. 4A-C; Hayward & Ryland 1979: 146, fig. 57

Colony encrusting, unilaminar. Autozooids oval to hexagonal, strongly convex, separated by deep grooves; 0.6-0.8 x 0.4-0.5 mm. Frontal wall finely granular, vitreous and glistening; areolae small and inconspicuous. Primary orifice semicircular; lyrula occupying about half proximal border, low, broad and truncated with a slightly concave distal edge. Proximally a broad, smooth shelf extends from the lyrula to the peristome. No condyles. Four distal oral spines (six in periancestrular autozooids), only the distal pair being developed in ovicelled individuals. Peristome well developed, raised, encircling the proximal and lateral borders of the orifice and produced proximally into a tall, tapered and pointed mucro. Ovicell prominent, slightly broader than long, recumbent on succeeding autozooid, surface finely granular. Ancestrula oval, 0.3 mm long; tatiform, the frontal membrane underlain by an extensive cryptocyst and surrounded by eleven spines.

A boreal-arctic species reported by Kluge (1962) from the Barents Sea and the White Sea, *E. klugei* is little known. Ryland (1963b) found it in the Bergen Fjord, and it occurs commonly around Faroe, at depths greater than 50 m. It perhaps extends further south along the edge of the continental shelf.

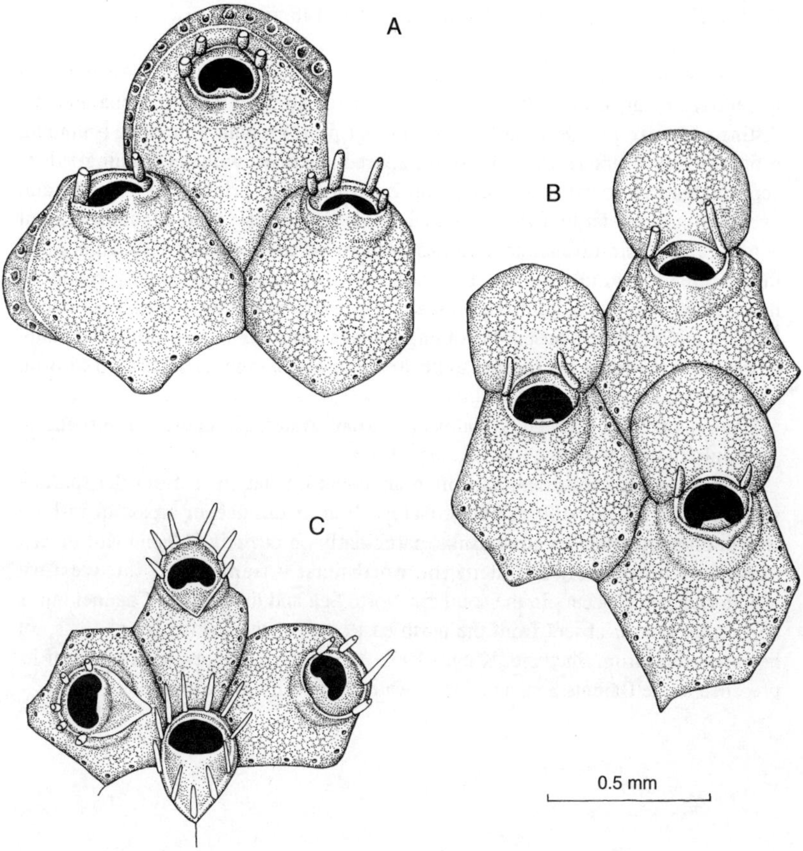

Fig. 40. *Escharella klugei*.
A, Autozooids at the colony margin. B, Ovicellate zooids. C, The ancestrula and periancestrular zooids.

Escharella variolosa (Johnston)
(Figs 41; 42B)

Lepralia variolosa Johnston 1838: 278, pl. 34, fig. 4
Mucronella variolosa : Hincks 1880: 366, pl. 51, figs 3-7
Escharella variolosa : Hayward & Ryland 1979: 148, fig. 58

Colony developing as an extensive flat, unilaminar sheet. Autozooids hexagonal or quadrate, often elongated, flat or slightly convex, separated by distinct sutures; 0.6-0.8 x 0.3-0.5 mm. Frontal wall granular, vitreous, becoming progressively coarser; areolae closely spaced, distinct, with intervening ridges accentuated by secondary calcification. Primary orifice orbicular, lyrula slender, occupying less than half the proximal border, typically anvil-shaped; without condyles, but with distinct articulating surfaces on either side of the lyrula. Two, rarely three, distal oral spines. Peristome low, thickened; proximally produced into a low, pointed mucro, occasionally bifid terminally. Ovicell broader than long, granular, prominent but becoming progressively immersed; ridged frontally where the peristome encroaches onto it. Embryos orange. Ancestrula tatiform, with nine spines.

Embryos present throughout the year in Manx waters (Eggleston, 1969) and in June-July in the Mediterranean (Gautier, 1962).

Encrusting hard substrata, and often abundant on flat shell, from the shallow sublittoral to at least 100 m but probably absent from deeper areas. In inshore waters often common on shell banks, particularly on *Ostrea, Chlamys* and *Pecten* valves. Widely distributed in the north-east Atlantic and the western Mediterranean, it occurs in the southern North Sea and the English Channel but is apparently rare or absent from the north-east coast of the British Isles; it has not been recorded from Shetland. Kluge (1962, 1975) showed that early reports of its presence in the Barents Sea, and Arctic waters generally, were erroneous.

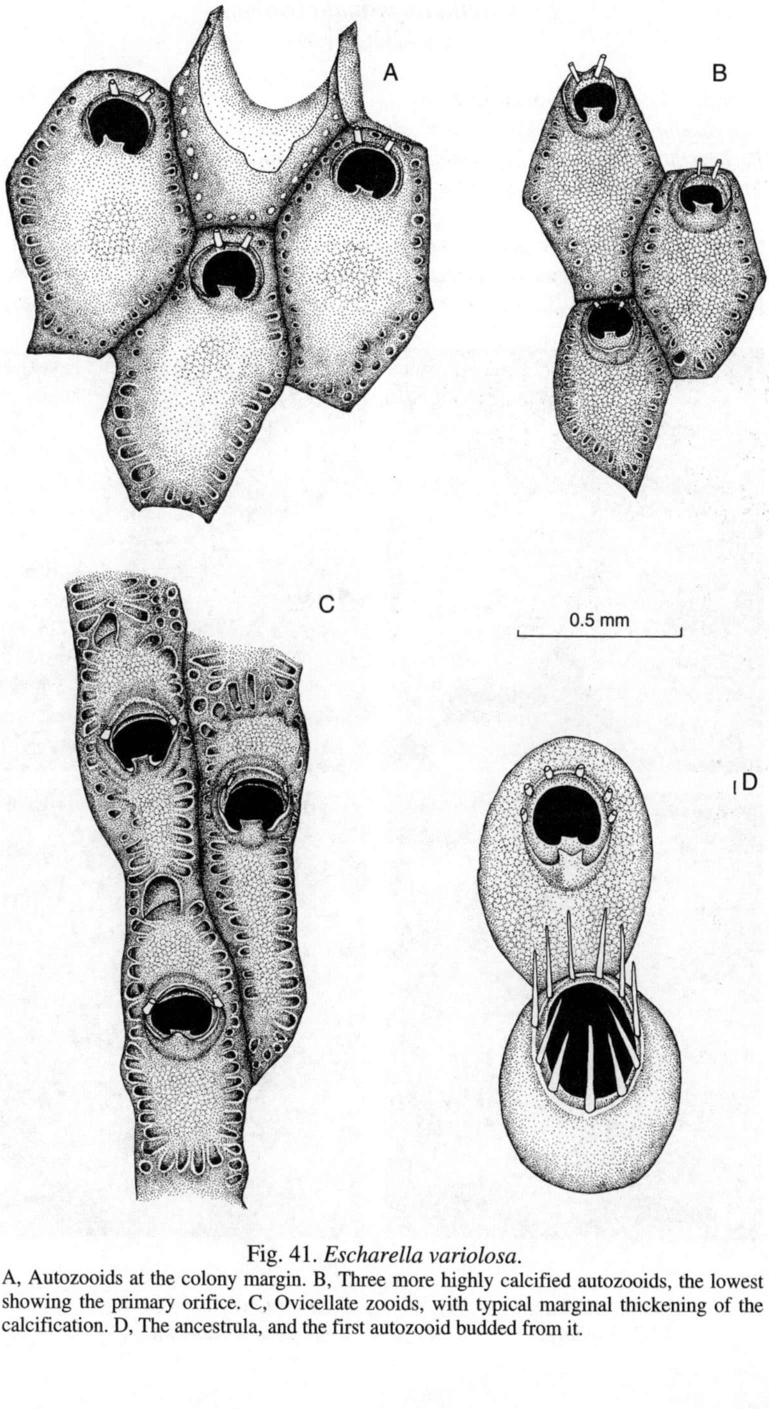

Fig. 41. *Escharella variolosa.*
A, Autozooids at the colony margin. B, Three more highly calcified autozooids, the lowest showing the primary orifice. C, Ovicellate zooids, with typical marginal thickening of the calcification. D, The ancestrula, and the first autozooid budded from it.

Escharella abyssicola (Norman)
(Figs 42C; 43; 44)

Lepralia abyssicola Norman 1869: 307
Mucronella abyssicola : Hincks 1880: 369, pl. 38, figs 1, 2
Escharella laqueata (not *E. abyssicola*): Marcus 1940: 230
Escharella abyssicola : Hayward & Ryland 1979: 150, figs 59, 60

Colony white, often extensive, with autozooids easily visible to the naked eye. Autozooids oval to hexagonal, often sharply tapered proximally: strongly convex, large, 0.8-1.3 x 0.4-0.7 mm, separated by deep grooves. Frontal wall finely

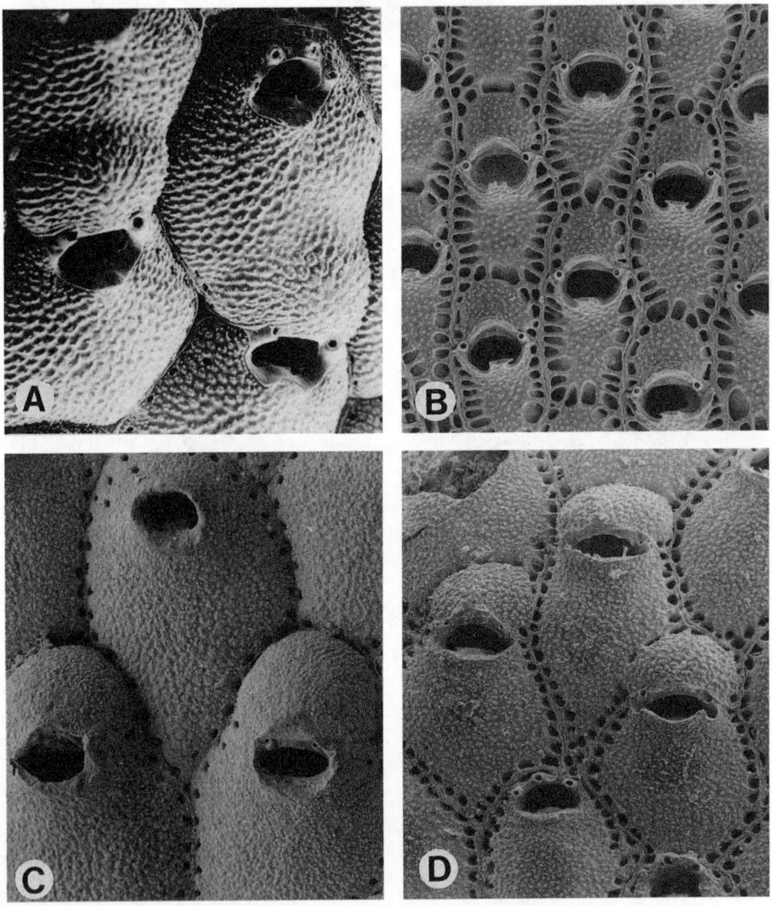

Fig. 42
A. *Escharella klugei*; x60. B, *Escharella variolosa*; x40.
C, *Escharella abyssicola*; x34. D, *Escharella laqueata*; x24.

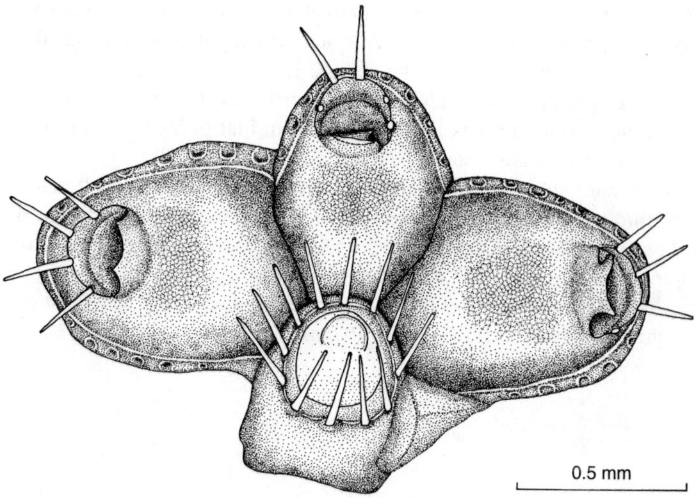

Fig. 43. *Escharella abyssicola*.
The ancestrula and periancestrular zooids

granular, vitreous, becoming coarser as calcification continues. Marginal pores small, inconspicuous, usually seen only in cleaned specimens. Primary orifice wider than long, elliptical; lyrula low and broad, variable in width but typically occupying one-half to three-quarters of the proximal border of the orifice. No condyles. Two or three short thick spines distally, usually lost or obscured by progressive calcification. Peristome well developed, forming a triangular or quadrate lip proximally, obscuring the orifice, and extending onto the frontal surface of the ovicell when present; in old autozooids it may form a complete tube enclosing the orifice. Ovicell globular, wider than long, partially immersed and lying between the two distally succeeding autozooids; often with a triangular ridge frontally as the peristome encroaches onto it. Ancestrula tatiform, elliptical, up to 0.6 mm long; with 12 or 13 tall, slender spines around the periphery. Periancestrular autozooids are smaller than later individuals and may have four or five oral spines.

The frontal calcification of *E. abyssicola* is substantial and becomes thicker and coarser with age. The orifice becomes enclosed in the tubular extension of the peristome, and the boundary between the autozooid and the ovicell is covered. This species is most easily confused with *E. laqueata*, which is of similar size, with a similarly structured orifice and the same tendency to heavy calcification. It is particularly unfortunate that Marcus (1940) transposed the names, and the identifications of Ryland (1963b) were based on Marcus. The two species may be distinguished by the marginal pores. In *E. abyssicola,* the vertical walls of the autozooid are practically non-existent; the frontal calcification extends almost to the substratum, where it joins the numerous, small, basal pore-chambers. The autozooid is consequently strongly convex and the marginal pores are small, often indistinct and may even appear to be absent. *E. laqueata* has substantial vertical walls with larger, more distinct pore-chambers. The frontal wall is scarcely convex, usually flat, and has large distinct marginal pores, interspersed with strong ridges which persist even in heavily calcified individuals.

Found on hard substrata, often stone, in offshore waters from well below the kelp zone to the continental slope and beyond. Its lower bathymetric limit is not known but is at least 350 m. Often found in association with *E. laqueata*. A boreal-arctic species, not circumpolar, *E. abyssicola* extends as far south as west Norway / Shetland, and reaches the Bay of Biscay in deep, cold water just off the edge of the continental shelf (Hayward, 1978a).

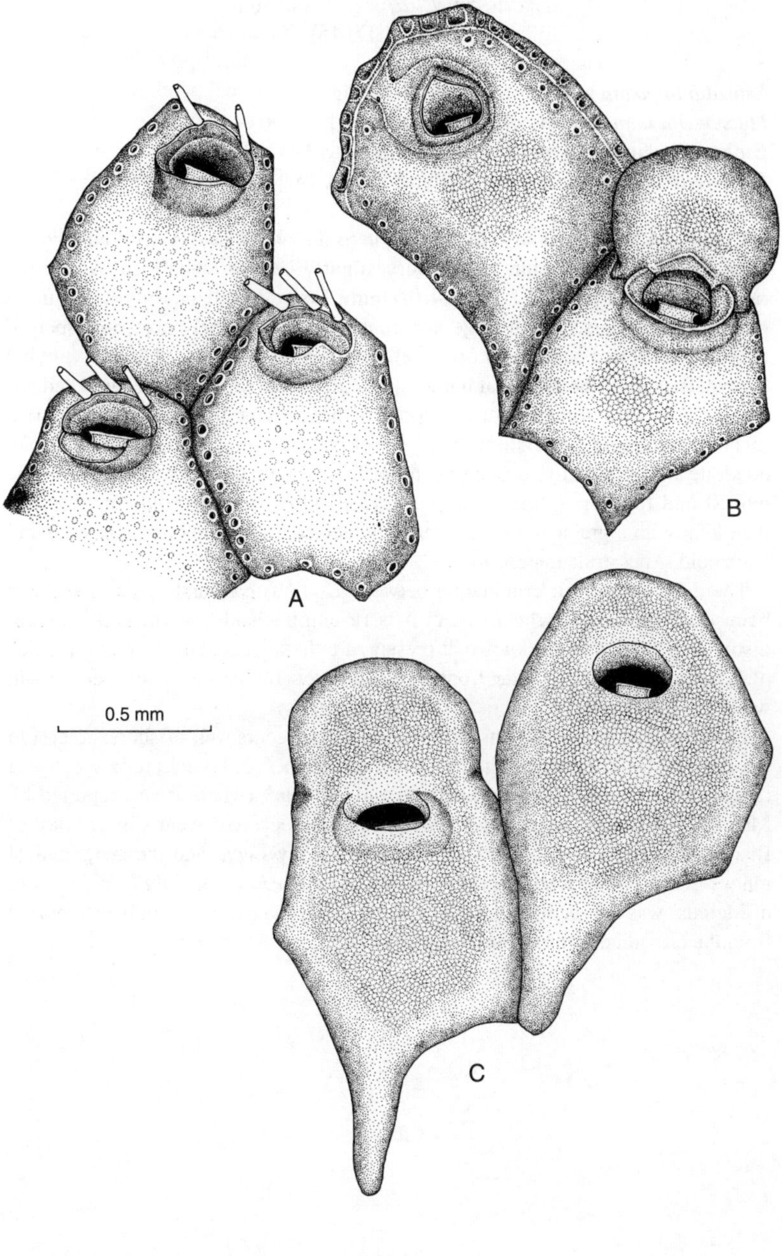

Fig. 44. *Escharella abyssicola*.
A, Autozooids at the colony margin. B, Autozooids in later ontogeny, one ovicellate. C, Old and heavily calcified autozooids (the left ovicellate) with deeply immersed orifices.

Escharella laqueata (Norman)
(Figs 42D; 45)

Lepralia laqueata Norman 1864: 85, pl. 10, fig. 5
Mucronella laqueata : Hincks 1880: 368, pl. 51, fig. 8
Escharella abyssicola (not *E. laqueata*): Marcus 1940: 231
Escharella laqueata : Hayward & Ryland 1979: 152, fig. 61

Colony white, autozooids clearly visible to the naked eye. Autozooids oval to hexagonal, often irregular, flat or very slightly convex, separated by narrow, raised ridges, large, 0.7-1.2 x 0.4-0.7 mm. Frontal wall granular, becoming progressively coarser, with large and conspicuous marginal pores, interspersed with thickened ridges. Primary orifice elliptical, wider than long; lyrula occupying the greater part of the proximal border, low and broad, with concave lateral edges, no condyles. Two or three short, stout oral spines present distally, lost or obscured in later autozooids. Peristome well developed, slightly flared, with a broad quadrate lip proximally, obscuring the orifice; eventually encroaching onto the ovicell and forming a complete tube. Ovicell globular and prominent, broader than long with a granular surface, partly immersed, not recumbent on succeeding autozooid. Ancestrula undescribed.

The difficulties in discriminating between *E. abyssicola* and this species have been discussed under the former. It is re-emphasised that the only certain distinguishing characters, known at present, are the large and conspicuous areolae of *E. laqueata*, and its flatter frontal surface, probably due to its deeper vertical walls and larger pore chambers.

Escharella laqueata encrusts hard substrata and occurs well offshore, at least to the edge of the continental shelf and perhaps deeper. A boreal-arctic species, it ranges south to west Norway, Shetland and Denmark (where it was reported by Marcus (1940) as *E. abyssicola)* and is found in deep cold waters in the Bay of Biscay. It is often found in association with *E. abyssicola* and the geographical ranges of the two species largely coincide, although *E. abyssicola* (=*E. laqueata* of Marcus) was not recorded from Denmark and *E. laqueata* has not been reported from the Canadian arctic or equivalent latitudes.

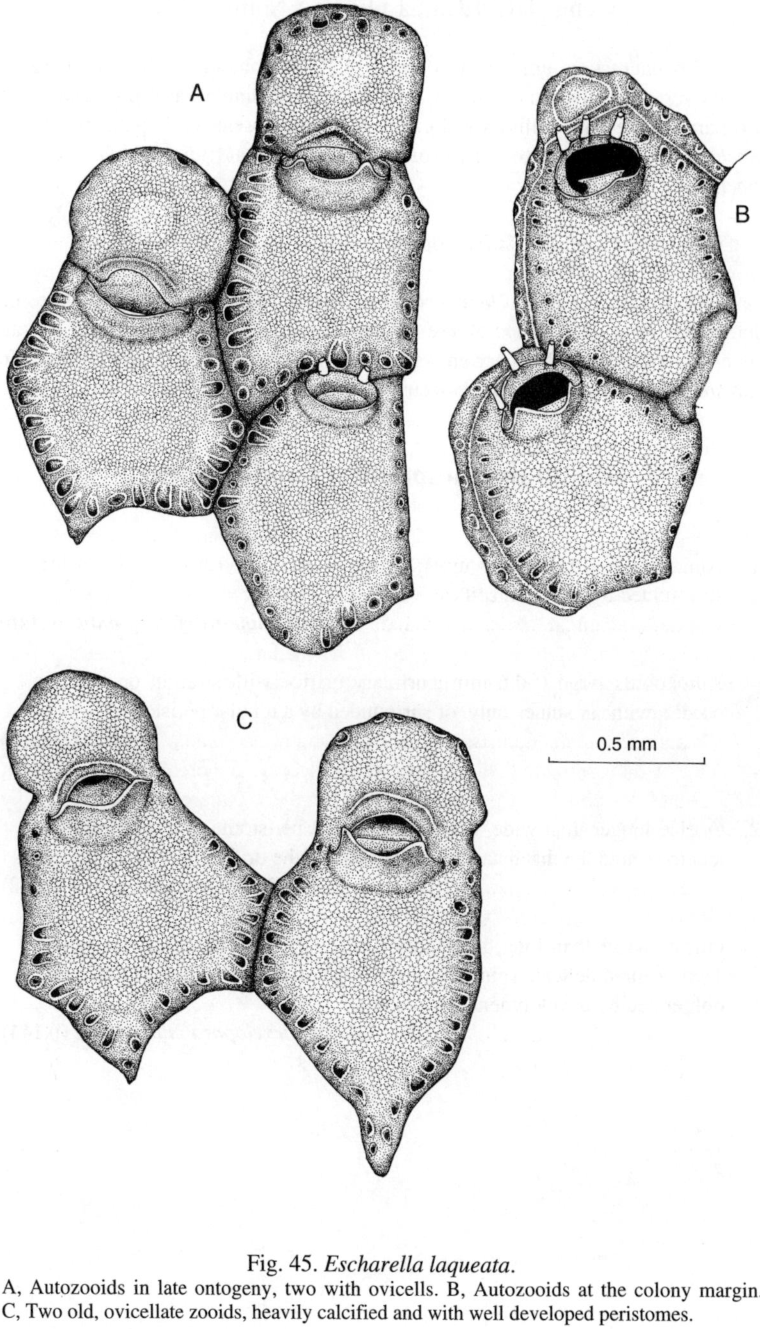

Fig. 45. *Escharella laqueata*.
A, Autozooids in late ontogeny, two with ovicells. B, Autozooids at the colony margin. C, Two old, ovicellate zooids, heavily calcified and with well developed peristomes.

Genus HEMICYCLOPORA Norman, 1894

Colony encrusting, unilaminar. Frontal shield umbonuloid with small marginal pores; vertical walls very reduced, with numerous minute basal pore chambers. Primary orifice with well-marked condyles; lyrula absent; oral spines present or absent. Ovicell hyperstomial, prominent, imperforate; closed by autozooid operculum. No avicularia.

Type species: *Lepralia polita* Norman, 1864

In the British sea area, *Hemicyclopora* species occur principally offshore on hard substrata. Three species are described here. In addition, *H. polita* var. *mucronata* Ryland (1963b), from west Norway and Faroe, was redescribed as *H. labrata* (Hayward, 1994); it may occur off Shetland.

Key to British species of *Hemicyclopora*

1. Autozooids small (<0.6 mm); primary orifice with convex border, surrounded by eight tall spines
.... ***Hemicyclopora multispinata*** (p. 146)

Autozooids large (>0.6 mm); primary orifice with straight or concave border, with six spines only, or surrounded by a tubular peristome
.... **2**

2. Orifice longer than wide, not obscured by a peristome. Six oral spines in an arc around the distal and lateral borders of the orifice
.... ***Hemicyclopora polita*** (p.142)

Orifice wider than long, hidden by a tall, tapering, cylindrical peristome. Two or three delicate spines on distal edge of orifice in earliest ontogeny, obliterated by development of peristome
.... ***Hemicyclopora microstoma*** (p.144)

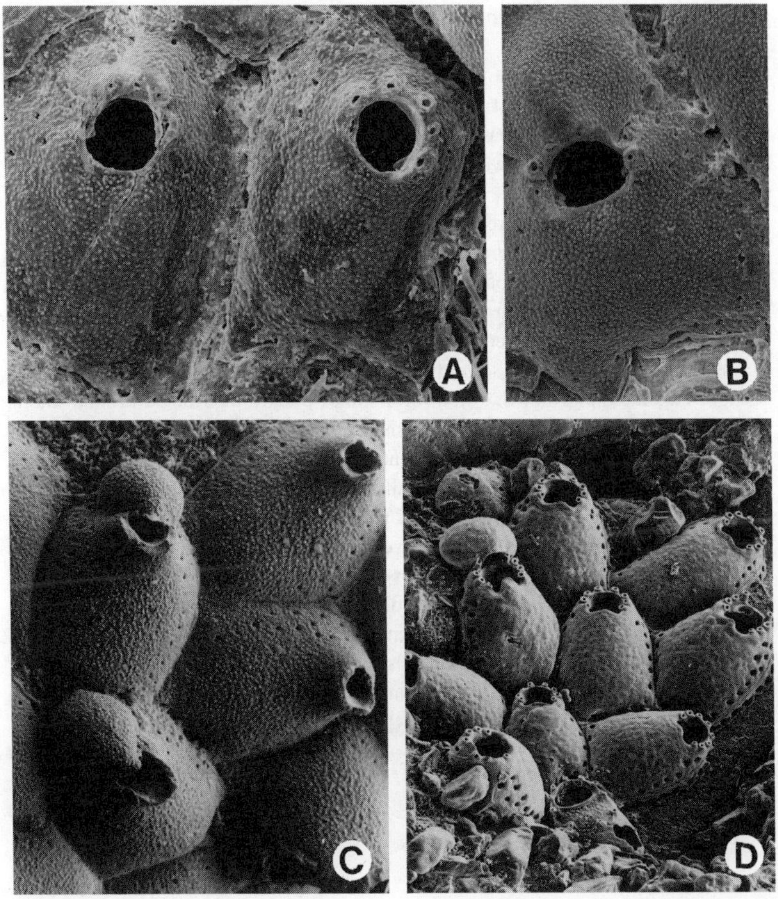

Fig. 46 A, B, *Hemicyclopora polita*.
A, Two autozooids; x50. B, An ovicellate zooid; x50.

C, *Hemicyclopora microstoma*; x40. D, *Hemicyclopora multispinata*; x35.

Hemicyclopora polita (Norman)
(Figs 46A, B; 47)

Lepralia polita Norman 1864: 87, pl. 11, fig. 1
Lepralia polita : Hincks 1880: 315, pl. 32, fig. 5
Hemicyclopora polita : Hayward & Ryland 1979: 156, fig. 62

Colonies developing spreading, irregular, often extensive sheets; thickly calcified, white. Autozooids oval to polygonal, large, strongly convex, widely separated by deep grooves, often with irregular orientation; 0.6-1.0 x 0.5-0.8 mm. Primary orifice longer than wide, with slightly convex proximal border, and conspicuous, basally deflected, triangular condyles; appearing bell-shaped. A low peristomial rim present, in some populations produced proximally as an erect, straight-edged, square lip. Six stout oral spines evenly distributed around lateral and distal rim of orifice, the bases persisting in later ontogeny; early astogenetic autozooids may bear eight spines. Frontal shield thick, smooth, steeply convex with a single or double series of small, rather inconspicuous marginal pores. Ovicell prominent, globular, slightly broader than long, but often irregular in shape; smooth surfaced, frequently developing a low frontal umbo; four spines persist in ovicelled autozooids. Embryos white. Ancestrula oval, tatiform, 0.7 mm long, with a semi-elliptical opesia bordered by eleven slender spines. Embryos present throughout the year in the Bergen Fjord (Ryland, 1963b).

On hard substrata, especially small stones, and often abundant on the coral *Lophelia*. Boreal-arctic, with circumpolar distribution, in the East Atlantic ranging south to Shetland, the Hebrides and the west coasts of Ireland, always well offshore, in deep water. It reaches the northern Bay of Biscay, where it occurs along the edge of the continental slope.

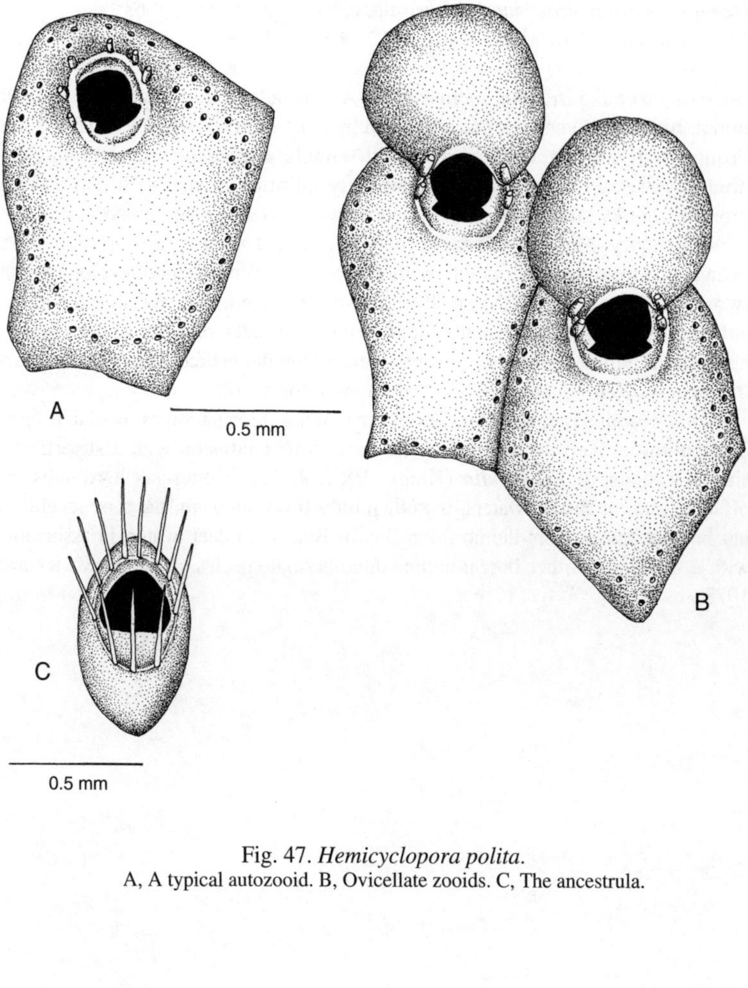

Fig. 47. *Hemicyclopora polita*.
A, A typical autozooid. B, Ovicellate zooids. C, The ancestrula.

Hemicyclopora microstoma (Norman)
(Figs 46C; 48)

Lepralia microstoma Norman 1864: 87, pl. 11, fig. 2
Mucronella microstoma : Hincks 1880: 370, pl. 38, figs 3, 4
Hemicyclopora microstoma : Hayward & Ryland 1979: 158, fig. 63
Not *Escharella microstoma* : Kluge 1962: 408; 1975: 494

Colony forming irregular incrustations. Autozooids oval, hexagonal or irregular in outline, strongly convex and separated by deep grooves; 0.6-0.9 x 0.5-0.7 mm. Frontal wall smooth, fine grained, with minute marginal pores in a single or double series. Primary orifice transversely elliptical, twice as broad as long, proximal border straight or shallowly concave, hidden by the peristome and seen with difficulty. No oral spines. Peristome initiated in two parts, proximal and distal, each forming an elevated, triangular process. With further development the two parts form a united, spout-like peristome, in both fertile and non-fertile autozooids. Ovicell prominent, tilted basally; broader than long, occasionally irregular in outline, narrowed towards the autozooidal orifice, finely granular and imperforate. Reproduction and ancestrula: no information.

H. microstoma is a little known species. What information is available on its distribution is likely to be unreliable owing to its confusion with a superficially similar species of *Escharella* (Kluge, 1962, 1975). It encrusts hard substrata offshore and, in British waters, is known only from Shetland, the type locality. It has been collected from the northern Bay of Biscay, in deep water, in association with a number of other boreal-arctic, stenothermal species (Hayward & Ryland, 1978).

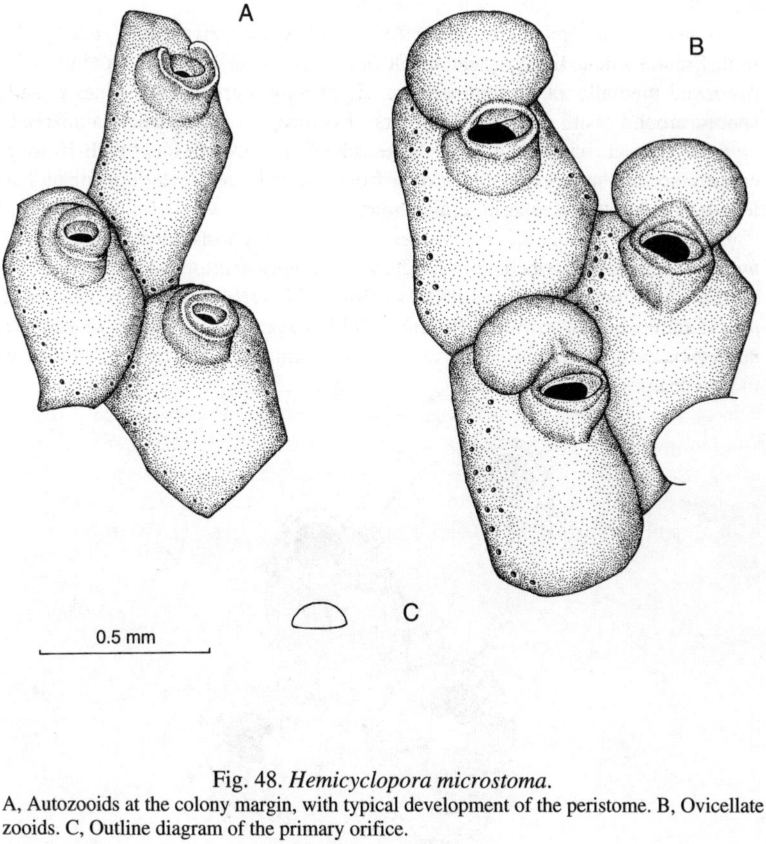

Fig. 48. *Hemicyclopora microstoma*.
A, Autozooids at the colony margin, with typical development of the peristome. B, Ovicellate zooids. C, Outline diagram of the primary orifice.

Hemicyclopora multispinata (Busk)
(Figs 46D; 49)

Lepralia multispinata Busk 1861: 78, pl. 32, figs 5, 6
Hemicyclopora multispinata : Norman 1909: 308 (part), pl. 42, fig. 6; Cook 1968: 216

Colony a small unilaminar patch. Autozooids oval, convex, 0.5-0.6 x 0.3-0.4 mm. Frontal shield finely granular. Orifice longer than wide, its proximal edge produced medially as a prominent lip. Eight long, curved oral spines equally spaced around distal and lateral borders of orifice, six in ovicellate autozooids. Ovicell globular, wider than long. Ancestrula oval, 0.35 mm long, with 10 short, equally spaced spines around an oval frontal membrane, underlain for half its length by a cryptocyst; opesia semicircular.

On shells and biogenic carbonates, offshore. A poorly known species, *H. multispinata* was described from Madeira and ranges south to the Azores (Cook, 1968). To the north, it reaches northern Biscay. The shallow water records of *H. polita* listed by Hayward & Ryland (1978) have been found to refer to *H. multispinata*. It is probably close to its northern limit off the western approaches to the Channel.

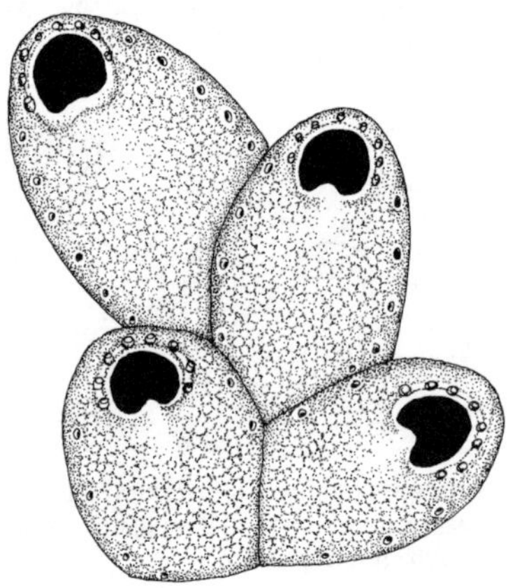

Fig. 49. *Hemicyclopora multispinata*

Fig. 50 (opposite).
A, B, *Neolagenipora collaris*.
A, Group of autozooids, two with ovicells; x110. B, The primary orifice; x400.

C, D, *Neolagenipora eximia*.
C, Ovicellate zooids; x40. D, Autozooids at the colony margin; x60.

Genus NEOLAGENIPORA Vigneaux, 1949

Colony encrusting. Frontal wall of autozooid an umbonuloid shield, with few marginal pores. Primary orifice lepralioid, with conspicuous condyles; oral spines present, peristome well developed. Ovicell hyperstomial, prominent, with scattered pores; not closed by autozooid operculum. No avicularia. Large basal pore chambers present.

Type species: *Lepralia collaris* Norman, 1867

Key to the British species of *Neolagenipora*

1. Proximal edge of orifice straight, or slightly concave
 .... ***Neolagenipora collaris*** (p. 148)

 Proximal edge of orifice convex, or distinctly peaked medially and appearing lyrulate
 .... ***Neolagenipora eximia*** (p. 150)

A third species, *N. rugosa* Hayward (1994), is presently known from Faroe.

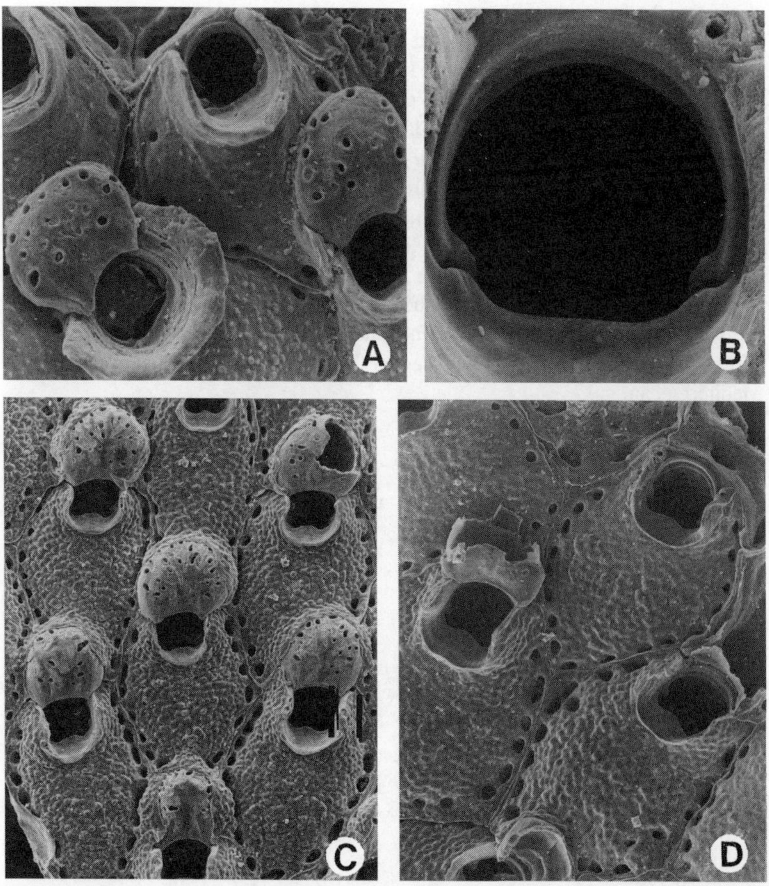

Neolagenipora collaris (Norman)
(Figs 50A, B; 51)

Lepralia collaris Norman 1867: 204
Phylactella collaris : Hincks 1880: 358, pl. 43, fig. 3
Phylactellipora collaris : Hayward & Ryland 1979: 162, fig. 65

Colony forming small rounded or irregular patches, rarely exceeding a few mm^2. Autozooids short, oval, strongly convex, separated by deep grooves; 0.38-0.5 x 0.2-0.3 mm. Frontal wall finely granular, smooth, with a few widely spaced, indistinct marginal pores. Primary orifice slightly wider than long, commonly 0.09 x 0.10 mm, proximal border shallowly concave between paired, blunt condyles. A pair of thin erect spines on distal edge of orifice in some, but not all, autozooids, lost in later ontogeny. Peristome well developed, thick, deep, with a widely flared rim, enclosing the proximal and lateral borders of the orifice. Pore chambers large and conspicuous, one distal and paired disto-lateral. Ovicell prominent, broader than long, thickly calcified, with 15-20 small round pores; developing a peripheral ooecial cover in later ontogeny. Embryos yellow. Ancestrula tatiform, with a smooth gymnocyst and a number of short spines encircling the frontal membrane.

Embryos recorded in June and August-December off the Isle of Man (Eggleston, 1969).

This inconspicuous species occurs offshore on the smallest shell substrata, most often on the inner surfaces of small lamellibranch valves. It occurs in association with *Callopora discreta, Phylactella labrosa* and other species characterised by small colony size. Mature colonies are tiny, rarely exceeding 100 autozooids, and colonies with as few as eight autozooids may bear ovicells. In the largest colonies, almost all autozooids bear ovicells. Bathymetric range unknown, but governed probably by distribution of suitable shell substrata; thus below kelp zone and perhaps to edge of the continental shelf. Described from Shetland, reported from the Antrim coast, Isle of Man and the Channel Islands, and apparently common off the western approaches to the Channel. Not reported from the North Sea or the Mediterranean.

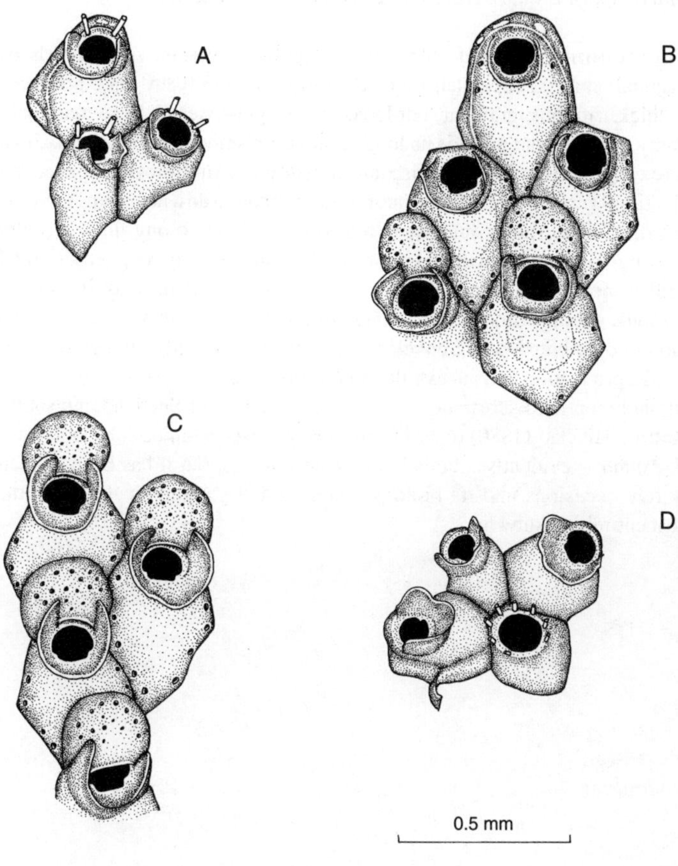

Fig. 51. *Neolagenipora collaris*.
A, Autozooids in early ontogeny, with oral spines. B, Autozooids at the colony margin, without oral spines. C, Ovicellate zooids. D, The ancestrula and periancestrular zooids.

Neolagenipora eximia (Hincks)
(Figs 50C, D; 52)

Lepralia eximia Hincks 1860b: 75, pl. 3, fig. 3
Phylactella eximia : Hincks 1880: 359, pl. 49, fig. 11
Phylactellipora eximia : Hayward & Ryland 1979: 164, fig. 66

Colony forming broad, coherent, encrusting sheets. Autozooids oval to hexagonal, convex, separated by distinct grooves; 0.5-0.8 x 0.3-0.5 mm. Frontal wall thick, vitreous, granular, bordered by a single series of large marginal pores. Primary orifice about as wide as long; distal edge smoothly arched, proximal edge convex, typically with a conspicuous rounded or bluntly tapered median projection, variably developed; condyles rounded, downcurved. Two or three small distal oral spines present in autozooids at the colony growing edge, not persisting in later ontogeny. Peristome low mid-proximally, erect and flared laterally, not developed distally. Ovicell recumbent on distally succeeding autozooid, slightly broader than long, smoothly calcified, with about 20 small, randomly distributed pores; enveloped distally and laterally by a granular ooecial cover. Reproduction and ancestrula: no information.

Offshore, on hard substrata. Occurs around Faroe, off Shetland and southwards to Antrim. Hincks' (1880) record from Cornwall has yet to be substantiated.

N. eximia is evidently a boreal cold water species, but it has been collected on very few occasions and its biology, ecology and geographical distribution are almost entirely unknown.

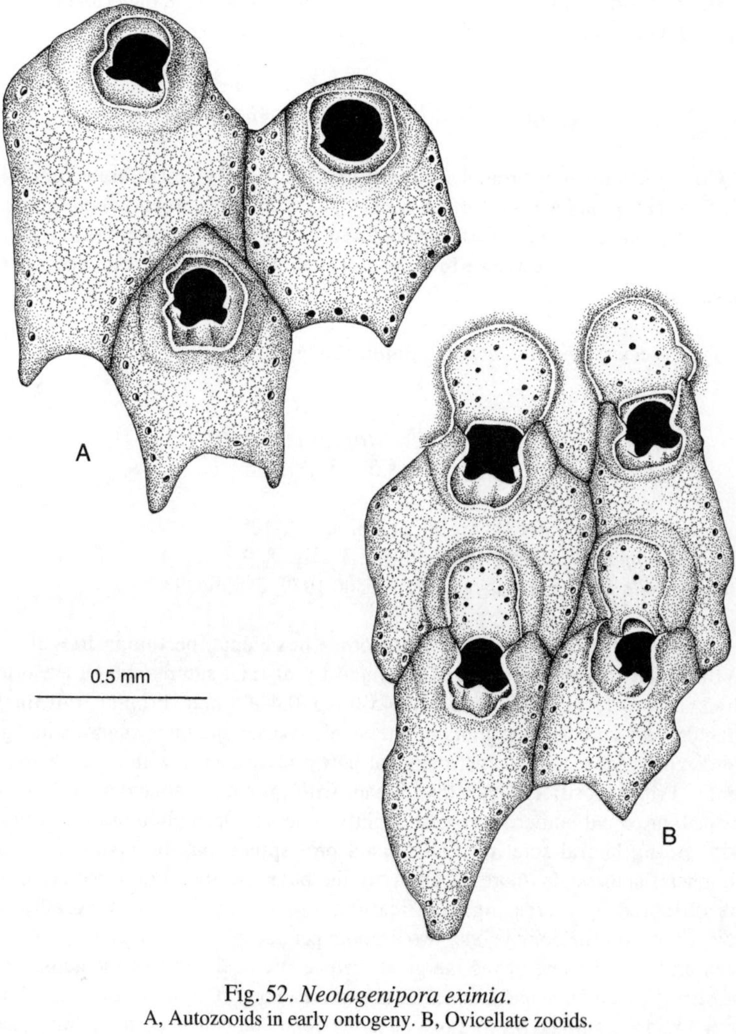

Fig. 52. *Neolagenipora eximia.*
A, Autozooids in early ontogeny. B, Ovicellate zooids.

Family LEPRALIELLIDAE Vigneaux, 1949

Colony encrusting; or erect, branching. Autozooids with umbonuloid frontal shield, bordered by marginal pores. Orifice bell-shaped or sinuate, often with well developed lateral condyles. Oral spines present or absent. Peristome variably developed. Adventitious avicularia present. Ovicell prominent, imperforate. Mural or basal pore chambers present.

Genus LEPRALIELLA Levinsen, 1916

Colony encrusting. Frontal wall with marginal pores, imperforate centrally. Orifice with prominent lateral condyles. Operculum with strong lateral sclerites. Ovicell prominent, imperforate, not closed by zooidal operculum. Oral spines present. Avicularia adventitious, usually small. Basal pore chambers well developed.

Type species: *Lepralia hippopus* Smitt, 1868

Lepraliella hippopus (Smitt)
(Figs 53; 57A)

Lepralia hippopus Smitt 1868b: 20, pl. 26, figs 99-105
Lepralia hippopus : Hincks 1880: 309, pl. 33, figs 8, 9
Hippoporella hippopus : Hayward & Ryland 1979: 218, fig. 93

Colony encrusting, irregular. Autozooids hexagonal, becoming irregular in outline, flat or slightly convex, separated by distinct sutures which are often thickened to form raised ridges; 0.45-0.6 x 0.4-0.5 mm. Frontal wall thick, granular and vitreous, becoming progressively coarser and more rugose with age; bordered by large and distinct marginal pores; occasionally with a small frontal umbo. Primary orifice longer than broad, with prominent condyles; horseshoe-shaped, proximal border straight or slightly concave. Operculum brown, distinct, with strong lateral sclerites. Two distal oral spines may be visible in some peripheral autozooids; more usually only the bases are seen briefly before these are obscured by increasing calcification. Avicularia elliptical, sporadically developed over the colony, one, two or more per autozooid but sometimes absent; each arises above one of the marginal areolae. Vertical walls of the autozooids substantial, with large and distinct basal pore chambers. Ovicell broader than long with a nodular surface bearing one or more frontal umbones; partially immersed, becoming covered by secondary calcification, and often indistinct.

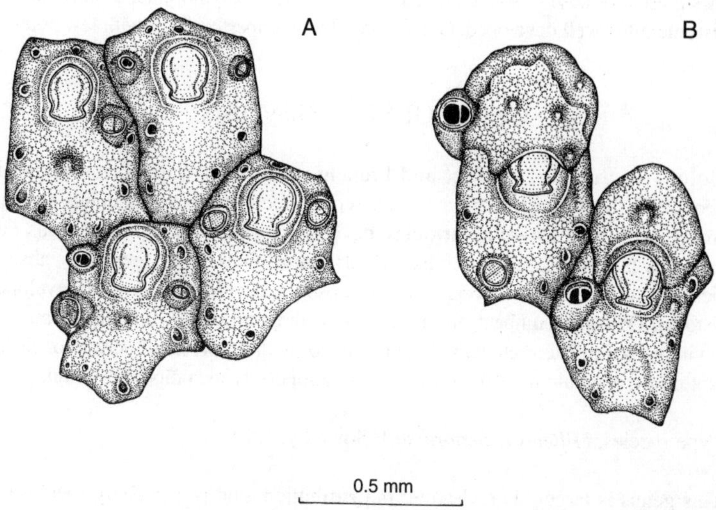

Fig. 53. *Lepraliella hippopus*.
A, Group of autozooids. B, Ovicellate zooids.

A common boreal-arctic species, circumpolar in distribution, *L. hippopus* is only doubtfully included in the British fauna. It was reported from off the Northumberland coast by Hincks (1880) but was not found by Eggleston (1975); it was not recorded from west Norway by Ryland (1963b). If it does occur, offshore from the north-east coasts of England and Scotland seem the most likely areas. Kluge (1975) reported it from 14-235 m.

Family BRYOCRYPTELLIDAE Vigneaux, 1949

Colony encrusting, unilaminar to multilaminar and nodular; or erect, branching, often robust. Autozooids with umbonuloid frontal shield bordered by marginal areolae. Primary orifice with or without a lyrula; condyles variably developed. Oral spines present or absent. Avicularia adventitious, often polymorphic, usually suboral. Peristome often well developed. Ovicell prominent, imperforate, or with few pores.

Genus PORELLA Gray, 1848

Colony encrusting, or erect and branching. Frontal wall of autozooid an umbonuloid shield, with large marginal septula recessed in conspicuous areolae. Primary orifice with lyrula variously developed, generally inconspicuous and sometimes lacking. Condyles indistinct or absent. Oral spines generally absent. Avicularia adventitious, suboral, often enclosed and hidden by a well developed peristome. Ovicell recumbent on distally succeeding autozooid, initially prominent, often immersed in later ontogeny; imperforate, or with a few, small pores; not closed by autozooid operculum. Vertical walls with multiporous and uniporous septula.

Type species: *Millepora compressa* J. Sowerby, 1805

This genus is largely boreal-arctic in distribution, and is especially speciose in the higher latitudes. Seven species are described here; at least four others occur around Faroe (Hayward, 1994) and might range south to Shetland, whilst many others occur further to the north (Kluge, 1962).

Key to the British species of *Porella*

1. Colony erect
.... **2**

Colony encrusting
.... **3**

2. Colony with cylindrical, tapering brunches; autozooids opening on one side of branch only. Peristome developing a round secondary orifice, with the avicularium usually still visible
.... ***Porella laevis*** (p.164)

Colony robust, with flat branches becoming irregularly rounded in later ontogeny; autozooid orifices open on all surfaces. Peristome developing a sinuate secondary orifice, the avicularium usually obscured
.... ***Porella compressa*** (p.158)

3. Avicularium diagnostic: large, the distal palate crossed by two longitudinal struts. Primary orifice twice as broad as long. Ovicell small, imperforate

.... ***Porella struma*** (p.170)

Avicularium small, the distal palate without longitudinal struts

.... **4**

4. Primary orifice semicircular, without a lyrula. Peristome well developed, often forming a complete tube in ovicellate autozooids; avicularium prominent, often umbonate. Ovicell smooth. Autozooids ~0.45 mm long. Northern only

.... ***Porella alba*** (p. 156)

Lyrula always present, though occasionally very small. Peristome often thickened, but usually incomplete distally in ovicellate autozooids

.... **5**

5. Lyrula very small (Fig. 62B). Calcification thin and delicate, areolae small and indistinct. Embryos pale yellow. Northern only

.... ***Porella patula*** (p.168)

Lyrula always distinct, quadrate; calcification thick and granular, areolae conspicuous

.... **6**

6. Autozooids large (>0.4mm long). Ovicell with a single central pore, occluded in later ontogeny. Often heavily calcified, the suboral avicularium immersed and obscured. Embryos red. Widespread and common

.... ***Porella concinna*** (p.160)

Autozooids small (<0.4mm long). Ovicell imperforate, always distinct. Calcification thickening to produce conspicuous interareolor struts between adjacent autozooids. Suboral avicularium usually prominent, umbonate. Embryos yellow. Southern and western coasts

.... ***Porella minuta*** (p.166)

Porella alba Nordgaard
(Figs 54; 57B)

Lepralia minuta Norman 1868: 308 (in part)
Porella minuta: Hincks 1880: 326 (in part), pl. 36, fig. 6; Kluge 1962: 455; 1975: 552
Porella alba Nordgaard 1906: 25, pl. 3, figs 43-46; Hayward & Ryland 1979: 122, fig. 46

Colonies forming irregular, spreading, unilaminar sheets. Autozooids small, oval to hexagonal, convex, separated by deep grooves; 0.3-0.45 x about 0.2 mm. Primary orifice broader than long, proximal border straight, without a lyrula; condyles minute, pointed, basally deflected. No oral spines. Peristome well developed, thin and raised, extending onto the ovicell of fertile autozooids and eventually forming a complete tube; with a stout, mid-proximal umbo. Avicularium oval, perpendicular to orifice and enclosed within peristome. Frontal shield finely granular, with large and conspicuous marginal pores; thickening steadily through ontogeny and developing raised ridges between the pores. Ovicell spherical, prominent, smoothly calcified and imperforate; partially obscured in later ontogeny by the development of the peristome.

Lower shore and shallow sublittoral, on a variety of substrata, including stones, shells and kelp holdfasts. This is an arctic-boreal species, with a circumboreal distribution. It is known from the Barents Sea, the White Sea, Spitzsbergen and northern Iceland. Dick & Ross (1988) recorded it as abundant on the lower shore at Kodiak Island, Alaska. *P. alba* is at the extreme southern limit of its range in Britain, and is rare here. It is known from a single specimen collected, by C. W. Peach, from the lower shore at Wick, which was included by Norman (1868) in his original account of *P. minuta*.

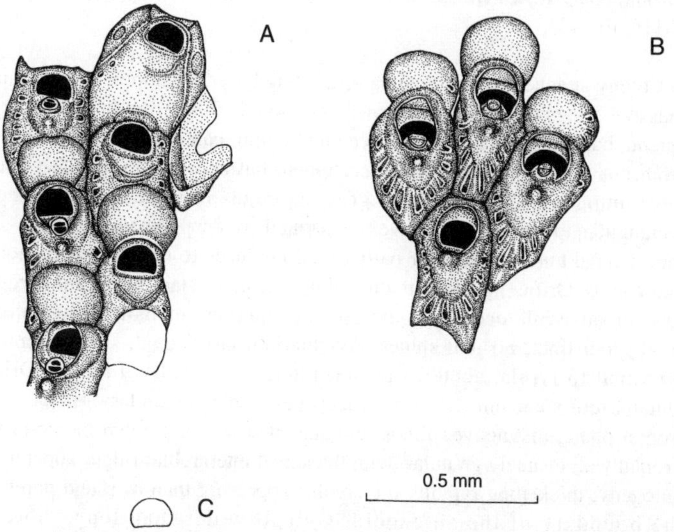

Fig. 54. *Porella alba.*
A, Autozooids at the colony margin, with the peristome in various stages of development. B, Autozooids in later ontogeny, three ovicellate, with typical tubular peristomes. C, Outline diagram of the primary orifice.

Porella compressa (J. Sowerby)
(Figs 55; 57C, D)

Millepora compressa J. Sowerby 1805: 83, pl. 41
Porella compressa : Hincks 1880: 116, pl. 45, figs 4-7; Hayward & Ryland 1979: 116, fig. 43

Colony attached by a spreading encrusting base, developing a thickly calcified, massive, irregularly branched growth; commonly 30-40 mm high, with equivalent spread, but often much larger, exceeding 90 mm; pale pinkish orange when living. Branches bilaminar, initially flat-sectioned, but becoming irregularly rounded in later ontogeny. Autozooids at growing edges in regular alternating series, rectangular to hexagonal, convex, separated by deep grooves; 0.4-0.6 x 0.2-0.3 mm; frontal budding in older parts of colony leads to random orientation of new autozooids. Orifice wider than long, almost semicircular, the proximal edge with a low, broad lyrula occupying almost three-quarters of its total width; condyles small, indistinct; no oral spines. Avicularium large, oval, situated immediately proximal to lyrula, acute to frontal plane, proximally directed. Orifice and avicularium soon immersed in a deep peristome; secondary orifice flush with frontal plane, distinctive, almost circular, with a small notch medio-proximally. Frontal wall coarsely granular, with thickened interareolar ridges apparent in early ontogeny; thickening rapidly, the areolae appearing then as round pores marking the boundary of the autozooid. Ovicell wider than long, more or less hemispherical, finely granular, imperforate; visible only in earliest ontogeny, soon completely immersed. Embryos pink to pinkish orange, recorded during December and January in western Norway (Ryland, 1963b).

Offshore, on coarse grounds, attached to stones, large shell debris or rock. Colonies are perhaps very long-lived. Growth continues through budding of new autozooids at branch tips and through frontal budding, giving rise to new laminae, on the proximal regions of the colony. Large colonies may be thickly encrusted by numerous other species of bryozoans. Occurs below the kelp belt and across much of the continental shelf. A boreal arctic, circumpolar, species. It is abundant around Faroe, occurs off all British coasts and ranges south to the Bay of Biscay. Absent from Mediterranean.

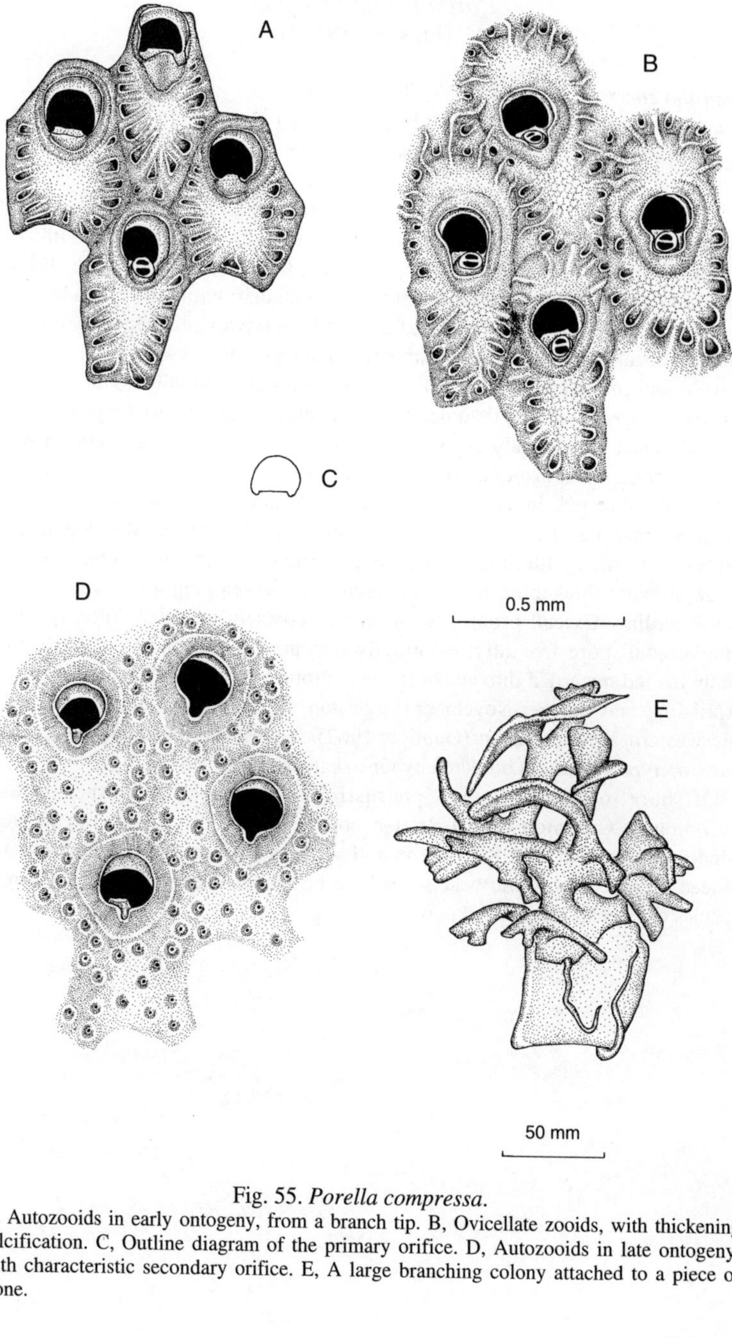

Fig. 55. *Porella compressa.*
A, Autozooids in early ontogeny, from a branch tip. B, Ovicellate zooids, with thickening calcification. C, Outline diagram of the primary orifice. D, Autozooids in late ontogeny, with characteristic secondary orifice. E, A large branching colony attached to a piece of stone.

Porella concinna (Busk)
(Figs 56; 58A, B)

Lepralia concinna Busk 1854: 67, pl. 99, figs 1-6
Porella concinna : Hincks 1880: 323, pl. 46, figs 1, 3-5, 7-8, 10-13
Porella concinna : Hayward & Ryland 1979: 118, fig. 44

Colony encrusting, unilaminar, developing extensive, spreading sheets. Autozooids oval to hexagonal, in later ontogeny rather irregular, with wavy outlines; convex, separated by deep grooves; 0.4-0.5 x 0.25-0.4 mm. Primary orifice as broad as long, appearing rather quadrangular; lyrula short, quadrangular, occupying about half proximal orifice width; condyles minute, indistinct; two slender distal oral spines present in early ontogeny only. A low, thick peristome developing in later ontogeny. Avicularium dimorphic: usually small, with oval rostrum lodged within peristome, no wider than lyrula, almost perpendicular to frontal plane; occasionally supplanted by a larger type, occupying almost whole frontal length of autozooid, with spatulate rostrum normal to frontal plane and proximally directed. In later ontogeny small adventitious avicularia may develop on lateral margins of autozooid. Frontal shield coarsely granular, developing a low umbo above the cystid of the suboral avicularium; bordered by a single series of large areolae; thickening in late ontogeny and developing a characteristically wavy outline. Ovicell prominent, spherical, coarsely granular; initially with a single small pore frontally, eventually obscured by secondary calcification. Embryos red, recorded throughout the year around the Isle of Man, with peaks in April-June and October-November (Eggleston, 1969); present January to April in the western Mediterranean (Gautier, 1962). Ancestrula oval, tatiform, with a narrow cryptocystal rim bordered by nine slender spines.

Offshore, on hard grounds, encrusting stones, shell and other organic carbonates. Common on all British coasts. Distributed from the western Mediterranean northwards to Faroe and western Norway. Its distribution, and indeed presence, in Arctic waters needs to be re-examined as boreal and Arctic species of *Porella* are poorly known.

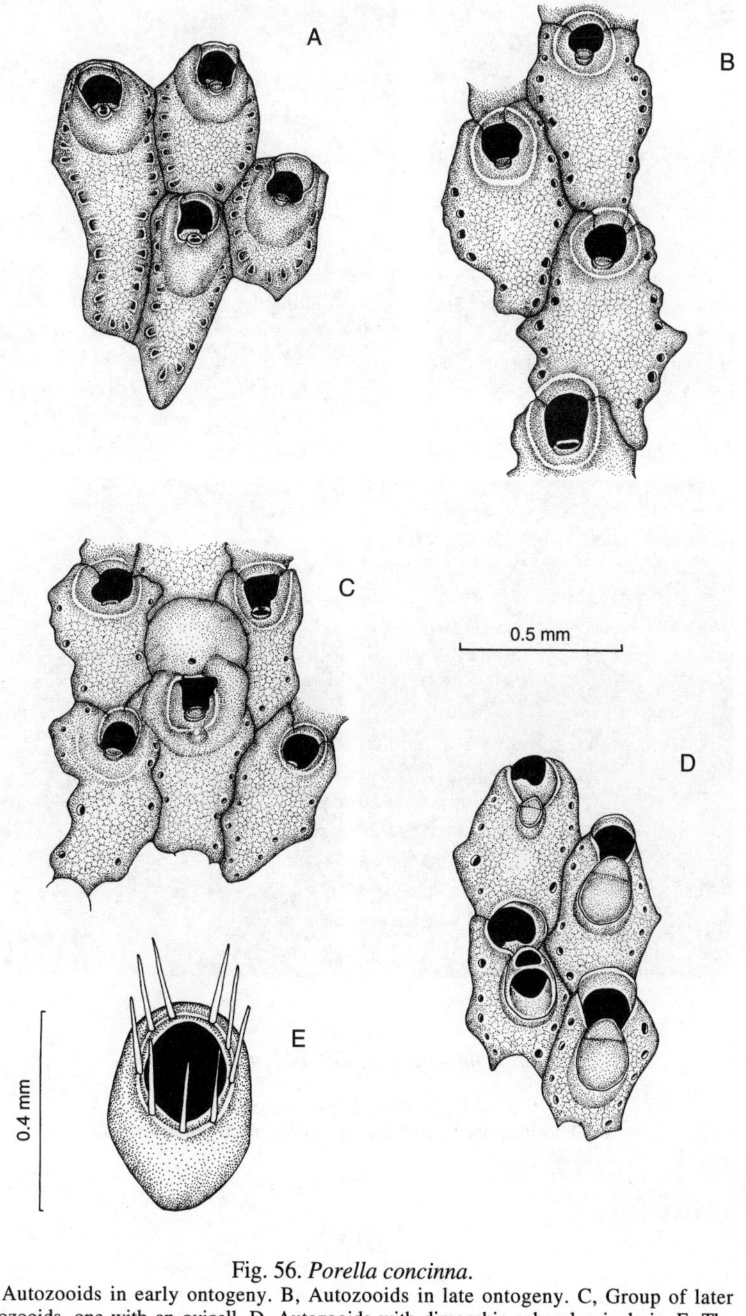

Fig. 56. *Porella concinna.*
A, Autozooids in early ontogeny. B, Autozooids in late ontogeny. C, Group of later autozooids, one with an ovicell. D, Autozooids with dimorphic suboral avicularia. E, The ancestrula.

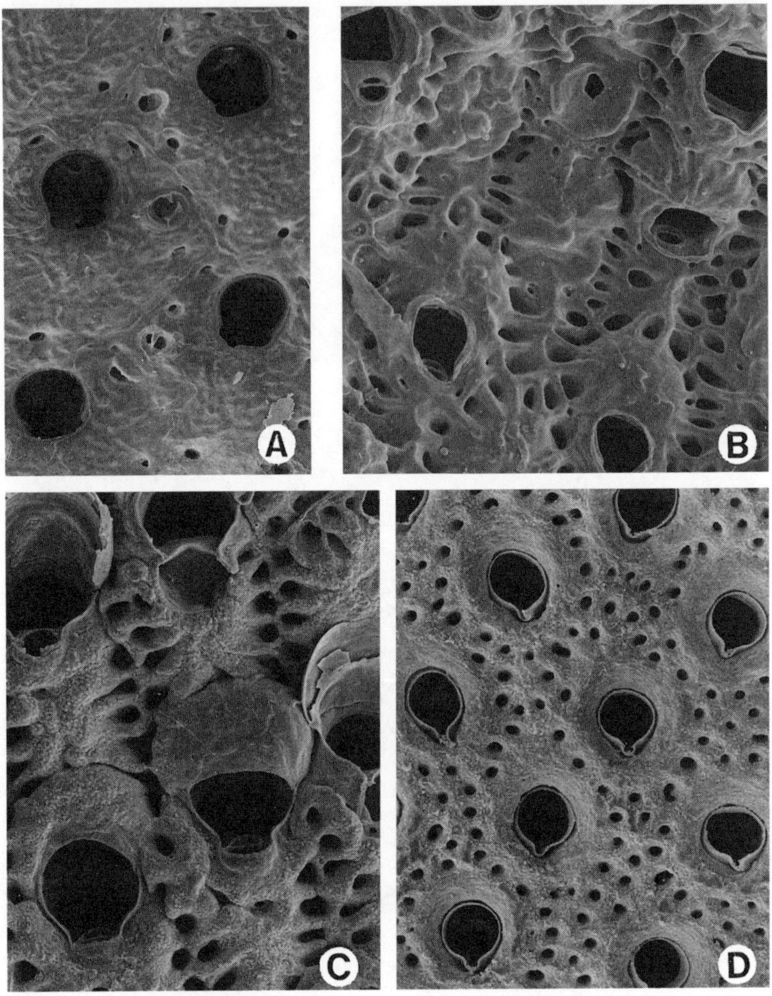

Fig. 57.
A, *Lepraliella hippopus*; x70. B, *Porella alba*; x60.

C, D, *Porella compressa*.
C, Ovicellate zooids; x80. D, Autozooids in late ontogeny; x50

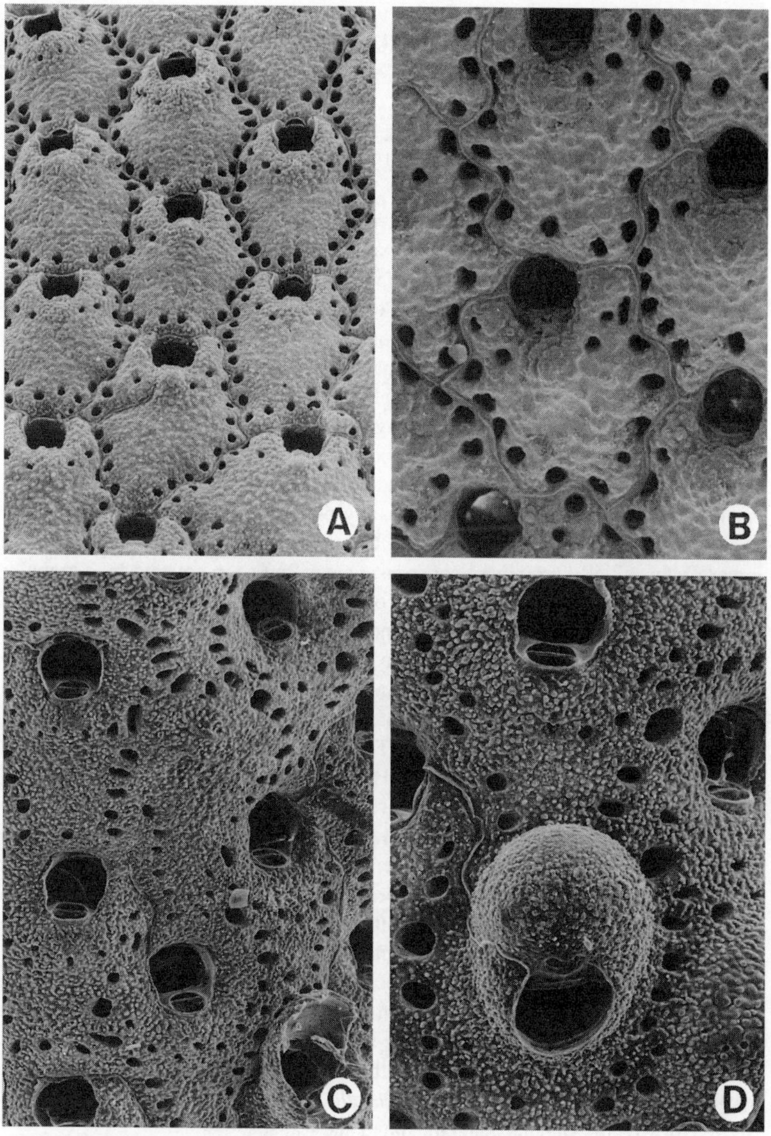

Fig. 58.
A, B, *Porella concinna*.
A, Group of autozooids (Martins Haven); x50. B, Detail (Faroe); x85.

C, D, *Porella laevis* (Faroe).
C, Group of autozooids; x45. D, Detail, with ovicellate zooid; x70.

Porella laevis (Fleming)
(Figs 58C, D; 59)

Cellepora laevis Fleming 1828: 532
Porella laevis : Hincks 1880: 334, pl. 47, figs 10, 11; Kluge 1962: 461; 1975: 559
Porelloides laevis : Hayward & Ryland 1979: 126, fig. 48

Colony attached by a spreading, encrusting base, developing an erect, rigid, irregularly branching, three dimensional form. Branches cylindrical, tapered, thickening steadily through ontogeny. Up to 40 mm high, yellow coloured in life. Autozooids in alternating whorls of four or five; oval to hexagonal, initially broad, convex, separated by distinct sutures. Frontal shield finely granular, bordered by a single series of relatively large marginal pores; a single line of smaller pores borders the peristome. Primary orifice as broad as long; proximal edge with a very short, straight-edged lyrula occupying most of its width; condyles minute, scarcely discernible. No oral spines except in ancestrula and periancestrular autozooids. Peristome low and thickened, most prominent laterally. Avicularia enclosed by peristome, situated immediately proximal to lyrula; rostrum broadly oval, perpendicular to plane of orifice. Ovicell more or less spherical, smooth, imperforate; initially prominent and conspicuous, but soon covered by a granular ooecial cover, becoming immersed and obscured. Ancestrula tatiform, the opesia with a narrow cryptocystal rim, surrounded by nine spines. Embryos present in August and December in western Norway (Ryland, 1963b).

The basal regions of the colony are thickened and strengthened by the continued calcification of the lowermost autozooids, rather than by frontal budding of new sheets of autozooids. The encrusting base continues to spread through the budding of new autozooids around its periphery, but thickens by steady frontal calcification, and thus remains essentially a single sheet. The surface of the colony thus alters considerably with age; autozooid orifices become deeply immersed, together with the avicularia, and the secondary orifice is more or less round. The orifices of many autozooids are obliterated around part of each branch periphery, defining apparent frontal and abfrontal surfaces.

Offshore, on hard substrata, frequently in association with the stone coral, *Lophelia*. Essentially a boreal-arctic species occurring off northwestern Britain at depths greater than 50 m. It is common around Faroe, occurs off Shetland and ranges south to the Bay of Biscay, where it was recorded at 150-1500 m (Hayward & Ryland, 1978), and to the Sea of Alboran, western Mediterranean, where it occurs at 175-200 m (Gautier, 1962). *Porella laevis* is distributed northwards to Arctic Norway, the Barents Sea, Greenland and Canada.

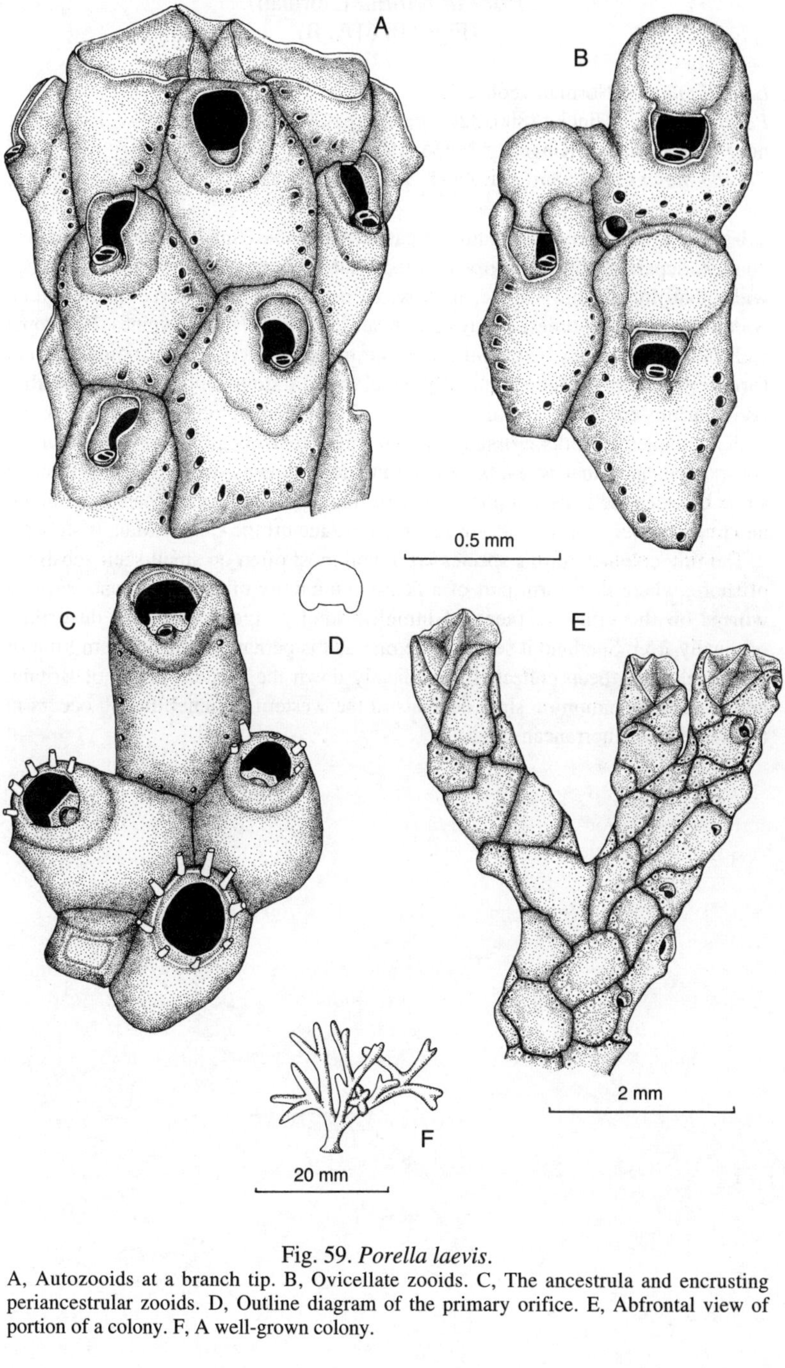

Fig. 59. *Porella laevis*.
A, Autozooids at a branch tip. B, Ovicellate zooids. C, The ancestrula and encrusting periancestrular zooids. D, Outline diagram of the primary orifice. E, Abfrontal view of portion of a colony. F, A well-grown colony.

Porella minuta (Norman)
(Figs 60; 61A, B)

Lepralia minuta Norman 1868: 308 (in part)
Porella minuta : Hincks 1880: 326 (in part), pl. 29, figs 1, 2
not *Porella minuta* : Kluge 1962: 455; 1975: 552
Porella minuta : Hayward & Ryland 1979: 120, fig. 45

Colony a small, round, unilaminar patch. Autozooids small, oval to hexagonal, convex, separated by deep grooves; 0.35-0.4 x about 0.2 mm. Primary orifice wider than long, with a broad, low, lyrula occupying about three-quarters total width of proximal border; condyles minute, indistinct. No oral spines. Peristome well developed, thick, extending onto the ovicell in fertile autozooids, but not forming a complete tube; typically developing a stout umbo mid-proximally. Avicularium enclosed within peristome, almost perpendicular to orifice, oval, with a complete, slender crossbar. Frontal shield coarsely granular, with small but conspicuous marginal pores, thickening in later ontogeny and developing distinct struts between the marginal pores. Ovicell spherical, prominent, coarsely granular and imperforate. Yellow embryos recorded in June off the Calf of Man, Irish Sea.

The tiny colonies of this species are found most often on small shell substrata, offshore, where they form part of a dense community of bryozoans and serpulid worms on the inner surfaces of lamellibranch valves. Although described originally from Shetland it seems to be rare and is perhaps at the northern limit of its range. It has been collected sporadically down the western coasts of Britain, and is not uncommon on shell deposits in the western Channel. It also occurs in the western Mediterranean.

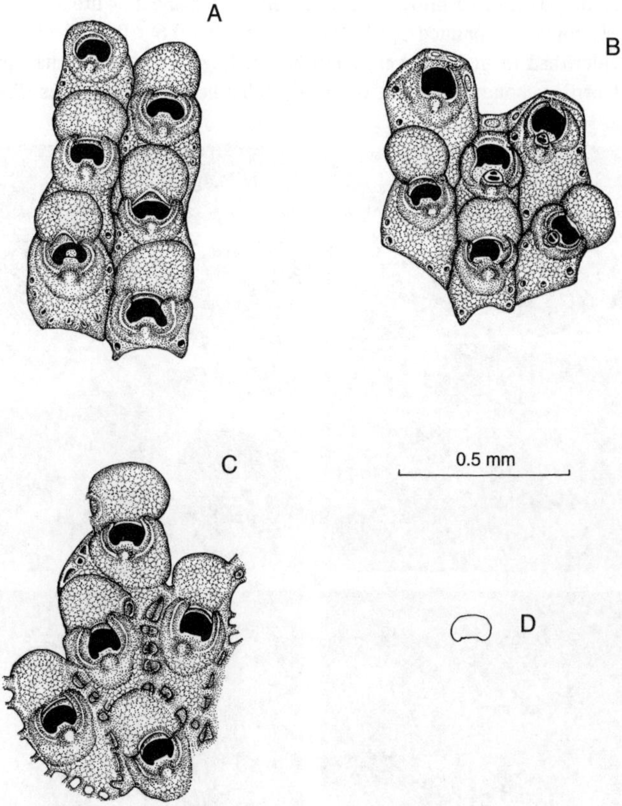

Fig. 60. *Porella minuta*.
A, Group of zooids, mostly ovicellate, showing the characteristic form of the peristome. B, Autozooids at the colony margin, with avicularia visible. C, Ovicellate zooids in late ontogeny. D, Outline diagram of the primary orifice.

Porella patula (M. Sars)
(Figs 61C; 62)

Lepralia patula M. Sars 1851: 148
Porella patula : Hayward & Ryland 1979: 124, fig. 47

Colony encrusting, forming a brittle, unilaminar sheet. Autozooids oval to hexagonal, convex, separated by distinct sutures; 0.4-0.6 x 0.3-0.4 mm. Primary orifice wider than long; lyrula very short, tapered, occupying about half proximal width of orifice; condyles small, rounded, indistinct; no oral spines. Peristome

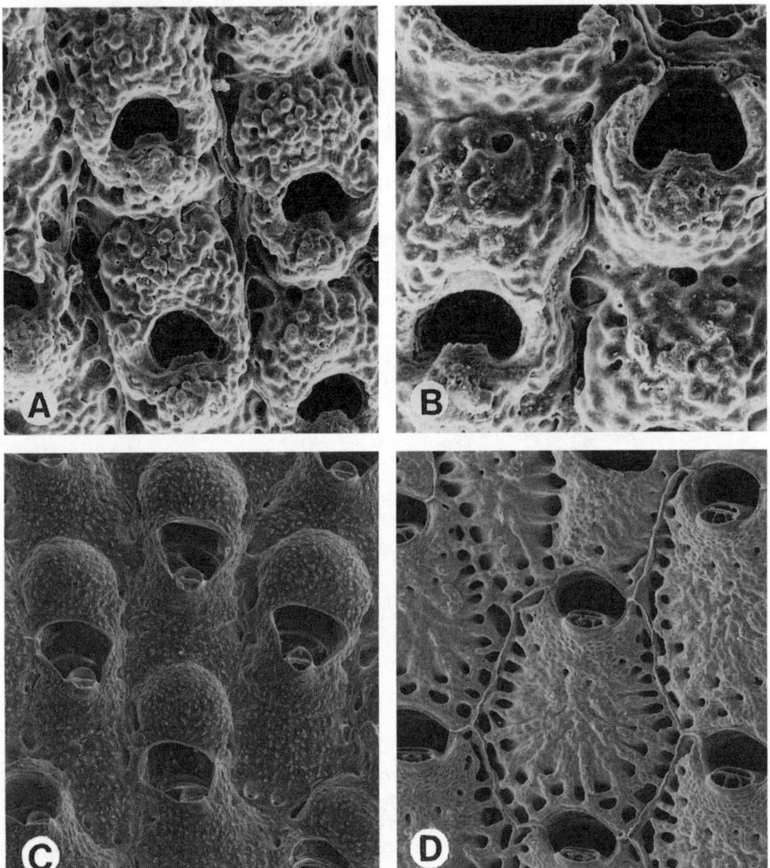

Fig. 61.
A, B, *Porella minuta*.
A, Group of ovicellate zooids; x90. B, Autozooids at the colony margin; x140.

C, *Porella patula*; x50. D, *Porella struma*; x45.

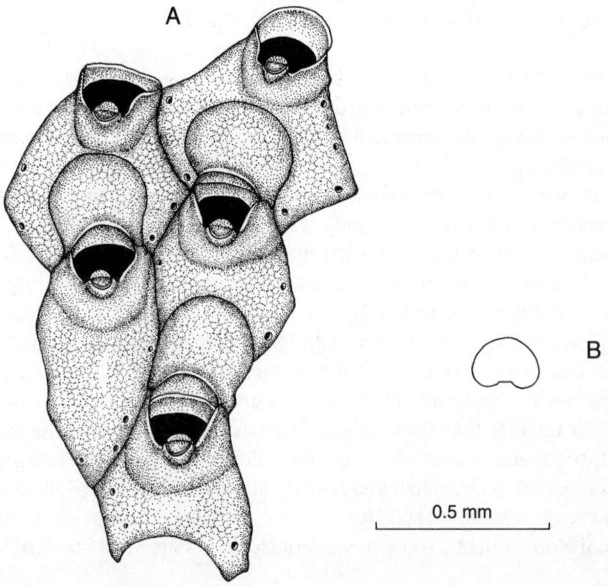

Fig. 62. *Porella patula*.
A, Group of autozooids, three with ovicells. B, Outline diagram of the primary orifice.

thin, raised, incorporating the suboral avicularium; rostrum only slightly acute to frontal plane, scarcely wider than the lyrula, tapered distally and appearing pear-shaped. Frontal shield finely granular, vitreous, bordered by few, widely spaced, small areolae; not developing pronounced interareolar struts. Ovicell broader than long, prominent, finely granular, imperforate; becoming partially immersed, but remaining conspicuous. Embryos pale yellow, recorded in March off Northumberland (Eggleston, 1969), and in August and October in western Norway (Ryland, 1963b).

Offshore, on hard substrata. A boreal-arctic species recorded from the Barents Sea, Greenland and Canada; in the eastern North Atlantic it ranges south to western Norway and north-east England, but is absent from the western coasts of Britain.

Porella struma (Norman)
(Figs 61D; 63)

Hemeschara struma Norman 1868: 221, pl. 7, figs 6-8
Porella struma : Hincks 1880: 329, pl. 39, figs 3-5
Porelloides struma : Hayward & Ryland 1979: 128, fig. 49

Colony encrusting, unilaminar, brittle; developing spreading, frilled sheets, often partly free of the substratum. Autozooids large, flat; quadrangular, hexagonal or polygonal, separated by distinct, thickened sutures. Primary orifice broader than long, proximal edge with a short, straight-edged lyrula occupying most of its width; condyles rounded, smooth and indistinct; two short distal oral spines present in early ontogeny only. Frontal shield coarsely granular, bordered by a single or double series of large areolae, with thickened struts developing between them in later ontogeny. Avicularian rostrum broadly oval, lodged within secondary orifice close to the lyrula, almost perpendicular to frontal plane; mandible semi-elliptical; crossbar complete, continuous with two stout struts crossing the distal palate; cystid voluminous, contributing to a proximal peristome, with a single line of small pores around its lower edge. Tentacles 17-18. Ovicell broader than long, slightly flattened frontally, imperforate, smooth surfaced, becoming immersed in granular calcification and obscured. Pale peach embryos reported in December and January in western Norway (Ryland, 1963b).

Offshore, on hard substrata. Boreal-arctic, perhaps circumpolar, distributed in the eastern North Atlantic south to western Norway, Faroe and Shetland.

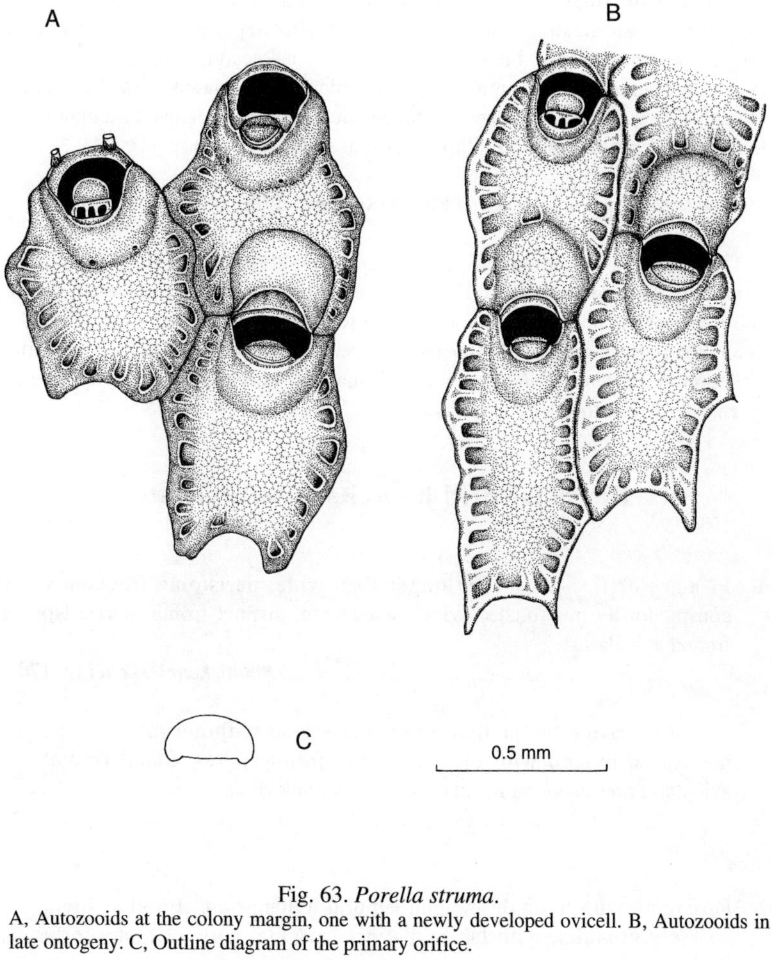

Fig. 63. *Porella struma*.
A, Autozooids at the colony margin, one with a newly developed ovicell. B, Autozooids in late ontogeny. C, Outline diagram of the primary orifice.

Genus PALMISKENEA Bishop & Hayward, 1989

Colony originating as an encrusting sheet, developing an erect, bilaminar, free branching form. Frontal shield of autozooids an umbonuloid shield; imperforate centrally, with small marginal pores, the frontal openings of which migrate medially as calcification thickens. Primary orifice lepralioid, with straight or convex proximal border; blunt proximo-lateral condyles present; no oral spines. Avicularia adventitious, monomorphic, typically associated with the orifice. Ovicell hyperstomial, recumbent, ectooecium with few, irregular foramina close to aperture. Vertical walls with tiny uniporous septula.

Type species: *Millepora skenei* Ellis & Solander, 1786

Three species are described here. Two others, *P. faroensis* Hayward and *P. aquilonia* Hayward have been described recently from Faroe (Hayward, 1994), and it is possible that either or both may be found to range further south along the outer continental shelf. *Palmiskenea* was only recently distinguished from the Celleporoidean genus *Palmicellaria* by Bishop & Hayward (1989). It appears to be entirely Arctic to subarctic in distribution.

Key to the British species of *Palmiskenea*

1. Primary orifice distinctly longer than wide; peristome frequently conspicuously mucronate; ovicell with large, distinct frontal pores. No frontal avicularia

.... ***Palmiskenea skenei*** (p. 178)

Primary orifice wider than long; peristome without conspicuous mucrones; ovicell with few, indistinct frontal pores. Small frontal avicularia present along the margins of older autozooids

.... **2**

2. Peristome with a single median suboral avicularium. Frontal shield coarsely granular, with large, distinct marginal pores. Colony form slender, dichotomously branching

.... ***Palmiskenea lorea*** (p. 174)

Peristome with one median suboral and paired lateral oral avicularia. Frontal shield finely shagreened, with small marginal pores. Colony broad, flat, lobed or indistinctly branched

.... ***Palmiskenea plana*** (p. 176)

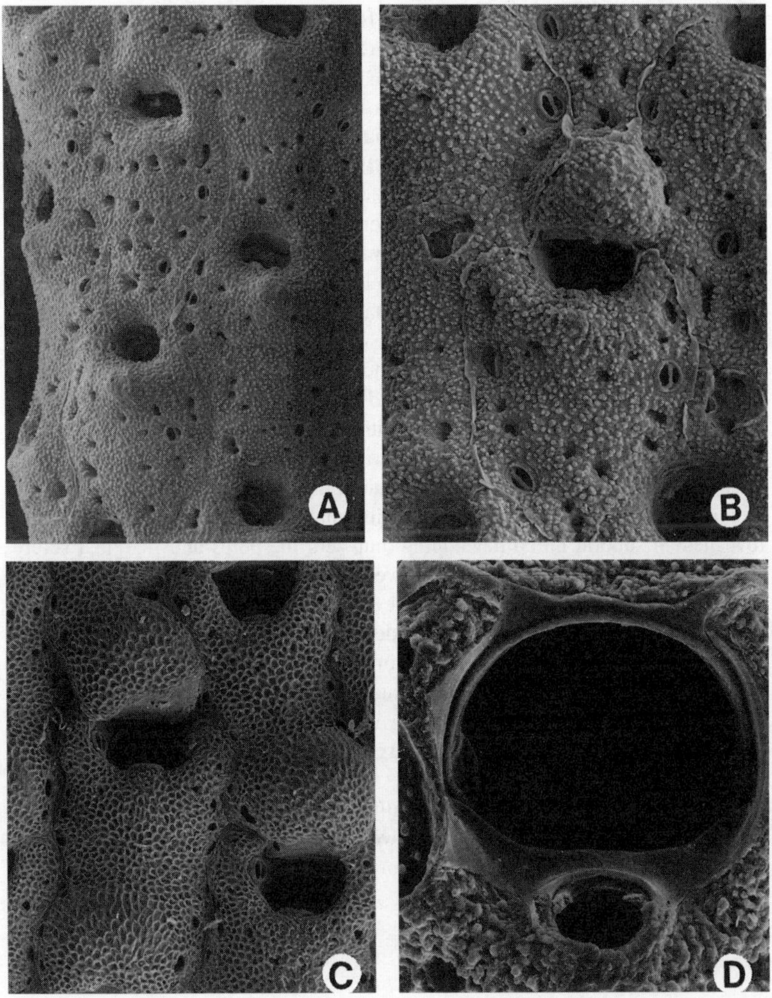

Fig. 64.
A, B, *Palmiskenea lorea*.
A, Autozooids in late ontogeny; x40. B, An ovicellate zooid; x65.

C, D, *Palmiskenea plana*.
C, A group of autozooids; x45. D, The primary orifice; x190.

Palmiskenea lorea (Alder)
(Figs 64A, B; 65)

Eschara lorea Alder 1864: 104, pl. 3, figs 5-7
Palmicellaria lorea : Hincks 1880: 382, pl. 52, figs 5, 6
Palmicellaria lorea : Hayward & Ryland 1979: 132, fig. 51

Colonies 30-40 mm high, with flattened, slender branches 1-2 mm wide, dividing dichotomously at irregular intervals to give a flattened, two-dimensional fan. Autozooids in about six alternating, longitudinal rows on each branch face; oval to rectangular, or irregularly polygonal, convex, separated by distinct sutures; 0.7-0.8 x 0.4-0.5 mm. Primary orifice wider than long, proximal edge very shallowly concave; condyles rounded, indistinct; peristome long and thickened, deepening markedly in later ontogeny and completely hiding the orifice; a single median suboral avicularium, enclosed within peristome, perpendicular to orifice, orientated along the disto-proximal axis; rostrum broadly oval; crossbar slender, complete, without a columella. Frontal shield finely granular, with a single series of comparatively large marginal pores; thickening greatly in later ontogeny, with a low median suboral umbo developing, but with the pores and marginal sutures remaining distinct. Additional, oval adventitious avicularia are budded from the marginal pores, frequently two, three or more per autozooid, with varying orientation. Ovicell relatively small, wider than long, with a few irregular frontal pores close to its aperture; becoming immersed in secondary calcification and quite obscured. In late ontogeny the frontal shields of the basalmost autozooids of the colony are grossly thickened, the primary orifice and ovicell become deeply immersed and quite hidden, but the marginal pores and frontal avicularia remain distinct.

Offshore, on coarse deposits. Originally described from Shetland, *P. lorea* has been reported sporadically from off the western coasts of Britain, but it is still rare and very little known, and nothing is known of its biology and ecology.

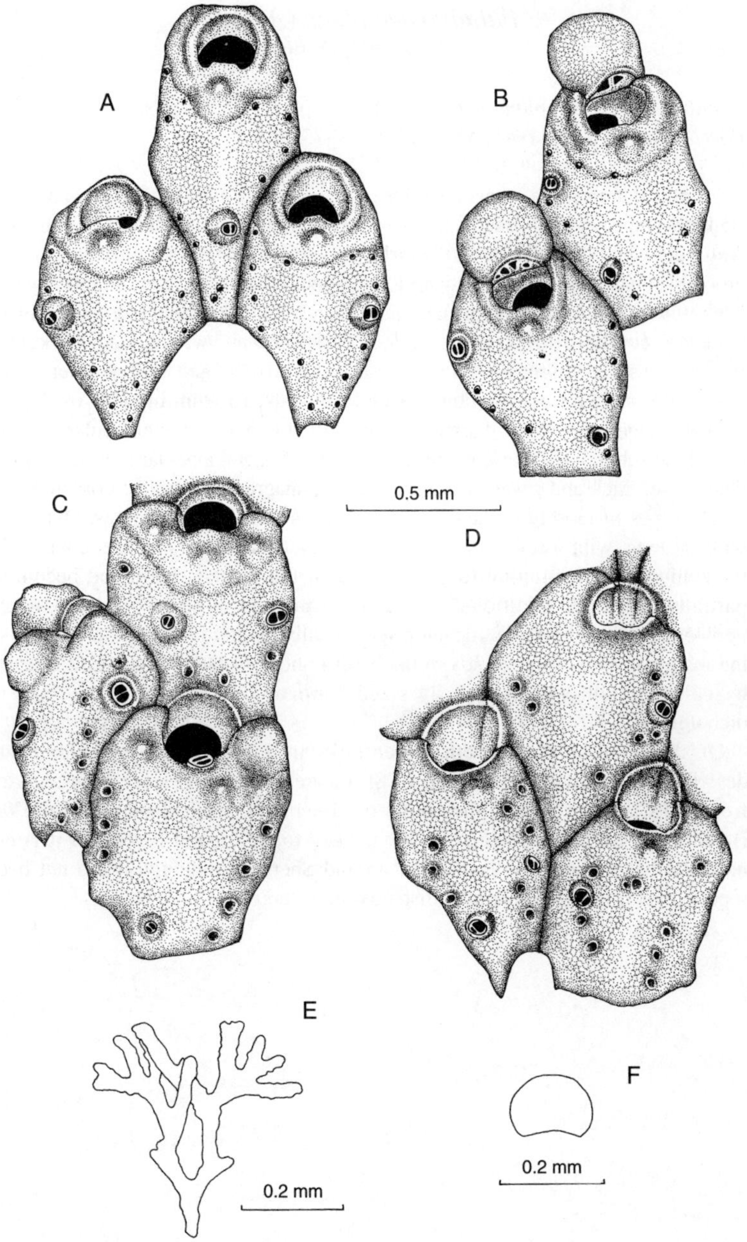

Fig. 65. *Palmiskenea lorea*.
A, Autozooids in early ontogeny. B, Ovicellate zooids. C, Autozooids in late ontogeny, with thickened peristomes. D, Autozooids from a colony base, with deeply immersed orifices. E, Outline of a typical colony. F, Outline diagram of the primary orifice.

Palmiskenea plana (Hincks)
(Figs 64C, D; 66)

Porella skenei forma *plana* Hincks 1888: 221, pl. 14, fig. 6
Porella plana : Kluge 1962: 460, fig. 307; 1975: 558

Colony flat and thin, irregularly lobed or branched, typically with a narrowed stalk developed from its encrusting base; exceeding 20 mm high and 15 mm wide. Autozooids rectangular to irregularly polygonal, convex, separated by deep grooves in early ontogeny, but with less distinct boundaries later; commonly 0.7-0.9 x 0.4-0.5 mm. Primary orifice wider than long, the proximal edge distinctly concave, condyles thick and rounded. Peristome a low, rounded thickening incorporating a single median suboral avicularium, and paired lateral oral avicularia, all with conspicuous, swollen cystids; rostrum broadly oval, with complete slender crossbar, lacking a columella, orientated perpendicular to plane of orifice; suboral avicularium about twice size of lateral avicularia. Frontal shield distinctive: thick and granular, with a regular, shagreened surface; bordered by a single series of closely spaced, comparatively small, marginal pores. Ovicell as wide as long, with a few indistinct perforations close to its aperture; covered by thickening calcification uniform with that of the frontal shield and becoming partially obscured. Additional avicularia, similar to those associated with the peristome but much smaller, develop sporadically in later ontogeny, at the sites of the marginal pores. Autozooids in the basal regions of larger colonies are sealed by calcification, and irregularly sized kenozooids are seen at the axils of dichotomies.

Offshore, on hard substrata. Essentially an Arctic species, *P. plana* was described originally from the Gulf of St. Lawrence, and has since been reported from the Canadian Arctic, the coasts of Greenland and Iceland, and from the Barents and Kara Seas. Recently, it has been found around Faroe and it is not unlikely that it occurs in deep water around Shetland, although it has not been reported yet from anywhere in British coastal waters.

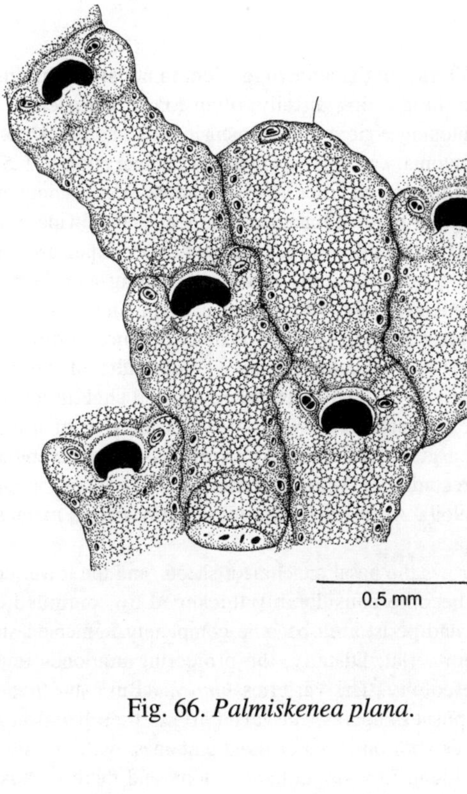

Fig. 66. *Palmiskenea plana*.

Palmiskenea skenei (Ellis & Solander)
(Figs 67; 68A, B)

Millepora skenei Ellis & Solander 1786: 135
Palmicellaria skenei : Hincks 1880: 379, pl. 52, figs 1-4; Hayward & Ryland 1979: 134, fig. 52

Colony up to 30 mm high, branching dichotomously at irregular intervals; branches flattened, broadening distally, often lobed or flabellate, occasionally anastomosed; developing a rigid three-dimensional structure. Autozooids oval to rectangular or hexagonal, convex, separated by distinct sutures; 0.5-0.7 x 0.3-0.4 mm. Frontal shield thickly calcified, finely granular, each autozooid with a few widely spaced marginal pores. Primary orifice longer than wide, proximal border gently convex, condyles small, rounded. Peristome developed as prominent lateral flanges and a stout, mid-proximal umbo; thickening and deepening in later ontogeny, largely obscuring the orifice. Avicularium suboral, medio-proximal, at the base of the umbo; acute to plane of orifice, proximally directed. Additional umbonate processes may develop on one or both sides of the peristome, also bearing small avicularia on their distal faces. Ovicell slightly longer than broad, flattened frontally, the ectooecium with a variable number of irregularly shaped pores; covered by a granular ooecial cover, immersed and eventually obscured. Embryos orange, recorded in April, September and December around the Isle of Man (Eggleston, 1969). Ancestrula tatiform, 0.3 mm, frontal membrane bordered by spines.

As the colony grows the basal attachment sheets, and the lowermost regions of the erect portion become considerably thickened by continued calcification; autozooid orifices and peristomes become completely immersed and the colony surface appears quite flat. Distally, the projecting umbones impart a bristly appearance to the colony. The varieties *bicornis* Busk and *tridens* Busk are distinguished by possessing respectively two or three peristomial umbones. However, autozooids with one, two or three umbones, with or without additional avicularia, can be found in a single large colony and there is thus no utility in retaining these taxa.

Offshore, below the kelp fringe, on hard substrata; extends at least to the edge of the continental shelf. *Palmiskenea skenei* is a boreal-arctic species, occurring off all British coasts, although most common off Scotland and Shetland. Ranges south to the Bay of Biscay but does not reach the Mediterranean.

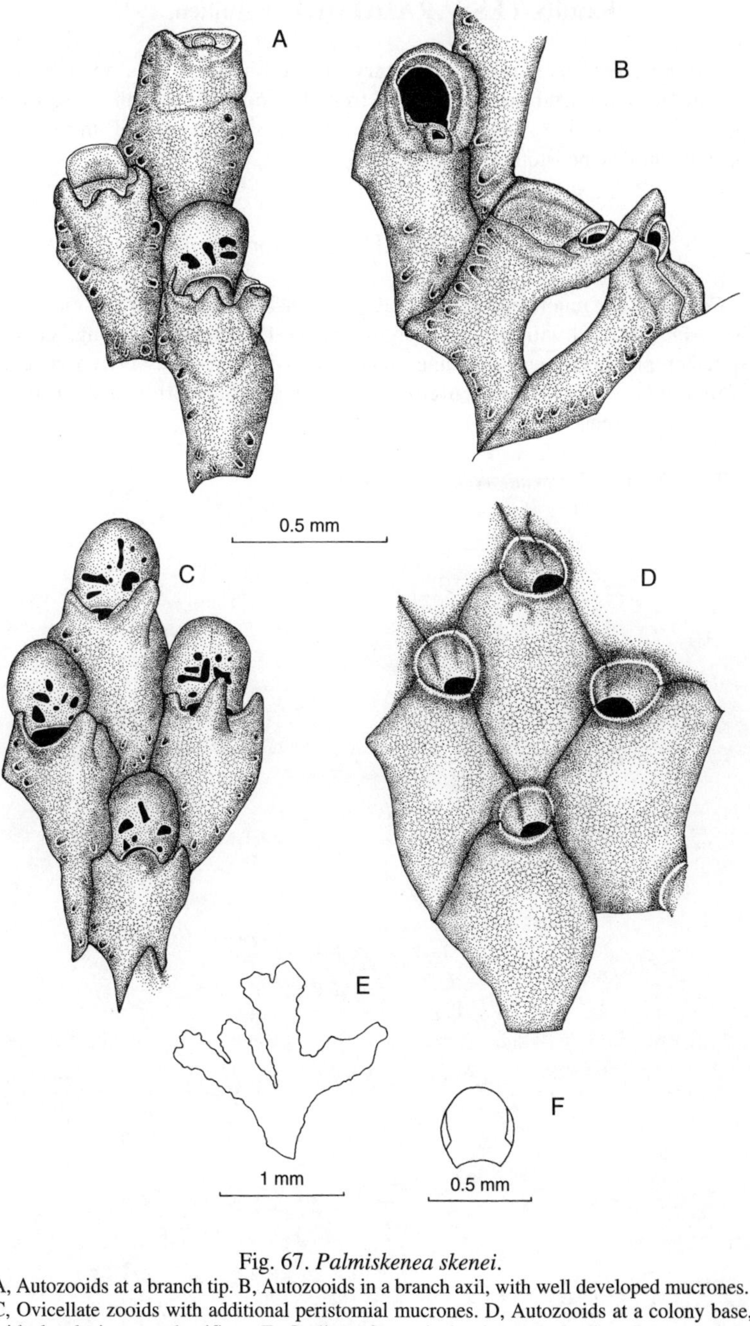

Fig. 67. *Palmiskenea skenei*.
A, Autozooids at a branch tip. B, Autozooids in a branch axil, with well developed mucrones. C, Ovicellate zooids with additional peristomial mucrones. D, Autozooids at a colony base, with deeply immersed orifices. E, Outline of a typical colony. F, Outline diagram of the primary orifice.

Family TESSARADOMIDAE Jullien, 1903

Colony erect or encrusting. Primary orifice obscured by a deep, tubular peristome; with a tubiform spiramen frontally, opening into the base of the peristome, above the zooidal operculum. Ovicell small, spherical, imperforate, opening into the peristome.

Genus TESSARADOMA Norman, 1869

Colony erect, unjointed, with cylindrical branches; attached by an encrusting base. Frontal wall with distinct marginal pores. Peristome projecting, smooth; spiramen prominent, not becoming immersed. Ovicells visible only on youngest autozooids, elsewhere obscured by continued calcification. Adventitious avicularia present.

Type species: *Tessaradoma boreale* (Busk)

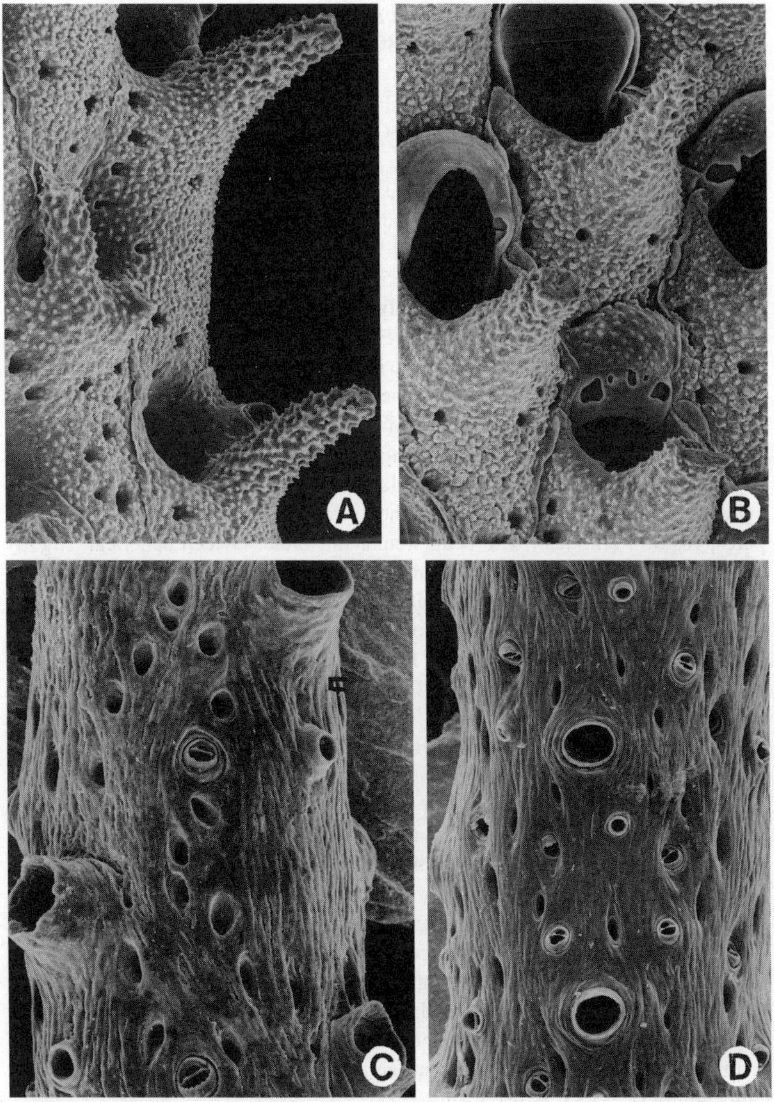

Fig. 68.
A, B, *Palmiskenea skenei*.
A, Autozooids in profile, showing suboral umbones; x80. B, Ovicellate zooids; x90.

C, D, *Tessaradoma boreale*.
C, Autozooids in early ontogeny; x110. D, Autozooids in later ontogeny, with immersed peristomes; x60.

Tessaradoma boreale (Busk)
(Figs 68C, D; 69)

Onchopora borealis Busk 1860c: 213, pl. 28, figs 6-7
Tessaradoma gracile : Norman 1869: 309; Marcus 1940: 222; Kluge 1962: 522; 1975: 634
Porina borealis : Hincks 1880: 229, pl. 31, figs 4-6
Tessaradoma boreale : Hayward & Ryland 1979: 242, fig. 104

Colony erect, rigid, spreading; up to 50 mm high, branching dichotomously at regular intervals. Always slender, rarely exceeding a thickness of 1 mm at the base, tapered towards the growing tips. Frontal wall finely granular, striated, with conspicuous marginal pores; spiramen immediately proximal to the peristome. Autozooids elongate, oval, arranged back-to-back in alternating pairs; 0.7-0.8 x 0.3 mm, broadening in the older parts of the colony. Primary orifice semicircular, not usually visible, obscured by the long, tubular peristome, projecting at a right angle to the autozooid, with an orbicular secondary orifice. Paired adventitious avicularia usually present lateral to the spiramen, developed on the marginal pores; rostrum elliptical, with semicircular mandible. Ovicell spherical, slightly broader than long, smooth and imperforate, opening into the peristome; conspicuous when newly completed, but soon obscured by the encroaching peristome and then completely immersed and hidden by thickening of the frontal calcification of adjacent autozooids. Embryos pure white, recorded in May, July and December in the Bergen fjords (Ryland, 1963b).

As the colony grows the basal regions are strengthened by continuous thickening of the now confluent frontal walls of the lowest autozooids. The original outlines of the autozooids are lost as the dividing walls are broken down, only the peristome and the spiramen remaining prominent. As the circumference of the stem grows, the openings of the marginal pores move laterally giving a striated effect to the frontal surface. Additional avicularia may be developed, and are often numerous.

Found attached to stones, shells, serpulid tubes, erect bryozoans, hydroids, and the stony coral *Lophelia*, in depths exceeding 50 m (Ryland, 1963b). Characteristic of deep fjords, the continental slope, and the outer shelf, with most British records coming from Shetland.

Widely distributed in the North and South Atlantic, at depths ranging from 70 m in the Arctic to 3700 m near the equator, being apparently restricted to temperatures of 2-13°C (Cheetham, 1972).

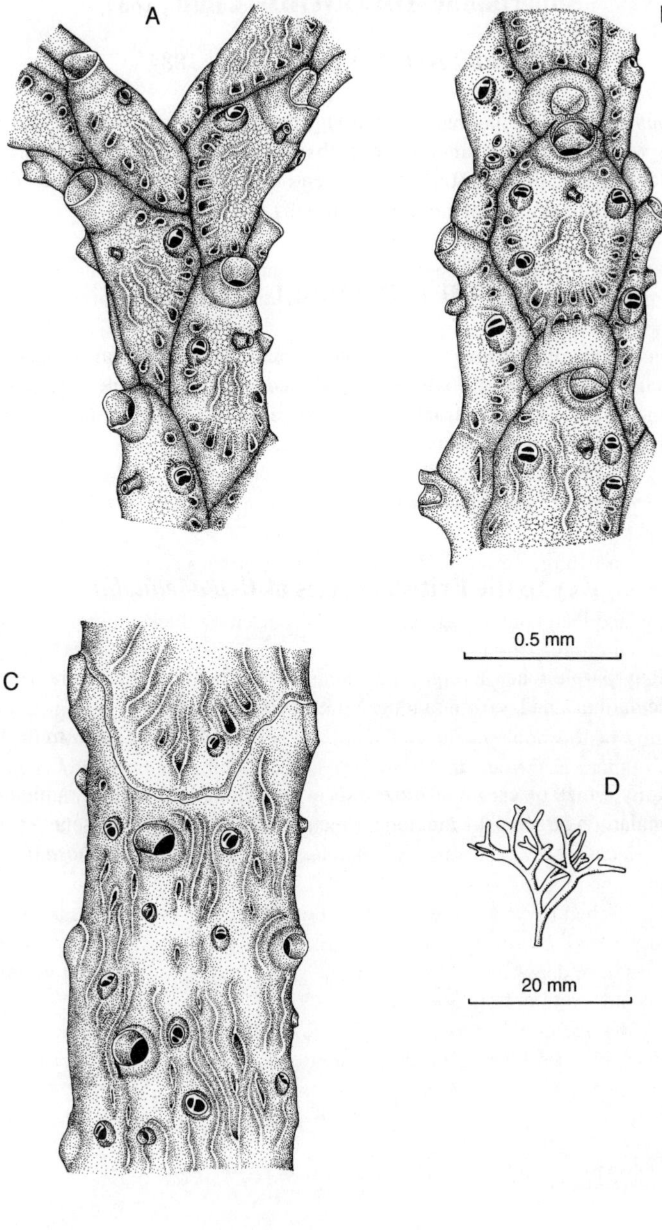

Fig. 69. *Tessaradoma boreale*.
A, Autozooids at a branch dichotomy, in early ontogeny. B, Ovicellate zooids. C, Autozooids in late ontogeny. D, Diagram of a typical colony.

Superfamily ADEONOIDEA Busk, 1884

Family ADEONIDAE Busk, 1884

Colony erect or encrusting. Frontal wall development umbonuloid. Primary orifice without a sinus, uncalcified; obscured by calcified peristome and secondary orifice. Marginal areolae present. Frontal spiramina evanescent or permanent, single or multiple. Avicularia usually present.

Genus REPTADEONELLA Busk, 1884

Colony encrusting. Primary orifice and operculum semicircular. Frontal wall with marginal pores and a single, or twinned, central spiramen. Avicularia adventitious, situated on the frontal wall proximal to the orifice. Embryos brooded in enlarged gonozooids. Basal pore chambers present.

Type species: *Lepralia violacea* Johnston, 1847

Key to the British species of *Reptadeonella*

1. Colony purple when living. Autozooids with a single frontal spiramen. Avicularium small, <0.1 mm long. No kenozooids visible
   .... ***Reptadeonella violacea*** (p. 186)

   Colony white or cream. Autozooids with twinned frontal spiramina. Avicularium larger, >0.1 mm long. Erect, tubular kenozooids present
   .... ***Reptadeonella insidiosa*** (p. 188)

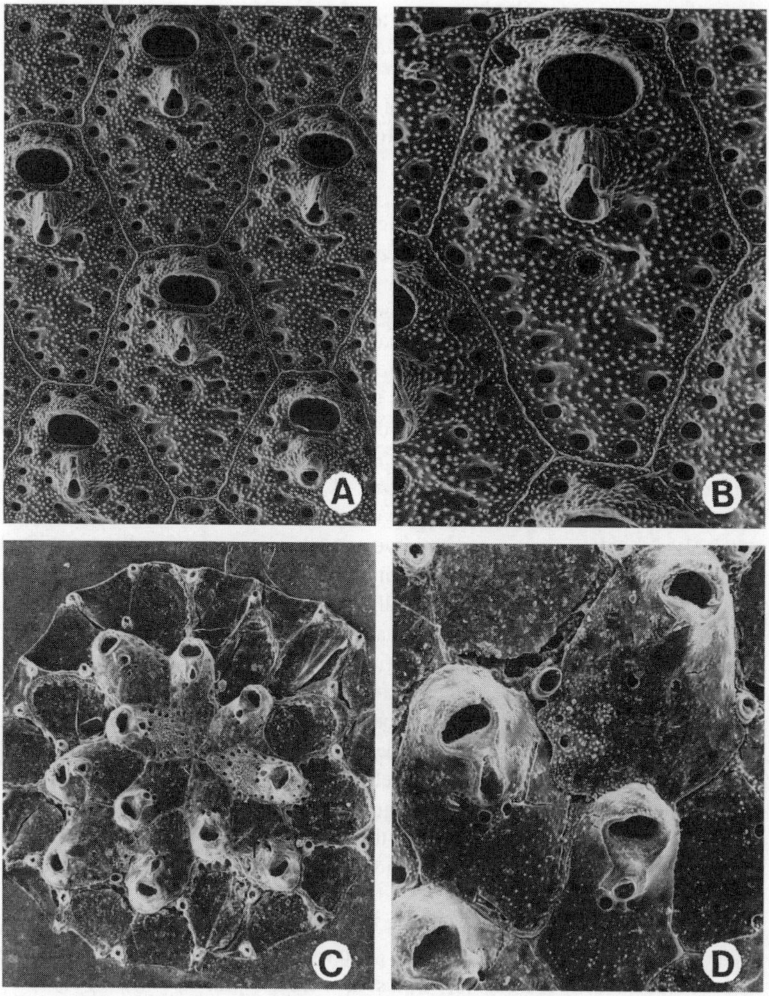

Fig. 70
A, B, *Reptadeonella violacea*.
A, Group of autozooids; x60. B, Detail; x110.

C, D, *Reptadeonella insidiosa*.
C, The ancestrula and periancestrular zooids; x31. D, Group of autozooids; x80.

Reptadeonella violacea (Johnston)
(Figs 70A, B; 71)

Lepralia violacea Johnston 1847: 325, pl. 57, fig. 9
Microporella violacea : Hincks 1880: 216, pl. 30, figs 1, 2, 4
Reptadeonella violacea : Hayward & Ryland 1979: 82, fig. 26

Colony forming broad, often extensive, incrustations, deep purple when living, with a paler edge of developing autozooids. Autozooids hexagonal, pyriform or lozenge shaped, clearly demarcated, flat or slightly concave with a more or less marked central depression; 0.5-0.68 x 0.2-0.34 mm. Frontal wall finely granular with a series of closely spaced marginal pores; a single round spiramen in the central depression. Primary orifice semicircular; secondary orifice surmounting a short peristome, arched distally; a short triangular avicularium set on the sloping proximal side, directed distally, rarely disto-laterally or transversely. Numerous small basal pore-chambers present, 14-16 in the distal half of each autozooid, each with a single communication pore. Gonozooids up to twice as broad as autozooids (0.5 mm) and more convex; the orifice wider, larger and frontal (rather than nearly terminal); frontal spiramen often transversely elongate, occasionally constricted across the middle; the avicularium may be absent. Embryos brown. Larva metamorphoses to form six, symmetrically grouped, ancestrular autozooids.

In newly budded autozooids the marginal pores are separated by slender interareolar ridges; as secondary calcification proceeds and the frontal surface becomes more rugose, the ridges are less evident. The tubular kenozooids characteristic of *R. insidiosa* do not appear. However, examination of a growing edge shows that small kenozooids are occasionally produced at the proximal end of the autozooid buds but do not develop further and are occluded as frontal calcification proceeds. The growing edge is paler than the rest of the colony and the fully developed autozooids receive their colouring from the accumulation of pigmented tissue in the coelomic spaces. An account of the ontogeny of *R. violacea* is given by Cook (1973).

Eggleston (1969) reported embryos in January, March to June and September in Manx waters; Gautier (1962), in the western Mediterranean, reported fertile autozooids from May to December with settlement in April.

R. violacea is present in the western English Channel and around the Channel Islands; it occurs on hard substrata offshore and may be abundant on shell banks. Reaches its northern limit at the Isle of Man; absent from the North Sea. A warm temperate species which is widespread in the Mediterranean and reported from the Cape Verde Islands, West Africa, the Gulf of Mexico and the Caribbean, Brazil, and the Pacific coasts of Mexico and North America.

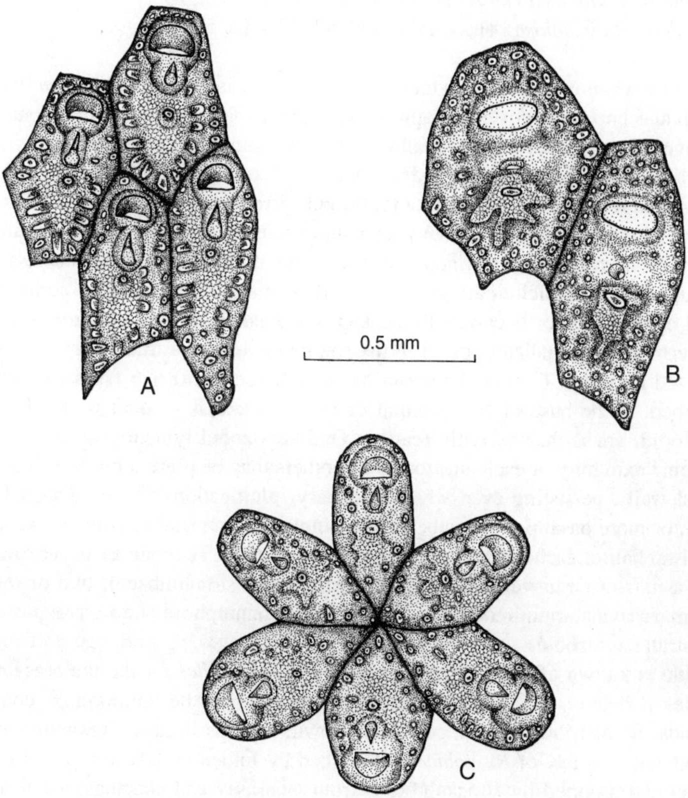

Fig. 71. *Reptadeonella violacea*.
A, Group of autozooids. B, Two gonozooids. C, The six ancestrular zooids.

Reptadeonella insidiosa (Jullien)
(Figs 70C, D; 72)

Adeonella insidiosa Jullien 1903: 54, pl. 6, fig. 4
Microporella violacea var. *a* : Hincks 1880: 216, pl. 30, fig. 3
Reptadeonella insidiosa : Hayward & Ryland 1979: 84, fig. 27

Colony white or yellowish-white, often extensive; in dried material the frontal membrane imparts a glistening appearance. Autozooids elongated, hexagonal or lozenge-shaped; separated by shallow, often indistinct grooves; flat proximally with a central depression, and rising towards the peristome; 0.5-0.6 x 0.24-0.34 mm. Frontal wall fine grained, smooth and vitreous; with a series of closely spaced marginal pores, distinct in young autozooids, less so when older. A pair of round, clearly delimited and non-coalescing spiramina in the central depression. Primary orifice semicircular; secondary orifice surmounting a raised peristome, distal edge arched or occasionally peaked. Avicularium on the proximal side of the peristome, immediately distal to the spiramina, elongate triangular; mandible directed distally. Tubular kenozooids, which seem to arise from the pore-chambers at the base of the terminal or proximo-lateral vertical walls of each autozooid, are a characteristic feature. One kenozooid typically occurs at the proximal extremity of each autozooid, and others may be present on the proximo-lateral walls, persisting even when secondary calcification is well advanced. A dozen or more basal pore-chambers, with single communication pores, present in the distal half of each autozooid. Gonozooid up to twice as broad as an autozooid, with a narrow, transversely elongated orifice; peristome absent, two or three spiramina, avicularium reduced in size. Larva metamorphoses into a group of six ancestrular autozooids.

Little is known of the biology or ecology of *R. insidiosa*. Like the preceding species it occurs offshore on hard substrata and may be common on coarse grounds. Its distribution is imperfectly known, perhaps because it was for long considered a variety of *R. violacea*. Described by Jullien (1903) from the Bay of Biscay and reported by Hincks (1880) from Guernsey and Hastings, it has also been collected from the Western Approaches and the Gulf of St. Malo.

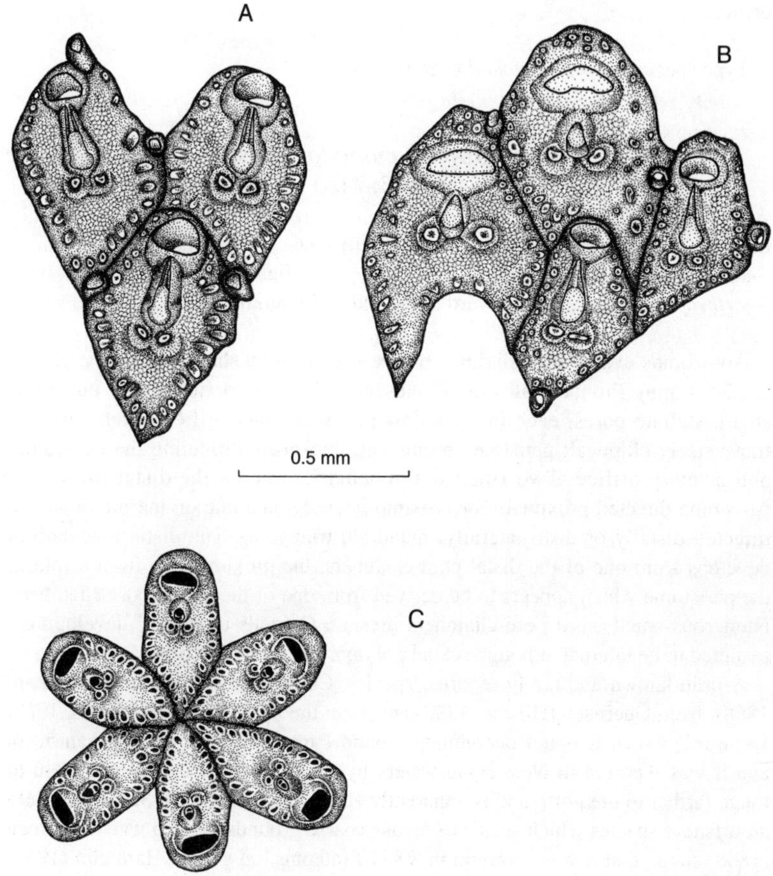

Fig. 72. *Reptadeonella insidiosa.*
A, Three autozooids. B, Two gonozooids and two autozooids. C, The six ancestrular zooids.

Genus ANARTHROPORA Smitt, 1868

Colony encrusting. Frontal shield perforated by numerous, scattered, stellate pores. Avicularia oral, associated with the peristome and semicircular secondary orifice.

Type species: *Lepralia monodon* Busk, 1860

Anarthropora monodon (Busk)
(Figs 73; 74A)

Lepralia monodon Busk 1860c: 213, pl. 29, figs 3,4
Anarthropora monodon : Hincks 1880: 233, pl. 33, figs 10, 11
Anarthropora monodon : Hayward & Ryland 1979: 86, fig. 28

Autozooids oval, flat or slightly convex, separated by shallow grooves; 0.6-0.7 x 0.3-0.4 mm. Frontal wall rugose, closely and evenly perforated by numerous small, stellate pores, each in a shallow pit. Secondary orifice semicircular or transversely elliptical; peristome prominent, thickened, obscuring the operculum and primary orifice. Two small oral avicularia: one on the distal rim of the peristome directed proximally or proximo-laterally, and one on the proximal rim directed distally or disto-laterally; mandible triangular. The distal avicularium develops from one of the distal pore-chambers, the proximal one from a fold in the peristome which appears to be derived from one of the lateral pore-chambers. Numerous small basal pore-chambers present. Ovicells unknown: development assumed to be internal although sexual polymorphs are undescribed.

A little known and rarely reported species. Collected from Shetland (Norman, 1868), from Guernsey (Hincks, 1880) and from the Aegean Sea (Harmelin, 1969; Hayward, 1974). It is not uncommon around Faroe in depths greater than 50 m and it was reported in Norwegian waters by Smitt (1868b). It does not seem to range farther to the north and is apparently absent from the North Sea. Essentially an offshore species which seems to favour coarse grounds in deep water. Norman (1868) stated that it was common in 80-170 fathoms, on gravel; Harmelin (1969) found it in depths of 150-1094 m in the Aegean, although Hayward (1974) showed it to be fairly common around the island of Chios in depths less than 100 m.

Fig. 73 (opposite, top). *Anarthropora monodon*.
A, A group of autozooids. B, Autozooids at the colony margin, showing pore chambers and development of the peristome.

Fig. 74 (opposite).
A, *Anarthropora monodon* (Shetland); x80. B, *Watersipora complanata* (Scilly); x70. C, *Cryptosula pallasiana* (Lewis); x40. D, *Hippoporina pertusa* (Mull); x60.

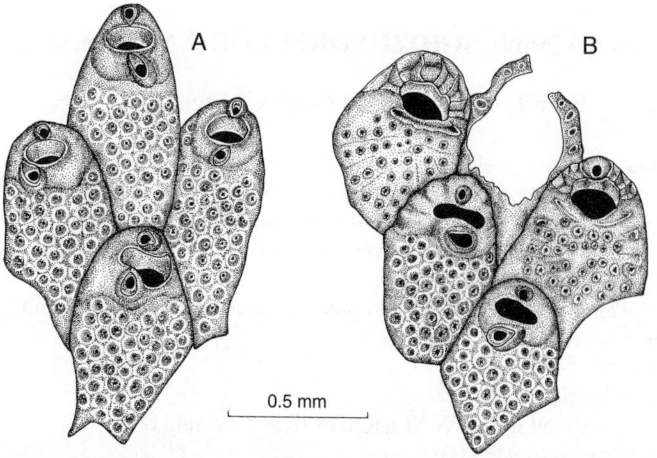
A
B
0.5 mm

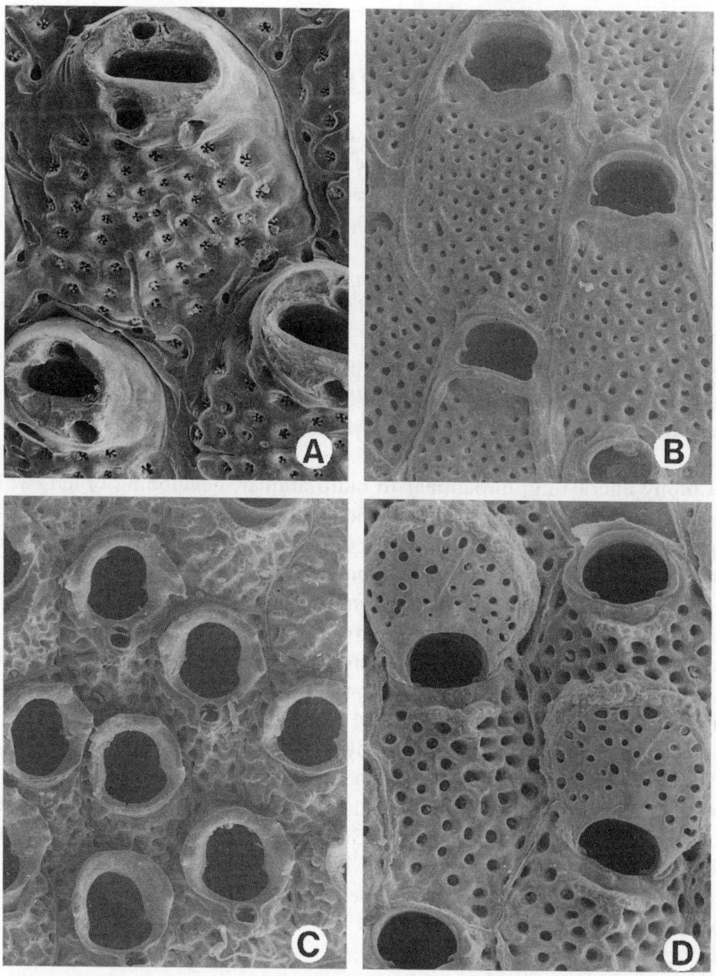
A
B
C
D

Superfamily SCHIZOPORELLOIDEA Jullien, 1883

Family WATERSIPORIDAE Vigneaux, 1949

Colony encrusting, sheet-like, multilaminar; or erect, bilaminar. Autozooids with evenly perforated cryptocystidean frontal shield, with marginal frontal septula. Vertical walls with large multiporous septula. Primary orifice bell-shaped or sinuate, with prominent condyles; no oral spines. No avicularia. Ovicells absent, embryos brooded internally.

D. P. Gordon (*in litt.*, Feb. 1998) proposed accommodating this family within the Smittinoidea.

Genus WATERSIPORA Neviani, 1895

Characters of the family

Type species: *Cellepora subovoidea* d'Orbigny, 1852

Gordon (1989) noted that *Cellepora subovoidea* d'Orbigny (*nomen novum* for *Cellepora ovoidea* Audouin) has no type specimen and that both it and the genus *Watersipora* can only be stabilised by the selection of a neotype specimen.

Watersipora complanata (Norman)
(Figs 74B; 75)

Lepralia complanata Norman 1864: 85, pl. 10, fig. 4
Micropora complanata: Hincks 1880: 175, pl. 23, figs 8, 9
Watersipora complanata: Hayward & Ryland 1979: 88, fig. 29

Colony encrusting, unilaminar to multilaminar, developing broad, white, spreading incrustations on rock and shell. Autozooids oval to hexagonal, flat, separated by narrow sutures; 0.6-0.9 x 0.3-0.5 mm. Primary orifice wider than long, more or less bell-shaped, with prominent condyles; proximal edge straight, or slightly convex; operculum pale brown, with a thickened, darker, distal sclerite, and two pale spots proximo-laterally. Frontal shield finely granular, regularly and closely punctured by numerous small pores; distal ends of lateral walls and border of orifice smooth, thickening in later ontogeny, developing a pair of knob-like umbones proximo-lateral to orifice, and typically a broadened triangular region mid-distally. Biology and ecology unknown.

Watersipora complanata was described from an unknown British locality. It was subsequently reported from the Adriatic (Hincks 1887) and from Morocco (Canu & Bassler 1925) and shown by Gautier (1962) to be widely distributed in the western Mediterranean. In Britain it is known only from Scilly, where it occurs on the lower shore, encrusting shells and the undersides of stones, on most islands in the group.

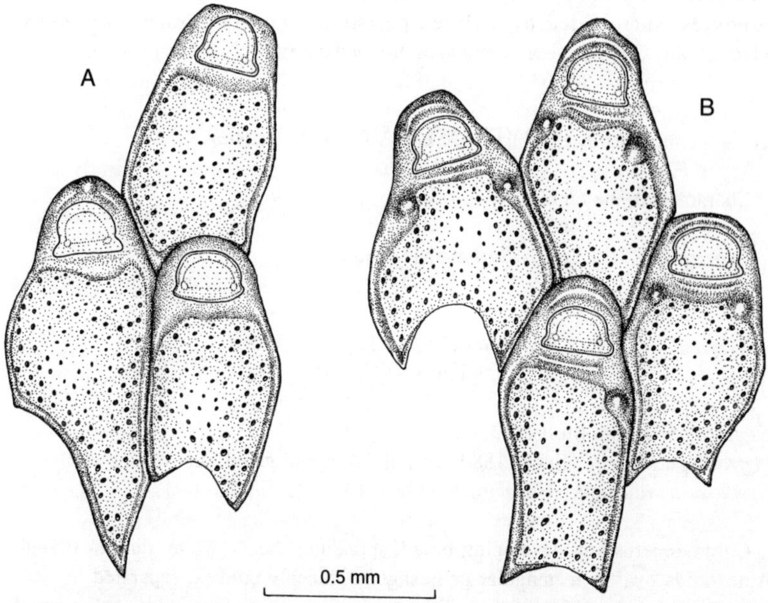

Fig. 75. *Watersipora complanata*.
A, Autozooids in early ontogeny. B, Autozooids in later ontogeny, with thickened orifice rims and prominent umbones.

Family CRYPTOSULIDAE Vigneaux, 1949

Frontal shield cryptocystidean, uniformly porous. Vertical walls with multiporous septula. Primary orifice rectangular to bell-shaped, with prominent condyles; surrounded by a flared peristome; no oral spines. Avicularia adventitious. Ovicells absent, embryos brooded internally.

Genus CRYPTOSULA Canu & Bassler, 1925

Characters of the family

Type species: *Eschara pallasiana* Moll, 1803

Cryptosula pallasiana (Moll)
(Figs 74C; 76)

Eschara pallasiana Moll 1803: 64, pl. 3, fig. 13
Lepralia pallasiana: Hincks 1880: 297, pl. 24, fig. 4; pl. 33, figs 1-3
Cryptosula pallasiana: Hayward & Ryland 1979: 90, fig. 30

Colony encrusting, developing broad, spreading sheets; white, pink or orange. Autozooids oval to rectangular or hexagonal, steeply convex, separated by deep grooves; 0.6-1.0 x 0.35-0.5 mm. Primary orifice longer than wide, bell-shaped, with concave proximal margin and prominent, downcurved condyles; large, occupying one-third to one-half total frontal length of autozooid, surrounded by an erect, flared peristome. Operculum deep brown, visible to the unaided eye, with a thickened marginal sclerite. Frontal shield thick, with small, regularly distributed pores, each bordered by calcification which thickens markedly in later ontogeny and becomes especially rugose; a stout umbo develops medioproximal to the orifice, in many autozooids incorporating a small avicularium. Rostrum of avicularium acute to plane of orifice, proximally directed, broadly oval. Polypide with 16-17 tentacles (Calvet, 1900). Embryos orange, conspicuous. Ancestrula 0.5 mm long, resembling later autozooids, developing a distal triplet of zooid buds. Embryos most abundant during the summer. Settlement recorded May to December in western Norway, with peak July and September (Ryland, 1963a). In Britain settlement continues through October.

Common on all British shores, from just below MTL and into the shallow sublittoral. On stones, shells and other hard substrata, as well as kelp holdfasts and *Himanthalia* buttons. Distributed from western Norway southwards to the Mediterranean and the Black Sea, and in the west Atlantic from Nova Scotia to North Carolina. A common fouling species, *C. pallasiana* has been recorded from docks and harbours in New Zealand (Gordon & Mawatari, 1992) and also from Pacific coasts of North America (Osburn, 1952).

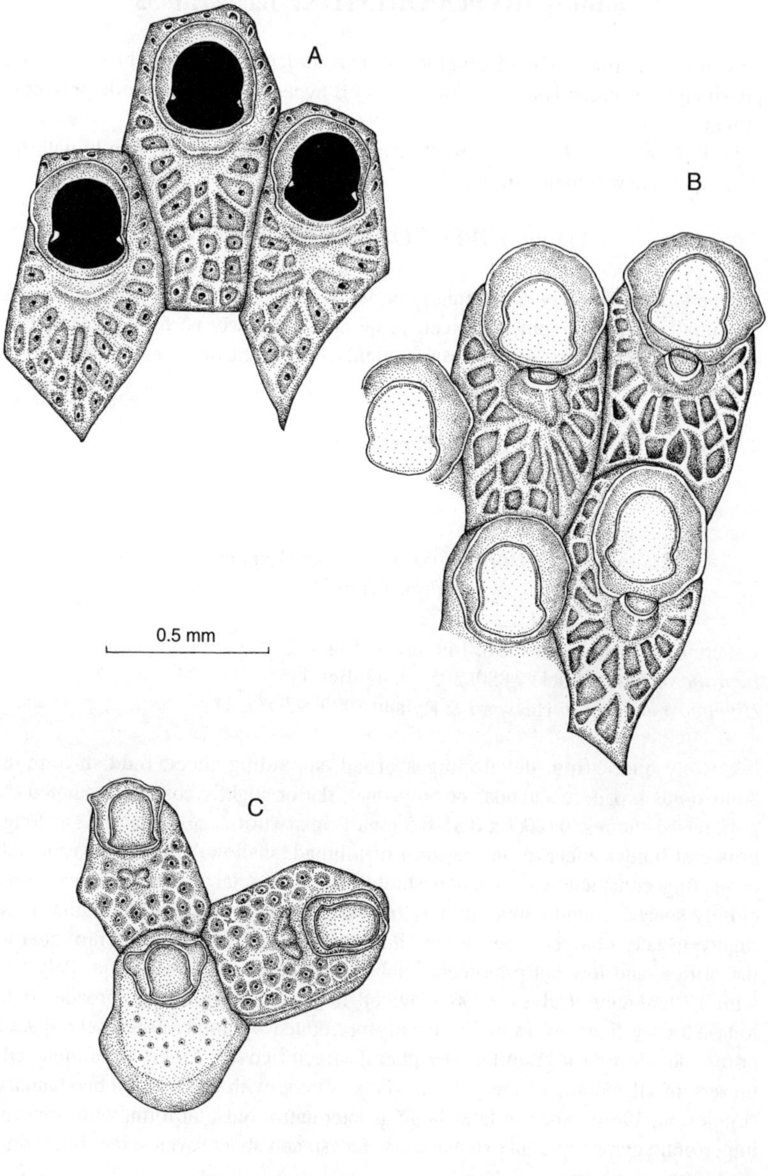

Fig. 76. *Cryptosula pallasiana.*
A, Autozooids with small peristomes, and lacking avicularia. B, Autozooids with deep, flared peristomes and suboral avicularia. C, The ancestrula and two of the periancestrular zooids.

Family HIPPOPORINIDAE Bassler, 1935

Autozooid frontal shield cryptocystidean. Orifice with straight or concave proximal border, condyles prominent. Ovicell hyperstomial. Avicularia present or absent.

D. P. Gordon (*in litt.,* Feb. 1998) proposed synonymising this family with the Bitectiporidae within the Smittinoidea.

Genus HIPPOPORINA Neviani, 1895

Frontal shield closely and regularly perforated. Primary orifice broadly concave proximally. No oral spines. Ovicell prominent, regularly perforated, closed by autozooid operculum. Adventitious avicularia present or absent. Multiporous septula present in vertical walls.

Type species: *Cellepora pertusa* Esper, 1796

Hippoporina pertusa (Esper)
(Figs 74D; 77)

Cellepora pertusa Esper 1796: 149, pl. 10, figs 1, 2
Lepralia pertusa: Hincks 1880: 305, pl. 43, figs 4, 5
Hippoporina pertusa: Hayward & Ryland 1979: 92, fig. 31

Colony encrusting, developing a broad, spreading sheet, reddish-orange. Autozooids broad, rectangular or polygonal, flat or slightly convex, separated by thin, raised sutures; 0.6-0.7 x 0.35-0.5 mm. Primary orifice about as wide as long; proximal border concave, in the form of a broad, shallow V; condyles rounded, projecting conspicuously. Frontal shield finely granular, evenly perforated by closely spaced, round pores; initially translucent, thickening and becoming more rugose in later ontogeny, developing a low, cruciform median umbo proximal to the orifice, and low but prominent knobs adjacent to it. No avicularia. Polypide with 18 tentacles (Calvet, 1900). Ovicell large and conspicuous, broader than long, slightly flattened frontally; evenly perforated by numerous closely spaced pores, developing a granular, peripheral ooecial cover. Embryos orange-red; present in all months of the year in Manx waters, with a peak October-January (Eggleston, 1969). Ancestrula as large as later autozooids; tatiform, with smooth, imperforate gymnocyst, an extensive cryptocyst, and about seven spines bordering the frontal membrane.

On hard substrata, especially large shells, offshore. Essentially a warm temperate species distributed in the eastern Atlantic from southern and western Britain to the Mediterranean and West Africa. In Britain, it ranges north to the Hebrides and is not uncommon on the shell banks in the western Channel. Elsewhere, recorded from the Gulf of Mexico, California and the Galapagos Islands.

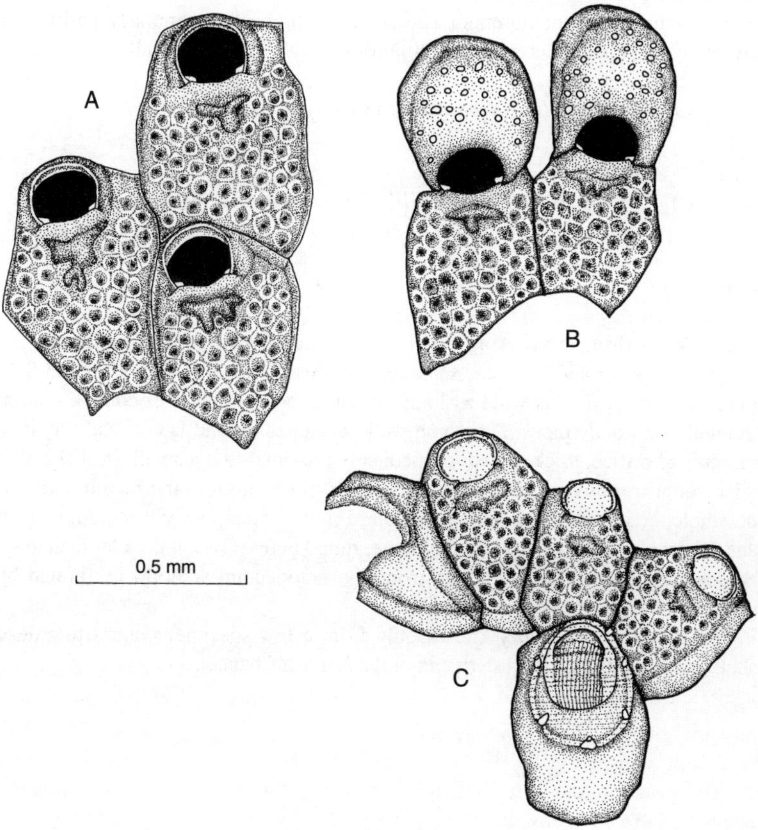

Fig. 77. *Hippoporina pertusa*.
A, Autozooids in early ontogeny. B, Ovicellate zooids. C, The ancestrula and periancestrular zooids.

Genus CODONELLINA Bassler, 1934

Colony encrusting. Autozooid frontal shield evenly perforated. Orifice with broad, deep poster and prominent condyles. No oral spines. Peristome typically well developed. Avicularia, when present, adventitious, suboral. Ovicell hyperstomial, with membranous ectooecium; entooecium regularly perforated; closed by autozooid operculum. Multiporous septula in vertical walls.

Type species: *Lepralia galeata* Busk, 1854

Codonellina lacunata Hayward & Hansen
(Fig. 78)

Codonellina lacunata Hayward & Hansen 1999

Colony a thin, encrusting, unilaminar sheet. Autozooids hexagonal to irregularly polygonal, convex, separated by distinct sutures; 0.5-0.7 x 0.3-0.48 mm. Primary orifice as wide as long, proximal border concave; condyles short, triangular, no oral spines. Peristome well developed around lateral and proximal margins of orifice, thick, flared; immediately proximal to it a small, oval chamber with membranous upper surface, resembling an avicularium but without mandible, condyles or originating septula. Frontal shield thinly calcified, smooth and glistening, evenly perforated by large, round pores. Ovicell broader than long; ectooecium entirely membranous frontally, entooecium smooth, perforated by about 20 small pores.

This species is presently known only from a few specimens encrusting dead shell, collected from the western end of the English Channel.

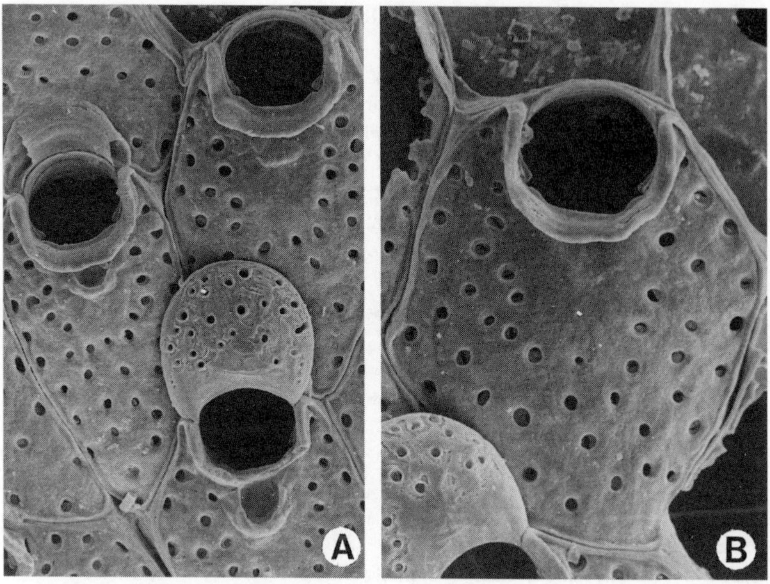

Fig. 78. *Codonellina lacunata* (The holotype specimen).
A, Autozooids and an ovicell; x100. B, Detail of an autozooid; x120.

Genus PENTAPORA Fischer, 1807

Colony attached by an encrusting base, developing an erect, three dimensional form of folded and anastomosed bilaminar plates, or narrow bifurcating branches; rigid, twiggy, honeycombed or tightly meandriform. Frontal shield cryptocystidean, evenly porous; vertical walls with large, multiporous septula. Primary orifice bell-shaped or rectangular, with prominent condyles. Adventitious avicularia present. Ovicell hyperstomial, perforate, closed by autozooid operculum.

Type species: *Millepora foliacea* Ellis & Solander, 1786 = *Eschara fascialis* Pallas, 1766

Pentapora fascialis (Pallas)
(Figs 79; 80; 81A, B)

Eschara fascialis Pallas 1766: 42
Millepora foliacea Ellis & Solander 1786: 133
Lepralia foliacea: Hincks 1880: 300, pl. 47, figs 1-4
Pentapora foliacea: Hayward & Ryland 1979: 94, figs 32, 33

Colony a rigid, domed structure resembling a coral head; built of lobed, folded and anastomosed bilaminar sheets of zooids, meandriform (like a brain coral) or with larger spaces between the plates and thus appearing coarsely honeycombed. Deep orange in life, fading to pale buff after death. Autozooids in regular quincuncial arrangement, oval to rectangular or hexagonal, flat or slightly convex, separated by distinct grooves; 0.7-0.9 x 0.3-0.5 mm. Primary orifice longer than wide, rectangular to semi-elliptical, the proximal margin straight; or slightly concave with a small mucro medially. Condyles small, downcurved. No oral spines. Operculum light brown, conspicuous, with distinct marginal sclerite. Frontal shield with numerous, closely spaced and regularly distributed pores; initially finely granular, becoming progressively more rugose in later ontogeny, with prominent ridges between the deeply sunk pores, and frequently a short umbo immediately proximal to the orifice. Avicularia sporadic, absent from many autozooids; suboral, most often small, with a short semi-elliptical mandible, slightly acute to frontal plane, less often enlarged, with broadly spatulate mandible normal to frontal plane. Polypide with 16-17 tentacles. Ovicell as wide as long, globular, smooth, with a few scattered frontal pores; rapidly immersed in thickening calcification and completely obscured.

The primary orifice becomes deeply immersed in later ontogeny and the rim of the secondary orifice may be produced laterally as a pair of stout processes (Fig. 79G). Reparative growth on old regions of the colony may be achieved through the frontal budding of new autozooids, often irregularly shaped and with varying orientation. However, strengthening of the colony base is otherwise achieved by continued thickening of autozooid frontal shields, which may exceed 1 mm thickness.

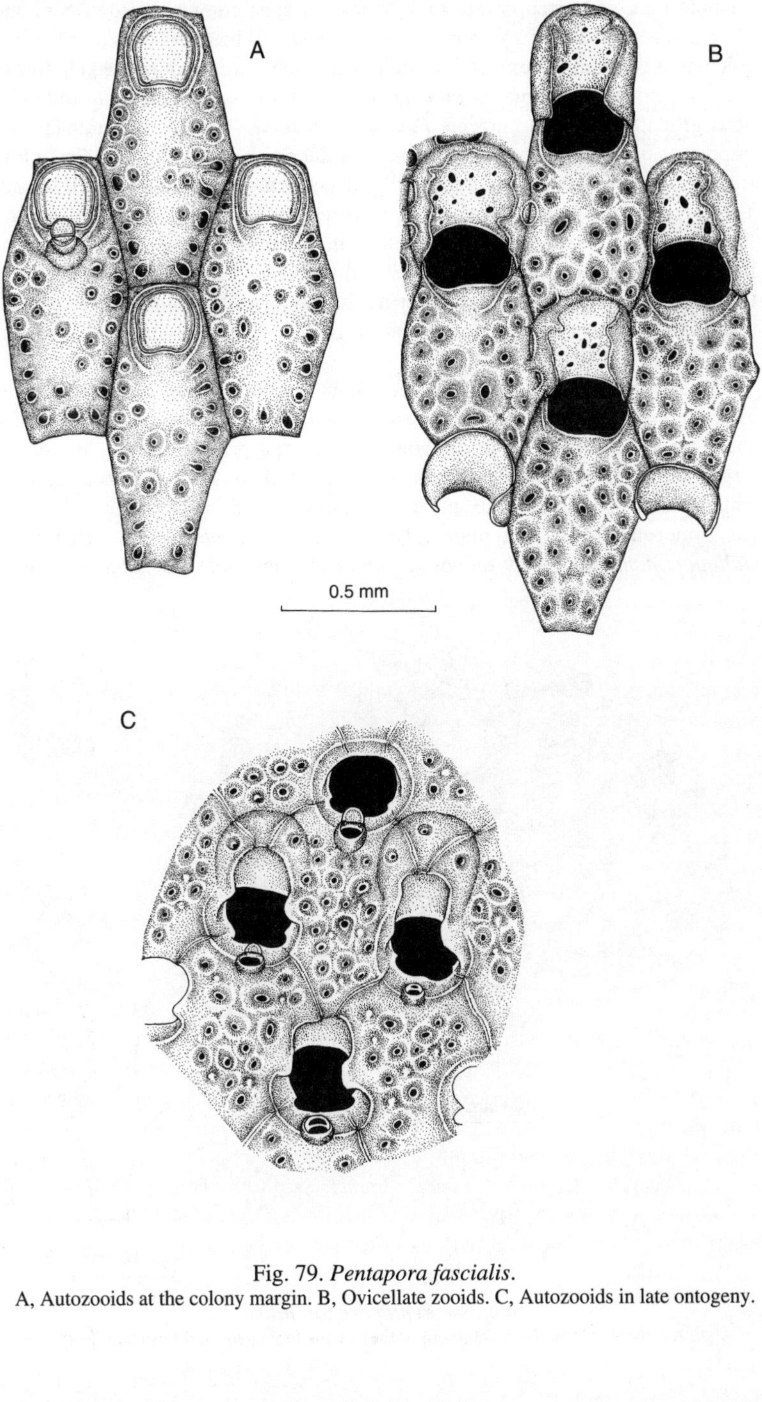

Fig. 79. *Pentapora fascialis*.
A, Autozooids at the colony margin. B, Ovicellate zooids. C, Autozooids in late ontogeny.

Sublittoral, on current-swept rocky bottoms, most common below 18 m and often abundant between 25-35 m. Usually attached to boulders or bedrock; small colonies occasionally found on kelp holdfasts at the lower limit of the kelp forest, and encrusting, unilaminar sheets occur rarely on unstable, hard substrata. On hard grounds with fast currents *Pentapora fascialis* may be a dominant and important part of the sessile benthos with densities of 1 colony per m^2 quite frequent. Colony size seems limited only by environment and may be substantial. The largest known seems to be that reported by Hincks (1880) from off the Eddystone, which "exceeded seven feet" (2 m) in circumference. Growth rate and longevity have still to be studied in detail, although Patzold, Ristedt & Wefer (1987), employing oxygen isotope analysis, estimated a minimum age of three years for a specimen from 20 m off Martins Haven, Pembrokeshire, which was 20 cm diameter and 10 cm high.

Pentapora is probably of major ecological importance to its environment, effectively mitigating the effects of strong bottom currents by providing habitat space for a broad range of invertebrate animals, and perhaps for fish as well. A large colony may shelter thousands of other animals, including polychaetes, amphipods, isopods, decapods and echinoderms, as well as providing settlement for numerous sessile organisms. The bryozoans *Amphiblestrum flemingii, Callopora dumerilii, Membraniporella nitida* and *Smittoidea reticulata* are almost

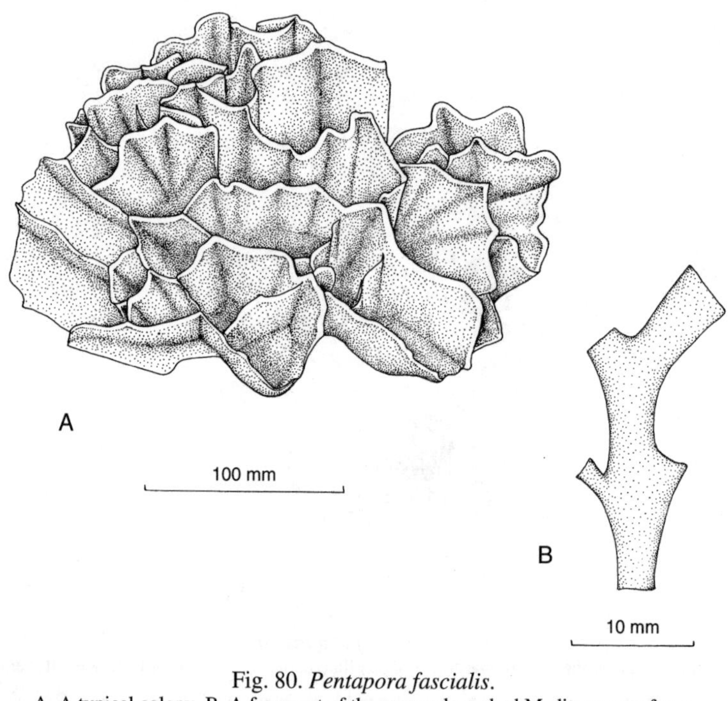

Fig. 80. *Pentapora fascialis*.
A, A typical colony, B, A fragment of the narrow branched Mediterranean form.

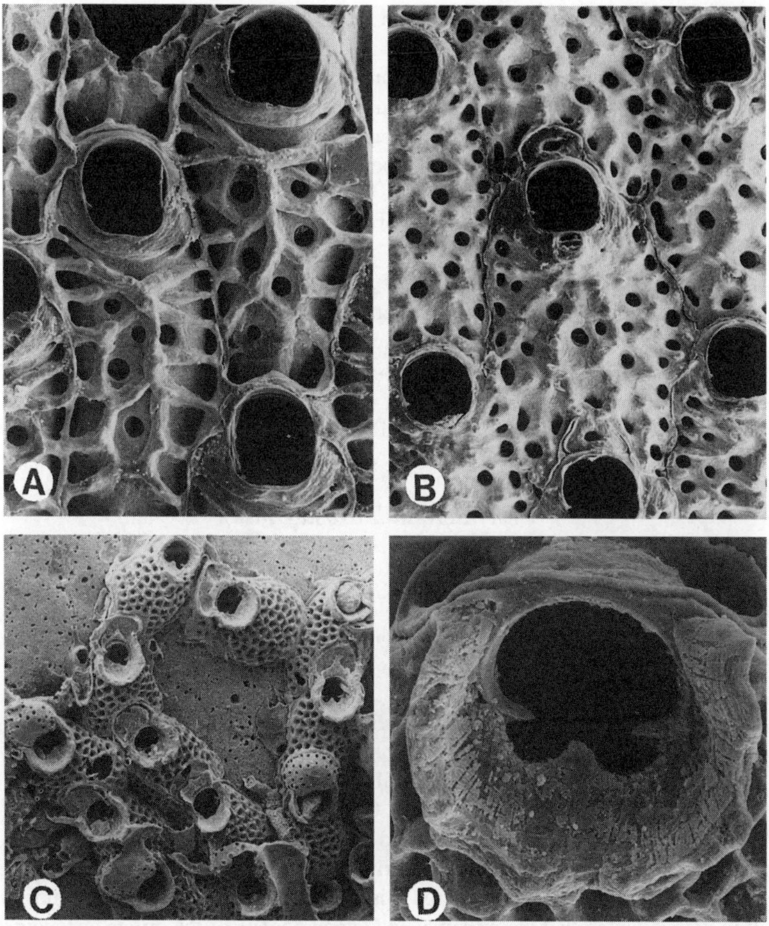

Fig. 81
A, B. *Pentapora fascialis*.
A, Autozooids at the colony margin; x70. B, Autozooids in later ontogeny, two with suboral avicularia; x50.

C, D, *Phylactella labrosa*
(specimen from Lundy). C, Portion of a colony; x33. D, The primary orifice; x260.

invariably found on the larger colonies of *Pentapora*. Similar bryozoan-dominated habitats in New Zealand have been recognised as important nursery grounds for fish and accordingly granted protection in law (Bradstock & Gordon 1983).

Pentapora fascialis occurs off rocky western coasts of Britain, from the Hebrides southwards, and along the central Channel. It is present in the western Mediterranean and the Adriatic in its narrow branched form.

Family TEUCHOPORIDAE Neviani, 1895

Autozooids with cryptocystidean frontal shield, evenly perforated. Primary orifice with or without lyrula and condyles, surrounded by a deep peristome. Ovicell prominent, perforate. Adventitious avicularia present or absent. Vertical walls with basal pore chambers or multiporous septula.

Genus PHYLACTELLA Hincks, 1879

Colony encrusting. Frontal shield of autozooids uniformly porous. Primary orifice with lyrula and condyles, bordered proximally and laterally by a deep, flaring peristome. No oral spines. No avicularia. Ovicell hyperstomial, prominent, perforate. Small basal pore chambers present.

Type species: *Lepralia labrosa* Busk, 1854

Phylactella labrosa (Busk)
(Figs 81C, D; 82)

Lepralia labrosa Busk 1854: 82, pl. 92, figs 1-3
Phylactella labrosa: Hincks 1880: 357, pl. 43, figs 1-2; Hayward & Ryland 1979: 160, fig. 64

Colony forming small circular or irregular patches; or frequently diffuse, in single or multiple linear series, producing fans or patches at irregular intervals. Often closely intergrown with other small ascophorines. Autozooids oval, convex, separated by deep grooves; 0.4-0.56 x 0.3-0.4 mm. Frontal shield evenly and closely punctured by large, thick-rimmed pores; thickening through ontogeny and appearing distinctly rugose. Primary orifice wider than long; condyles sharply pointed, conspicuous, lyrula tapered, distinct. Peristome prominent, forming a deep, flared cup around the proximal and lateral borders of the orifice, extending onto the ovicell of brooding autozooids. Ovicell recumbent on distally succeeding autozooid, slightly wider than long, smooth surfaced, with about 20 variably sized, randomly distributed pores. Ancestrula tatiform, 0.25 mm long, with smooth, imperforate gymnocyst; opesia oval, bordered by a broad cryptocyst and with eight peripheral spines. Embryos reported in August off the Isle of Man (Eggleston 1969).

This species is characteristic of the encrusting epifauna of small shell substrata. It occurs in the concavity of valves of *Venerupis, Tapes* and other Veneracea, often in association with *Neolagenipora collaris, Trypostega venusta, Escharina johnstoni, Puellina* spp., and other cheilostomates which typically produce ovicells at small sizes. It occurs along the south and western coasts of the British Isles, on offshore shell gravels, from Sussex and Cornwall, northwards to the Shetlands. It remains only rarely reported and its broader geographical distribution is generally unknown.

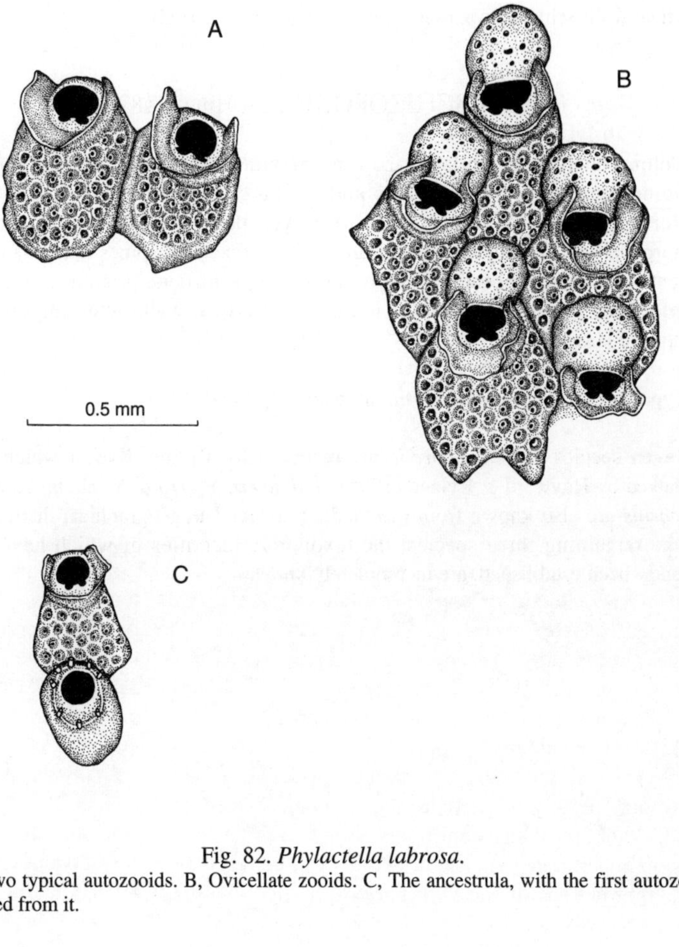

Fig. 82. *Phylactella labrosa*.
A, Two typical autozooids. B, Ovicellate zooids. C, The ancestrula, with the first autozooid budded from it.

Family SCHIZOPORELLIDAE Jullien, 1903

Autozooids with cryptocystidean frontal shield, evenly perforated, or with marginal pores only. Primary orifice with defined anter and poster, with well defined condyles; poster often narrowed to constitute a sinus. Oral spines present or absent. Avicularia adventitious and/or vicarious, or absent. Ovicell present. Vertical walls with multiporous septula or basal pore chambers.

Genus SCHIZOPORELLA Hincks 1877

Colony encrusting, unilaminar to multilaminar, sheet-like or mamillate, or developing partially erect plates and mounds. Autozooids with an evenly perforated, cryptocystidean frontal shield. Avicularia adventitious, occasionally dimorphic, typically adjacent to the autozooid orifice. Oral spines impermanent or absent. Ovicell recumbent on distally succeeding autozooid, prominent, spherical or globular, perforate; ribbed or umbonate. Vertical walls with uniporous or multiporous septula.

Type species: *Lepralia unicornis* Johnston, in Wood, 1844

Seven species of *Schizoporella* are recorded for Britain, five of which were reviewed by Hayward & Ryland (1995). *S. dunkeri, S. errata, S. magnifica* and *S. unicornis* are also known from the Mediterranean. The geographical distribution of the remaining three species, the taxonomic identities of which have only recently been established, are incompletely known.

Key to the British species of *Schizoporella*

1. Sinus broadly and shallowly U-shaped, wider than deep
.... **2**

Sinus narrow; as deep as, or deeper than, its greatest width
.... **3**

2. Orifice wider than long, appearing semicircular; sinus occupying half proximal border. Frontal shield finely perforate. Avicularia typically paired
.... ***Schizoporella unicornis*** (p. 220)

Orifice as wide as long, appearing almost circular; sinus occupying more than half proximal border. Frontal shield coarsely perforate. Avicularium single
.... ***Schizoporella errata*** (p. 212)

3. Sinus very narrow, slit-like. Avicularium often single, with long, slender mandible; crossbar without a columella
.... ***Schizoporella dunkeri*** (p. 210)

Sinus U- or V-shaped. Avicularia usually paired, mandible short, acute, triangular; crossbar with or without a columella
.... **4**

4. Sinus deeper than wide. Avicularia situated distal to the sinus, usually paired and typically directed distally
.... ***Schizoporella magnifica*** (p. 216)

Sinus not deeper than wide. Avicularia situated more or less level with sinus and directed disto-laterally
.... **5**

5. Crossbar of avicularium with a thick columella on its distal face. Sinus V-shaped, condyles with acute corners
.... ***Schizoporella cornualis*** (p. 208)

Crossbar of avicularium without a columella. Sinus V- or U-shaped
.... **6**

6. Sinus U-shaped, almost parallel-sided. Condyles thick, rounded, with faceted surface
.... ***Schizoporella hesperia*** (p. 214)

Sinus V-shaped; condyles broad and short, flat-surfaced
.... ***Schizoporella patula*** (p. 218)

Schizoporella cornualis Hayward & Ryland
(Figs 83; 84A, B)

Schizoporella longirostris Hayward & Ryland 1979: 173, fig. 70
Schizoporella cornualis Hayward & Ryland 1995: 46, pl. Vc, d

Colony a broad, thick, unilaminar to multilaminar sheet, frequently mamillate, white. Autozooids broad, rectangular to irregularly polygonal, convex, separated by distinct, thickened sutures; 0.58-0.69 x 0.43-0.59 mm. Primary orifice as wide as long, anter D-shaped, poster a shallow V-shape, constituting about one-quarter total orifice length; condyles broad, deep and flat, with acute corners. No oral spines. Frontal shield finely granular, densely perforated, the marginal series of pores distinctly larger than the rest; typically with a well developed median umbo. Avicularia single or paired (rarely, three), lateral to orifice, acute to frontal plane, disto-laterally directed; rostrum abruptly tapered, its distal tip acute; crossbar complete, with a stout, conspicuous columella. Ovicell as wide as long, globular, with small pores around its edges and sides, coarsely granular and imperforate frontally, with a prominent umbo.

On stones and shells, offshore to at least 100 m. Presently known from around Scilly and off the western end of the English Channel. This species has only recently been distinguished from the Mediterranean *Schizoporella longirostris* (=*S. dunkeri* (Reuss)).

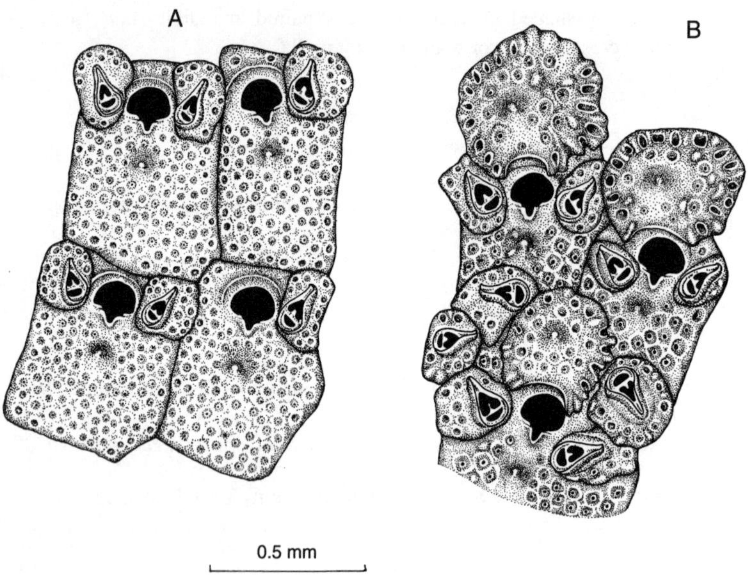

Fig. 83. *Schizoporella cornualis*.
A, Four typical autozooids. B, Ovicellate zooids, with additional frontal avicularia.

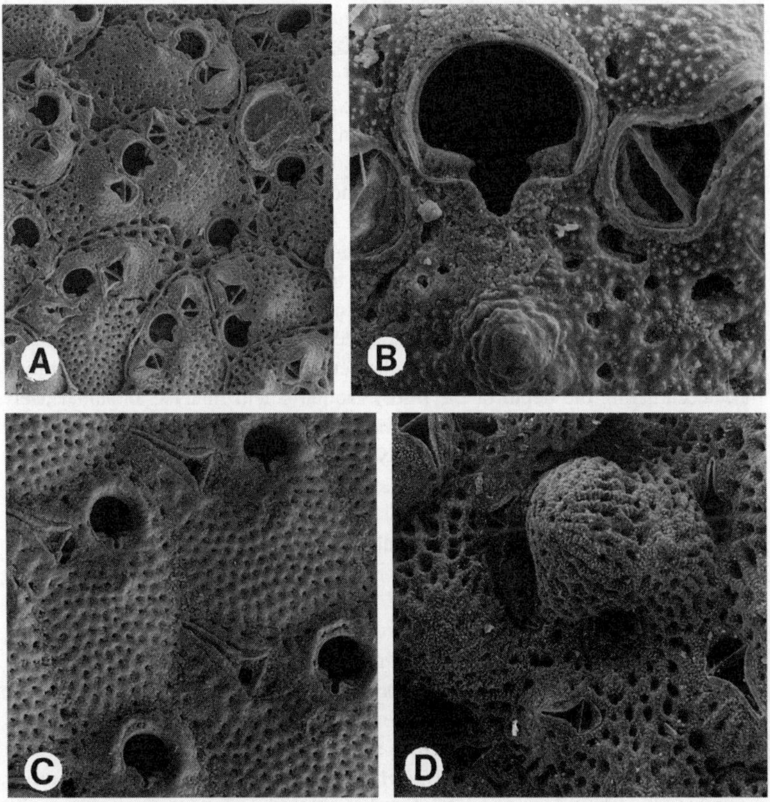

Fig. 84.
A, B, *Schizoporella cornualis* (The holotype specimen).
A, Group of autozooids; x32. B, The primary orifice; x180.

C, D, *Schizoporella dunkeri*.
C, Group of autozooids (Scilly); x50. D, Ovicellate zooids (Aegean); x50.

Schizoporella dunkeri (Reuss)
(Figs 84C, D; 85)

Cellepora dunkeri Reuss 1848: 90, pl. 10, fig. 27
Schizoporella unicornis form *longirostris* Hincks 1886: 266, pl. 10, fig. 2
not *Schizoporella dunkeri*: Hayward & Ryland 1979: 172, fig. 69
Schizoporella longirostris: Hayward & Ryland 1979: 173 (part)
Schizoporella dunkeri : Hayward & Ryland 1995: 39, pls. I, II

Colonies forming extensive, unilaminar to multilaminar sheets. Autozooids large, rectangular to polygonal, convex, separated by deep grooves; 0.5-1.0 x 0.35-0.6 mm. Primary orifice about as wide as long; anter D-shaped, poster a narrow, parallel-sided slit, with a circular terminal portion; condyles broad and flat, accentuating the depth of the sinus. Frontal shield finely granular, densely perforated by small round pores; developing a stout, median suboral umbo and appearing very rugose in later ontogeny. Avicularia single or paired, lateral to sinus, directed disto-laterally; cystid swollen, voluminous, rostrum abruptly narrowed distal to the crossbar, its distal half slender and parallel-sided; crossbar without a columella. Ovicell recumbent on distally succeeding autozooid, but globose and prominent; surface finely granular, with scattered pores, mostly around the margins; becoming more rugose in later ontogeny, frequently developing a small frontal umbo. Embryos reddish orange, present February-May in the western Mediterranean (Gautier 1962), in July at Scilly (Hayward 1976).

Sublittoral, on stones and shells. Recorded from 20-60 m in the Mediterranean, 12-52m around Scilly. It has only recently been established that the Miocene species *Schizoporella dunkeri* (Reuss) is conspecific with the living Adriatic species *Schizoporella longirostris* (Hincks), which thus becomes a junior synonym. *S. dunkeri*, as newly defined, is abundant in the Aegean and the Adriatic, and in British waters is known with certainty from Scilly. Western Mediterranean records of "*S. longirostris*" may represent *S. dunkeri* (Reuss), or may include one or more other species. The material figured by Zabala & Maluquer (1986) from the Catalan coast shows the characteristic sinus of *S. dunkeri*, and the slender avicularium, with no columella, and may be assigned with confidence to the species. The material of *S. longirostris* from the western English Channel referred to by Hayward & Ryland (1979) is now regarded as a separate species, *S. cornualis* (q.v.) and it is possible that some western Mediterranean records of *S. longirostris* may also refer to this species.

Records of *S. dunkeri* from the Mediterranean present a great problem. The records from the British sea area have been assigned to two species: *S. patula* (figured by Ryland 1968: 539, fig. 2) occurs in northern Britain, ranging north to Faroe and western Norway, and is probably the species variously termed by previous authors '*S. unicornis var. ansata*' or '*S. unicornis* deep water form'; *S. hesperia* (q.v.) occurs around southwestern Britain and possibly ranges south to the Mediterranean.

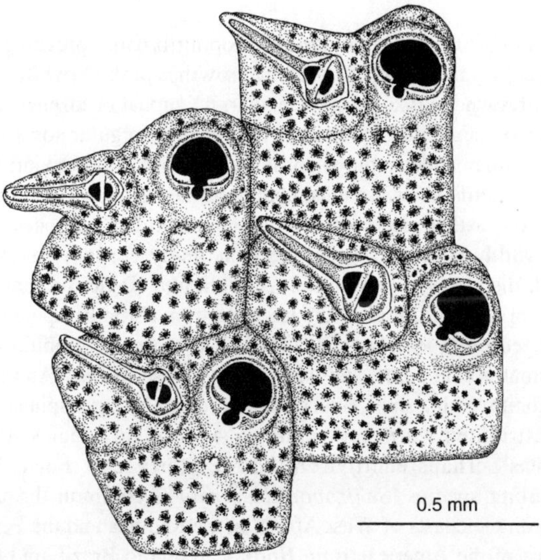

Fig. 85. *Schizoporella dunkeri.*

Schizoporella errata (Waters)
(Figs 86; 87A, B)

Lepralia errata Waters 1878: 11, pl. 1, fig. 9
Schizopodrella violacea Canu & Bassler 1930: 40, pl. 4, figs 1-14
Schizoporella unicornis Marcus 1940: 237, fig. 121
Schizoporella errata: Hayward & Ryland 1979: 170, fig. 68; Gordon & Mawatari 1992: 31, pl. 9, fig. B

Colony unilaminar or multilaminar, developing broad, spreading sheets, or erect, foliaceous, branched or cylindrical growths; pinkish-white, to reddish brown or violet-brown. Autozooids rectangular, polygonal or lozenge-shaped, flat or slightly convex, separated by shallow grooves; in regular rows and evenly spaced in unilaminar sheets, but disordered without common orientation in frontally budded multilaminar growths; commonly 0.4-0.7 x 0.3-0.5 mm. Primary orifice slightly longer than wide; sinus a deep rounded U-shape occupying about half proximal width of orifice, its corners distinctly cusped; condyles small, bluntly pointed, distinct. No oral spines. Avicularium single, proximo-lateral to orifice, on right or left; cystid relatively large, rostrum abruptly tapered to narrow, blunt tip, directed obliquely laterally. Ovicell prominent, globular, regularly perforated by small round pores, and with a very rugose surface. Ancestrula oval, smooth, imperforate, with a short, oval opesia bordered by eight spines.

A widely distributed warm temperate to subtropical species. Its original distribution was perhaps centred on the Mediterranean, but it is now an established fouling species. *Schizoporella errata* is common throughout the Mediterranean, on the coasts of West Africa, in the Red Sea and the Persian Gulf, on the east coast of the Americas from North Carolina to Brazil, and throughout the Caribbean, and on the Pacific coast of the U.S.A. Its occurrence in New Zealand has recently been documented by Gordon & Mawatari (1992). Its often massive colonies, which may exceed 150 mm diameter are conspicuous among the fouling communities of docks and harbours, and other man-made installations. There is no known permanent population in British waters.

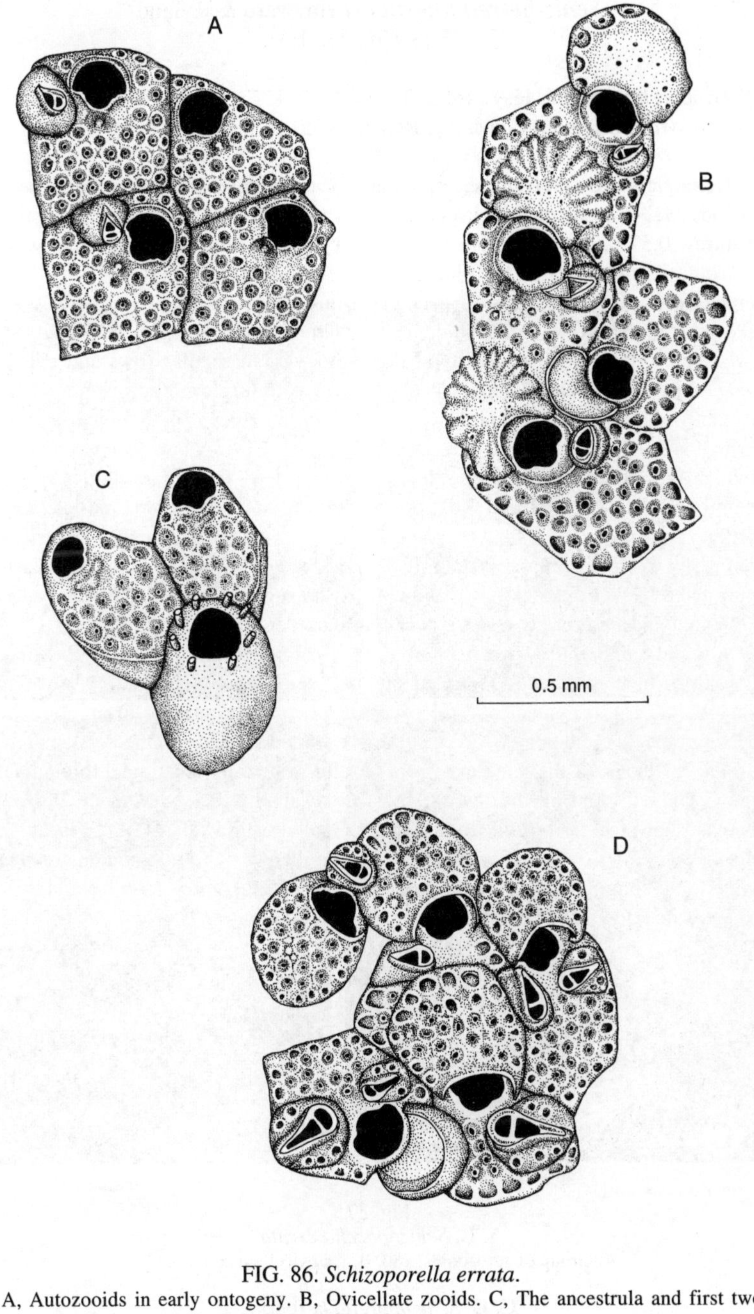

FIG. 86. *Schizoporella errata*.
A, Autozooids in early ontogeny. B, Ovicellate zooids. C, The ancestrula and first two zooids budded from it. D, Ovicellate zooids in later ontogeny.

Schizoporella hesperia Hayward & Ryland
(Figs 87C, D; 88)

Schizoporella dunkeri : Hayward & Ryland 1979: 172 (in part), fig. 69
Schizoporella hesperia Hayward & Ryland 1995: 44, pl. V, figs a, b

Colony a thick, spreading sheet, unilaminar to multilaminar, white. Autozooids broad, oval to hexagonal or irregularly polygonal; convex, separated by distinct sutures; 0.57-0.65 x 0.36-0.46 mm. Primary orifice slightly longer than wide,

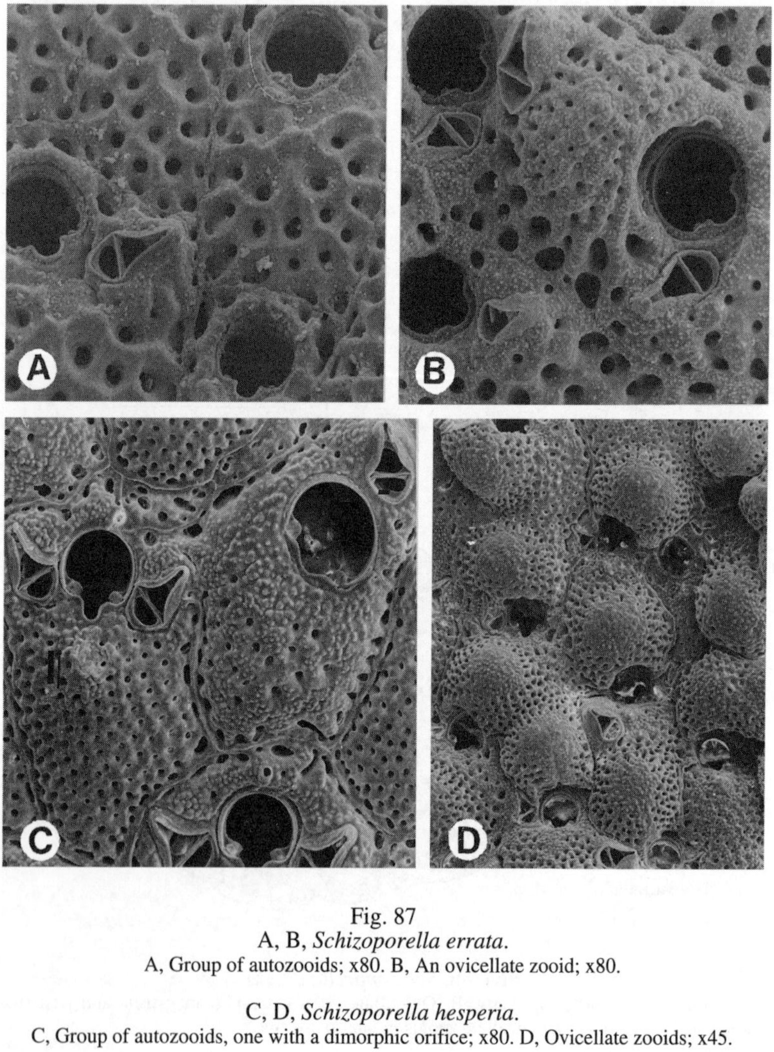

Fig. 87
A, B, *Schizoporella errata.*
A, Group of autozooids; x80. B, An ovicellate zooid; x80.

C, D, *Schizoporella hesperia.*
C, Group of autozooids, one with a dimorphic orifice; x80. D, Ovicellate zooids; x45.

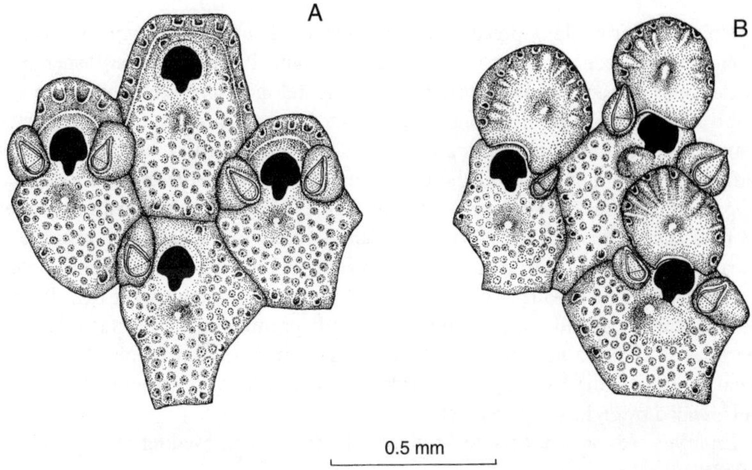

Fig. 88. *Schizoporella hesperia.*

anter D-shaped, poster one-third length of anter, U-shaped, parallel-sided, occupying about one-third width of proximal border; condyles prominent, rounded, thick with faceted frontal surface. No oral spines. Frontal shield finely granular, densely perforated by small round pores, with a median umbo developing in later ontogeny.

Avicularia single or paired (rarely, triple), proximal portion situated level with sinus, tip of rostrum level with distal rim of orifice; rostrum slightly acute to frontal plane, disto-laterally directed, abruptly tapered and slightly hooked distally; crossbar slender, complete, lacking a columella. Ovicell as wide as long, densely perforated marginally, the central area imperforate, with a pronounced umbo. Dimorphic zooids occur sporadically, with large, almost semicircular orifice, a shallow, U-shaped sinus occupying most of proximal border; avicularium lacking, or distal to orifice.

Formerly confused with '*S. dunkeri*' (= *S. cornualis* pt., *S. patula* pt.), this species was described from material collected off Lundy Island, Bristol Channel. It is known with certainty also from Gulland Rock, Padstow, from Guernsey and from the Hurd Deep, English Channel, but its wider distribution remains to be established.

Schizoporella magnifica Hincks
(Figs 89; 90A)

Schizoporella magnifica Hincks 1886: 268, pl. 10, fig. 1
Schizoporella magnifica: Hayward & Ryland 1979: 174, fig. 71

Colony an irregular, spreading sheet. Autozooids oval to hexagonal, convex, separated by deep grooves; 0.5-0.6 x 0.3-0.4 mm. Primary orifice longer than wide; sinus a deep, narrow U-shape, occupying half proximal width of orifice and equivalent to one-third of its length, accentuated by conspicuous, rounded, faceted condyles; delicate oral spines present on periancestrular autozooids, and infrequently on autozooids at the colony margin. Frontal shield finely granular, appearing vitreous, densely perforated by numerous small pores; marginal series of pores larger than the rest. Avicularia paired, rarely single, lateral to orifice and clearly distal to the sinus; rostra directed distally, with acute triangular mandibles. Ovicell longer than wide, narrowed towards aperture and appearing egg-shaped, recumbent on distally succeeding autozooid; prominent, the surface densely perforated and distinctly rugose, developing a small frontal umbo. Ancestrula oval, with a short oval opesia, bordered by a narrow cryptocystal rim and surrounded by eight slender spines.

Embryos red, present May to September in the western Mediterranean (Gautier 1962).

This species occurs throughout the Mediterranean, on hard substrata, from 5-80 m depth. It has been reported from Skomer, West Wales (Ryland, 1968), but not from any other British locality.

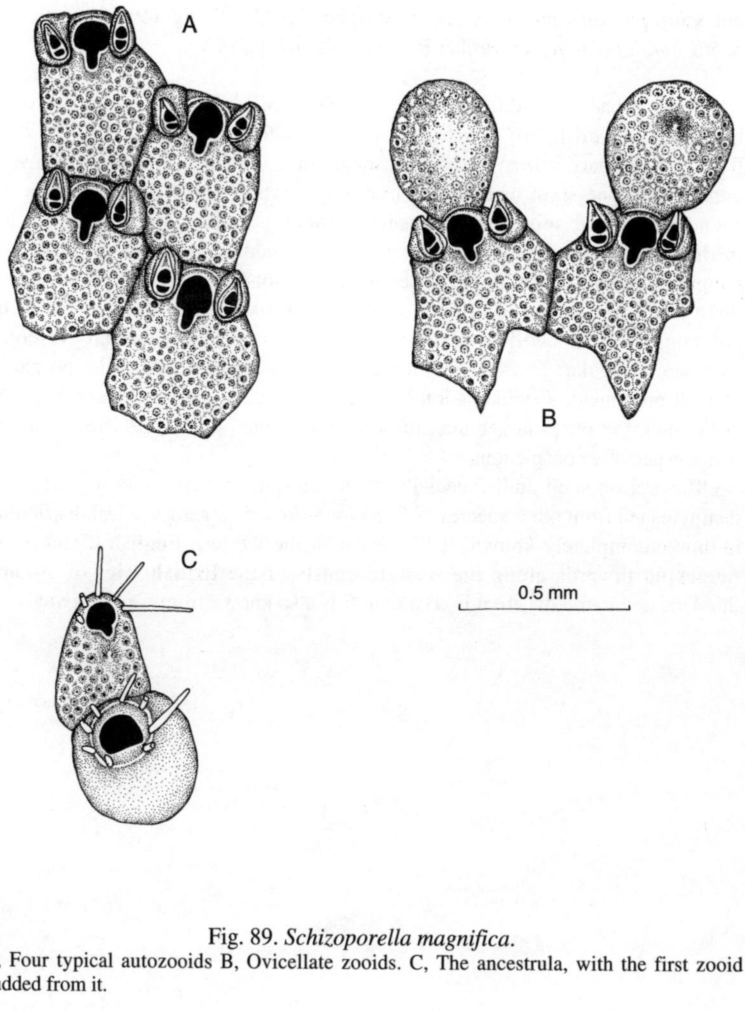

Fig. 89. *Schizoporella magnifica*.
A, Four typical autozooids B, Ovicellate zooids. C, The ancestrula, with the first zooid budded from it.

Schizoporella patula Hayward & Ryland
(Figs 90C, D)

Schizoporella dunkeri : Ryland 1968: 538, fig. 2
not *Schizoporella dunkeri* : Hayward & Ryland 1979: 172, fig. 69 (=*S. hesperia*)
Schizoporella patula Hayward & Ryland 1995: 42, pl. IV

Colony broad, spreading, thickly calcified, unilaminar. Autozooids broadly hexagonal, flat or slightly convex, separated by indistinct sutures; 0.57-0.73 x 0.4-0.65 mm. Primary orifice as wide as long, anter D-shaped, poster a shallow V, constituting one-sixth orifice length; condyles short and wide, flat, thick and rounded. No oral spines. Frontal shield finely granular, typically with a low median umbo, closely and evenly perforated by numerous tiny pores, appearing stippled; bordered by a single series of larger, slit-like marginal pores. One to three avicularia, lateral to orifice, or more proximally along the margins of the autozooid; cystid swollen, rostrum acute to frontal plane, laterally directed, elongate triangular; crossbar slender, complete, without a columella; no palate. Ovicell prominent, as wide as long, umbonate; surface coarsely granular, evenly perforated, the pores larger than those of the frontal shield, the basal marginal series especially conspicuous.

Offshore, on shell and, especially, rock. This species has only recently been distinguished from other species of *Schizoporella* and its geographical distribution is thus incompletely known. It is present in the western English Channel and ranges northwards along the western coasts of the British Isles to Orkney, Shetland and Faroe, where it is common. It is also known from west Norway.

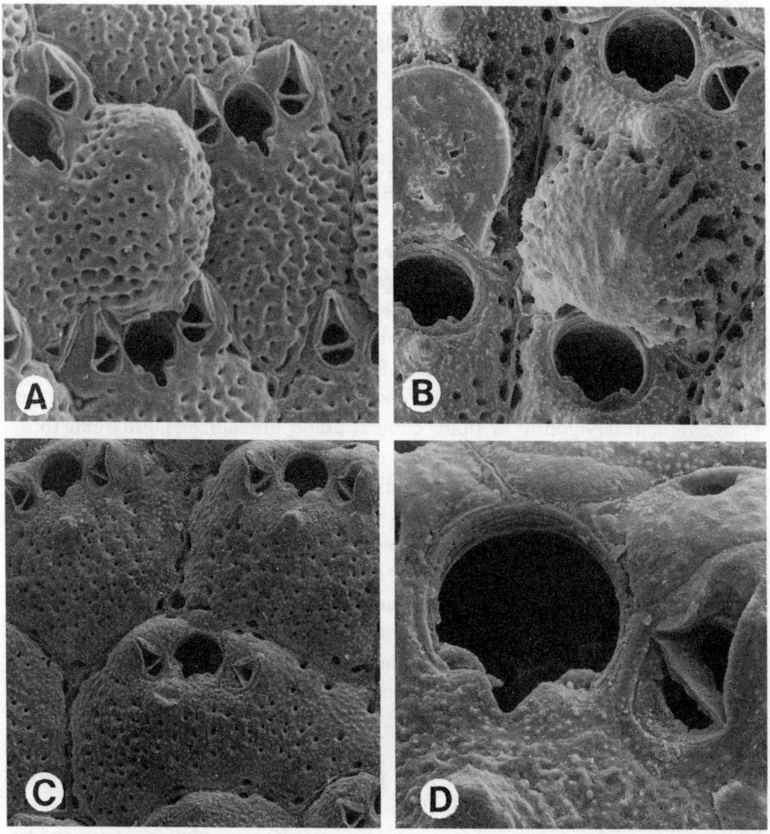

Fig. 90.
A. *Schizoporella magnifica* (Aegean); x80. B, *Schizoporella unicornis* (Shetland); x62. *Schizoporella patula* (Faroe), C. Group of autozooids; x50. D, The primary orifice; x210.

Schizoporella unicornis (Johnston in Wood)
(Figs 90B; 91)

Lepralia unicornis Johnston in Wood 1844: 19
Schizoporella unicornis: Hincks 1880: 288, pl. 35, figs 1, 2, 4, 5; Hayward & Ryland 1979: 168, fig. 67
not *Schizoporella unicornis*: Marcus 1940: 237.

Colonies developing extensive, unilaminar or multilaminar sheets, white or pinkish. Autozooids rectangular to polygonal, convex, separated by deep grooves; commonly 0.4-0.7 x 0.3-0.5 mm. Primary orifice wider than long, proximal edge with a broad, deep, V-shaped sinus occupying about half its total width; condyles small, indistinct, with acute corners. Frontal shield finely granular, densely perforated by numerous small, round pores; thicker and distinctly rugose in later ontogeny, and developing a stout, conical umbo immediately proximal to the sinus. Avicularia single or paired, lateral to the orifice on a level with the sinus; rostrum acute triangular, acute to frontal plane, disto-laterally directed; crossbar slender, without a columella. Polypide with 21-24 tentacles (Calvet 1900). Ovicell recumbent on distally succeeding autozooid, but prominent, globular, wider than long, with a highly arched aperture; frontal surface smooth, imperforate, developing a small umbo in later ontogeny, margins with scattered small pores and strong vertical fluting. Ancestrula oval, tatiform, 0.25 mm long; frontal shield vitreous, imperforate, with eight spines surrounding a semicircular opesia; budding a distal triplet of autozooids. Embryos present throughout the year, with a peak from July to September.

Lower shore and shallow sublittoral, on a wide range of substrata, including stones, shell, *Laminaria* holdfasts and the undersides of *Himanthalia* buttons. Widespread and common on all British coasts. Distributed in the northeast Atlantic region from western Norway and Faroe south to northwest Africa and the western Mediterranean. Records from outside this range should be considered doubtful.

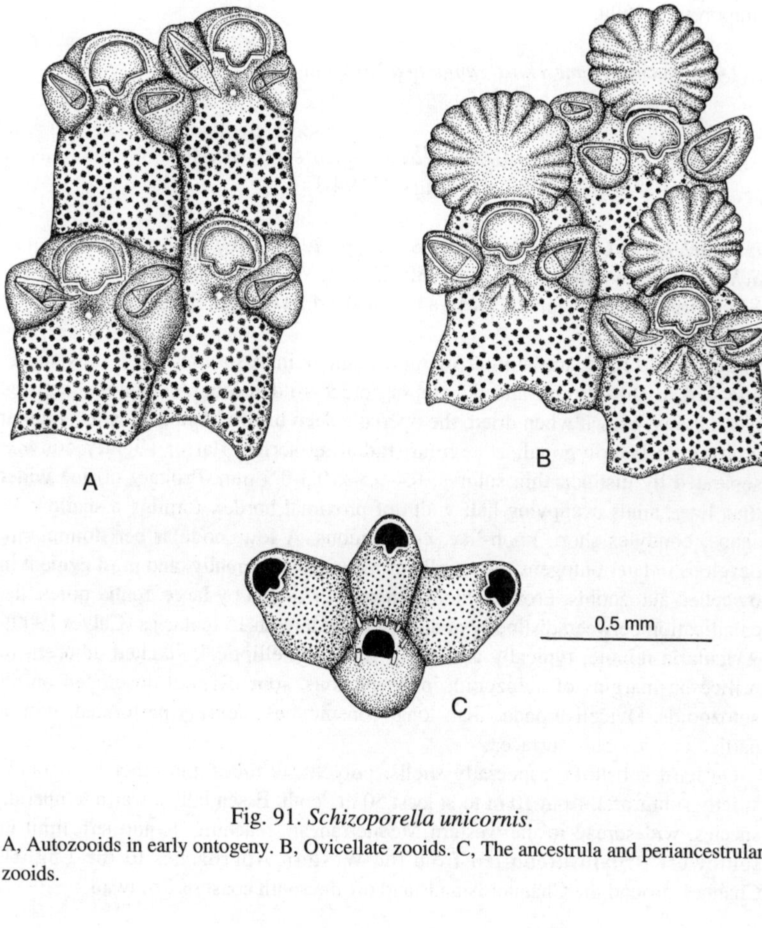

Fig. 91. *Schizoporella unicornis*.
A, Autozooids in early ontogeny. B, Ovicellate zooids. C, The ancestrula and periancestrular zooids.

Genus SCHIZOBRACHIELLA Canu & Bassler, 1920

Colony encrusting, or forming erect, folded plates. Frontal shield cryptocystidean, densely perforated. Primary orifice sinuate, with prominent condyles. Oral spines absent. Ovicell hyperstomial, prominent, closed by autozooid operculum; densely perforated, rugose. Avicularia adventitious, small, budded from frontal septula adjacent to orifice. Vertical walls deep, with small uniporous septula.

Type species: *Hemeschara sanguinea* Norman, 1868

Schizobrachiella sanguinea (Norman)
(Figs 92; 94B)

Hemeschara sanguinea Norman 1868: 222, pl. 7, figs 9-11
Schizoporella sanguinea: Hincks 1880: 252, pl. 39, figs 6, 7
Schizobrachiella sanguinea: Hayward & Ryland 1979: 188, fig. 78

Colony a broad, unilaminar or multilaminar incrustation, often only partly attached to substratum, and developing erect, folded lobes or plates; deep red when living, brown when dried, the opercula deep brown. Autozooids rectangular or irregularly polygonal, in regular, radiating series, flat or slightly convex, separated by distinct, thin sutures; 0.4-0.8 x 0.3-0.5 mm. Primary orifice wider than long; sinus occupying half width of proximal border, forming a shallow V-shape; condyles short, knob-like, conspicuous. A low, nodular peristomial rim develops in later ontogeny, especially pronounced proximally, and most evident in ovicelled autozooids. Frontal shield densely perforated by large, round pores, the calcification between distinctly nodular. Polypide with 15 tentacles (Calvet 1900). Avicularia minute, typically about 0.1 mm long, elliptical; situated adjacent to orifice on margins of autozooid, inconspicuous, sporadic, not developed on all autozooids. Ovicell broader than long, conspicuous; densely perforated, with a particularly knobbly surface.

On hard substrata, especially shells, polychaete tubes and other bryozoans; strictly sublittoral, from 10 m to at least 50 m depth. Essentially a warm temperate species, widespread in the western Mediterranean, reaching its northern limit in southwest Britain. Reported from the Western Approaches to the English Channel, around the Channel Islands and off the south coast of Cornwall.

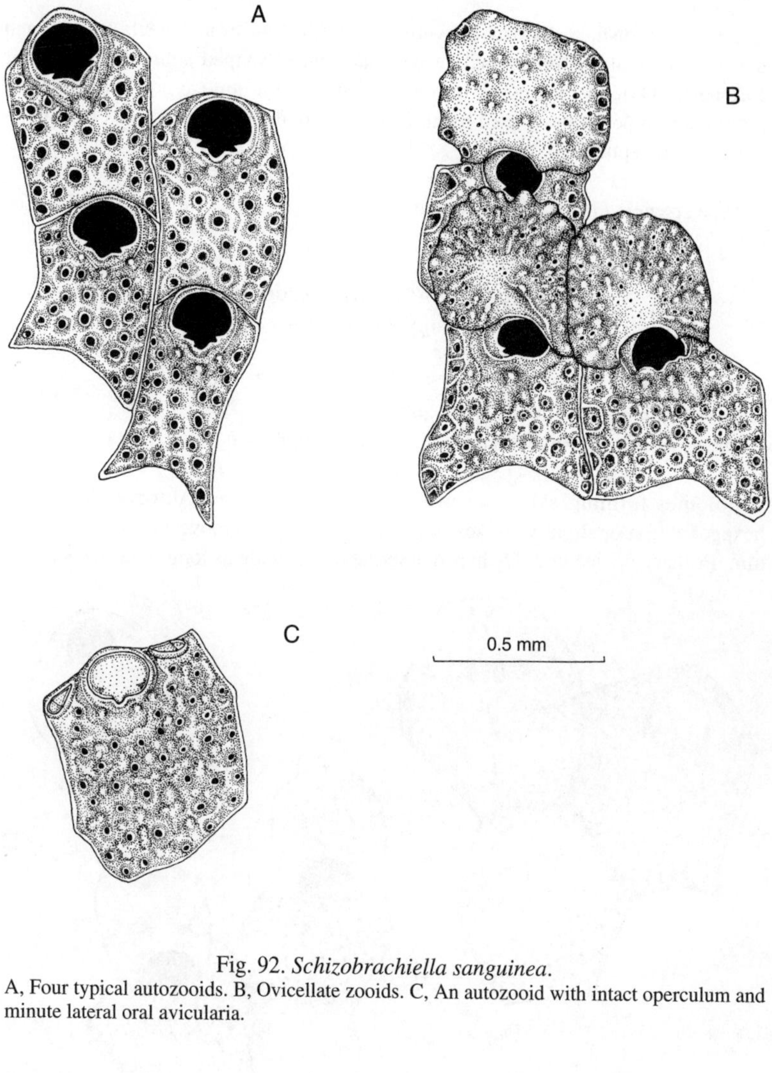

Fig. 92. *Schizobrachiella sanguinea*.
A, Four typical autozooids. B, Ovicellate zooids. C, An autozooid with intact operculum and minute lateral oral avicularia.

Genus ARTHROPOMA Levinsen, 1909

Colony encrusting. Autozooids with cryptocystidean frontal shield, evenly and densely perforate. Primary orifice with narrow, U-shaped sinus. No spines or avicularia. Ovicell recumbent on distally succeeding autozooid; hyperstomial, prominent, imperforate; closed by autozooid operculum. Vertical walls with large multiporous septula.

Type species: *Flustra cecilii* Audouin, 1826

Arthropoma cecilii (Audouin)
(Figs 93; 94A)

Flustra cecilii Audouin 1826: 239; Savigny pl. 8, fig. 3
Schizoporella cecilii: Hincks 1880: 269, pl. 43, fig. 6
Arthropoma cecilii: Hayward & Ryland 1979: 190, fig. 79

Colonies forming extensive, white or silvery patches. Autozooids oval to hexagonal, flat or slightly convex, separated by shallow grooves; 0.7-0.8 x 0.4-0.6 mm. Primary orifice with D-shaped anter, about as wide as long, with the straight

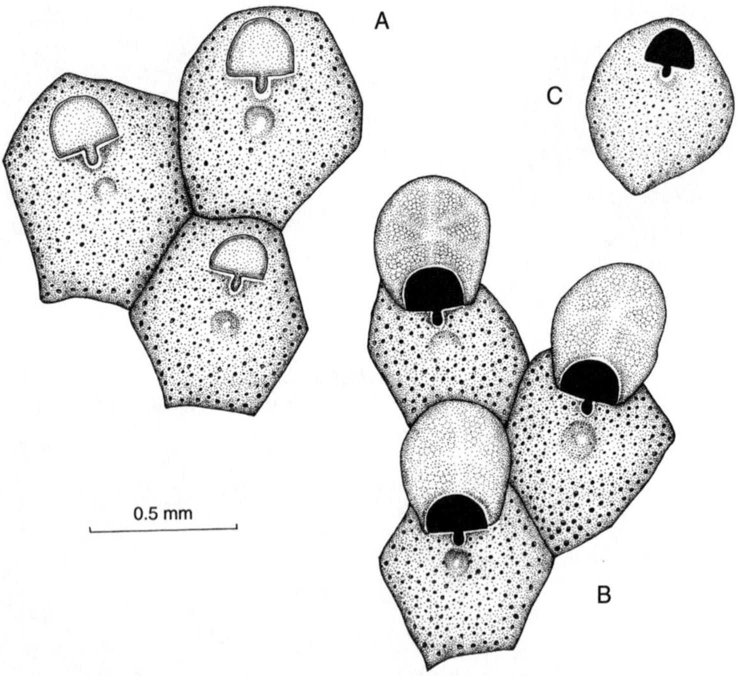

Fig. 93. *Arthropoma cecilii*.
A, A group of autozooids. B, Ovicellate zooids. C, The ancestrula.

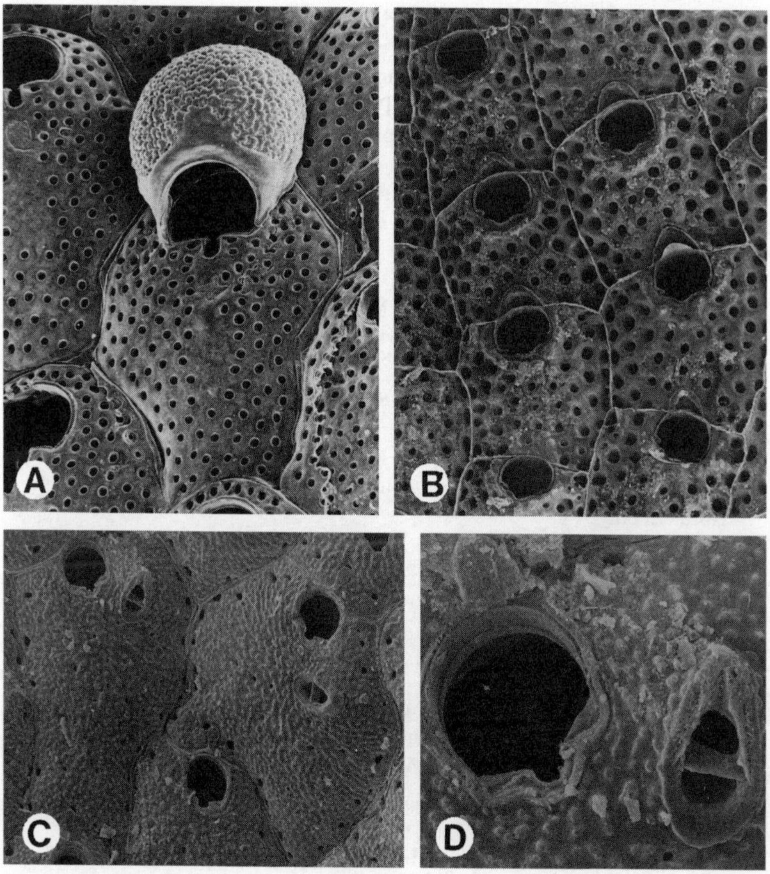

Fig. 94.
A, *Arthropoma cecilii*; x60. B, *Schizobrachiella sanguinea* (specimen from Naples); x40. C, D, *Escharina alderi*. C, Group of autozooids; x50. D, The primary orifice; x160.

proximal border indented by a narrow, U-shaped poster; operculum thick, deep brown, with a conspicuous marginal sclerite. Frontal shield smooth, densely perforated by very small round pores; a low umbo sometimes present proximal to the sinus in later ontogeny. Ovicell prominent, globular, about as wide as long; dense white, with a granular, or faintly tessellate surface; occasionally asymmetrical, with one side extending proximal to the orifice. Embryos and larvae present June to December in the western Mediterranean (Gautier 1962); ancestrula identical to later autozooids but smaller.

Offshore, on hard substrata, especially shell and other biogenic carbonates; recorded from 30-80 m in the Mediterranean. This is a warm temperate to tropical species, with a circumtropical distribution. It occurs throughout the Mediterranean and in the E. Atlantic reaches the northern limit of its range around the Channel Islands, the western mouth of the Channel and off south Cornwall.

Genus ESCHARINA Milne-Edwards, 1836

Colony encrusting, unilaminar. Autozooids with a cryptocystidean frontal shield, with few marginal perforations. Primary orifice sinuate, with well marked condyles, typically encircled by a conspicuous peristome. Oral spines present or absent. Adventitious avicularia present, with acute or setiform mandibles. Basal pore chambers present. Ovicell hyperstomial, imperforate, typically umbonate; recumbent on substratum or on distally succeeding autozooids, closed by zooidal operculum.

Type species: *Eschara vulgaris* var. *a* Moll, 1803

Key to the British species of *Escharina*

1. Autozooids in single linear series; sometimes forming small contiguous groups but not broad sheets. Avicularia small, latero-frontal, rare. Northern only
.... ***Escharina alderi*** (p. 228)

Autozooids budded in regular encrusting sheets
.... **2**

2. Primary orifice surrounded by a deep, flared peristome. No oral spines. Ovicell and frontal shield with one to many conical umbones. Avicularia with acute triangular mandibles, infrequent
.... ***Escharina johnstoni*** (p. 234)

Without a marked peristome. Oral spines present, umbones infrequent, avicularia with setiform mandibles
.... **3**

3. Paired avicularia lateral to orifice; cystid small, mandible whip-like
.... ***Escharina dutertrei haywardi*** (p. 230)

Paired or single avicularia proximal to orifice
.... **4**

4. Avicularia typically paired; rostrum acute, mandible rarely longer than an autozooid. Ovicell uniformly granular
.... ***Escharina vulgaris*** (p. 236)

Avicularia typically single; rostrum oval, mandible often exceeding length of autozooid. Ovicell with a prominent transverse ridge
.... ***Escharina hyndmanni*** (p. 232)

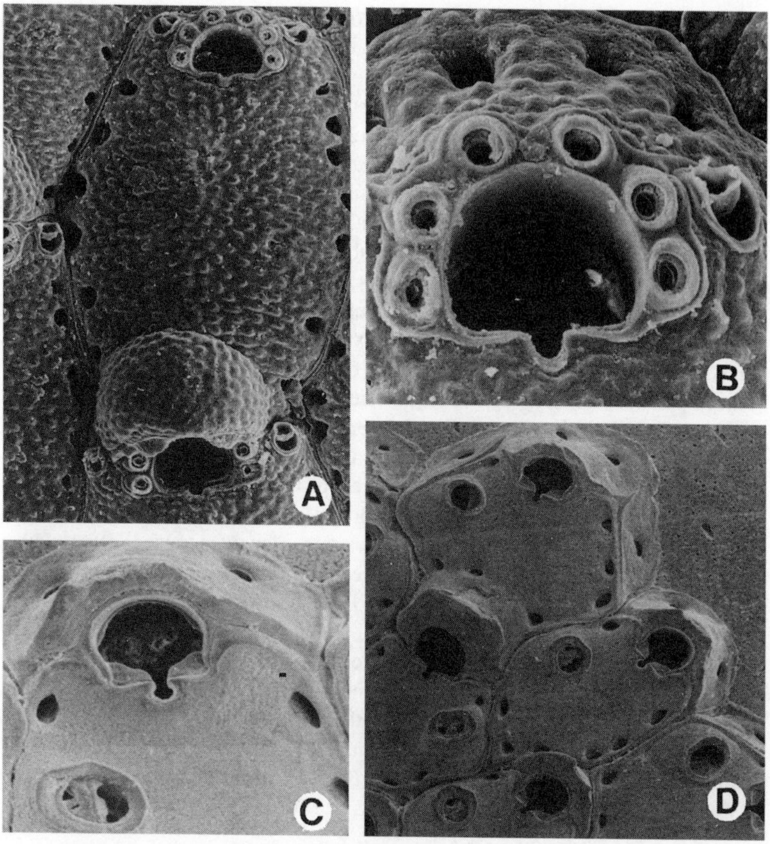

Fig. 95.
A, B, *Escharina dutertrei haywardi* (Shetland).
A, Group of autozooids, with an ovicell; x70. B, The primary orifice; x210.

C, D, *Escharina hyndmanni* (Biscay).
C, Detail of an autozooid; x110. D, The colony margin, including an ovicellate zooid; x50.

Escharina alderi (Busk)
(Figs 94C, D; 96)

Alysidota alderi Busk 1856b: 311, pl. 9, figs 6, 7
Schizoporella alderi: Hincks 1880: 243, pl. 36, figs 9, 10
Escharina alderi: Hayward & Ryland 1979: 200, fig. 84

Colony composed of autozooids in continuous, irregularly bifurcating, linear series, forming a fine tracery over the substratum, occasionally converging to form a loose, disordered sheet. Contiguous autozooids, usually no more than 12-20, may rarely be found in a normal quincuncial arrangement. Autozooids oval, elongate, flattened and only slightly convex; 0.5-0.7 x 0.4-0.5 mm. Frontal shield thick, fine grained, perforated by a number of irregularly scattered, inconspicuous, minute pores, frequently obscured in older autozooids; marginal pores small, indistinct, best seen in cleaned material. Orifice subcircular, slightly longer than broad, proximal margin with a narrow, U-shaped sinus. A thin rim may be developed around the anter but there is no prominent peristome. Oral spines absent. A low, conical umbo may be developed proximal to the sinus. Avicularia rare, lateral to the orifice, often absent; rostrum triangular, raised, often almost perpendicular to autozooid, commonly directed proximally or proximo-laterally. Mandible elongate, triangular and pointed. Basal pore chambers numerous, small, tubular, entirely enclosed within the lateral walls, all around the periphery of the autozooid. Ovicell prominent, recumbent on the succeeding autozooid, globular with a finely granular surface often raised into a central umbo.

A cold temperate species, restricted to the north-east Atlantic. At Shetland it occurs offshore on hard substrata (Norman, 1868), and has been reported as far south as the Hebrides (Ryland, 1963a); it has been collected from Bergen, Norway (Ryland, 1963b) and ranges into the Barents Sea (Kluge, 1975).

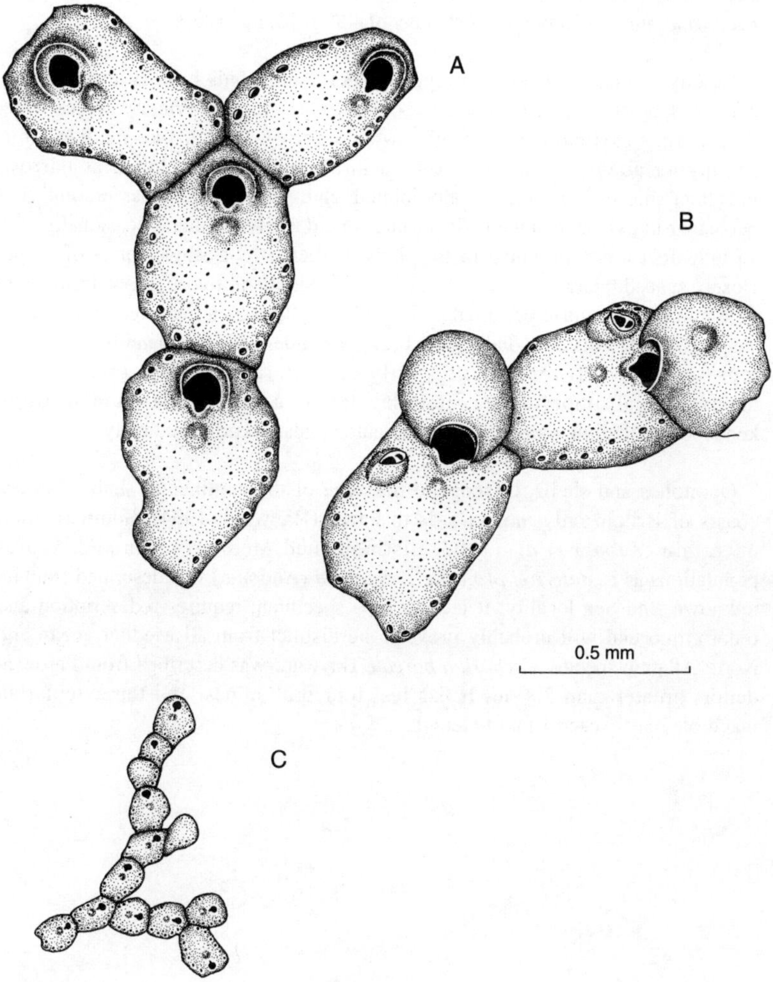

Fig. 96. *Escharina alderi*.
A, Four autozooids at the bifurcation of a linear series. B, Ovicellate zooids. C, Part of a colony showing typical growth habit.

Escharina dutertrei haywardi Zabala, Maluquer & Harmelin
(Figs 95A, B; 97)

Mastigophora dutertrei: Hincks 1880: 279, pl. 37, figs 1, 2
Escharina dutertrei: Hayward & Ryland 1979: 198, fig. 83
Escharina dutertrei haywardi Zabala *et al.* 1993: 73, fig. 12d-f

Colony a broad, spreading, irregular sheet. Autozooids broadly oval, slightly convex, separated by distinct sutures; 0.6-1.0 x 0.4-65 mm. Primary orifice wider than long; proximal edge straight, with a narrow, notch-like median sinus occupying about one-sixth its total width; condyles developed as narrow, indistinct shelves; six stout, basally jointed, club-shaped oral spines present, four persisting in ovicelled autozooids. Frontal shield finely nodular, occasionally with an indistinct umbo proximal to the sinus; bordered by a single series of round, closely spaced, marginal pores. Avicularia paired, lateral to orifice; rostrum about 0.06 mm long, rounded proximally, truncate distally, appearing more or less shoe-shaped; directed disto-medially, supporting a slender setiform mandible 0.35-0.45 mm long. Ovicell recumbent, partly on distally succeeding autozooid; hemispherical, imperforate, often developing a small frontal umbo. Blind kenozooids, of similar size and shape to autozooids, occur sporadically.

On stones and shells, offshore to the edge of the continental shelf. Western coasts of Britain only; northwards to Faroe (Hayward, 1994). Southern limit uncertain; Zabala *et al.* (1993) distinguished Mediterranean and Azores populations as *E. dutertrei protecta. E dutertrei* (Audouin) was described from an unknown Red Sea locality; it lacks a type specimen, requires redescription and redefinition and will probably prove to be distinct from all Mediterranean and North Atlantic species. *Escharina boreale* Hayward was described from Faroe, at depths greater than 200 m. It has just four oral spines, and the avicularian mandible may exceed 1 mm in length.

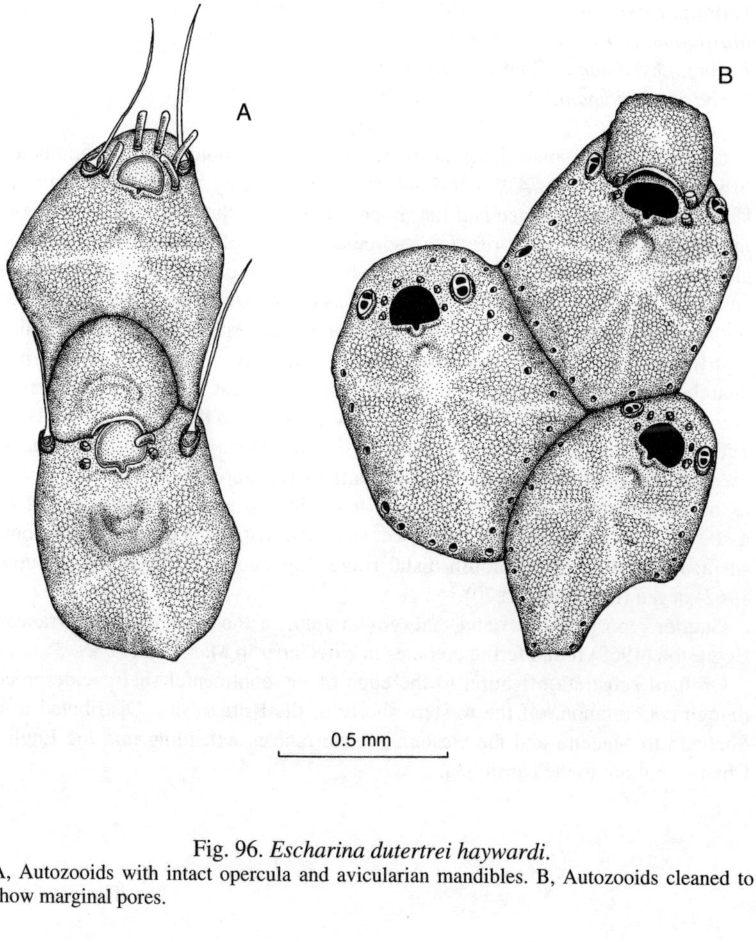

Fig. 96. *Escharina dutertrei haywardi*.
A, Autozooids with intact opercula and avicularian mandibles. B, Autozooids cleaned to show marginal pores.

Escharina hyndmanni (Johnston)
(Figs 97C, D; 98)

Lepralia hyndmanni Johnston 1847: 306, pl. 54, fig. 6
Mastigophora hyndmanni: Hincks 1880: 281, pl. 37, figs 3-4
Herentia hyndmanni: Gautier 1962: 159
Escharina hyndmanni: Hayward & Ryland 1979: 196, fig. 82

Colony forming small, irregular incrustations. Autozooids oval to rhomboidal, broad and flat, 0.6-0.74 x 0.36-0.44 mm; separated by raised sutures. Frontal shield smooth, fine grained and flat; imperforate centrally with a series of distinct marginal pores. Primary orifice semicircular, broader than long; sinus short and slender, slit-like, or like an inverted keyhole. Three or four very long, slender spines on the distal edge of the orifice; lost as the autozooid ages and usually seen only on peripheral autozooids. Distal border of the orifice rapidly thickened by additional calcification, apparently originating from the distal, basal pore chambers, forming a raised and curved rim above that of the primary orifice. Avicularium interzooidal, on the right or left, proximal to the orifice and replacing one of the lateral, basal pore chambers. Frontal surface flat and smooth, not clearly demarcated from the frontal wall of the autozooid with which it is associated; rostrum orbicular, setiform mandible up to twice the length of the associated autozooid. Ovicell immersed, spherical, with a curved, flattened frontal surface delimited by a smooth distal ridge. Embryos orange-yellow (Gautier, 1962) or red (Eggleston, 1970).

Gautier reported larvae and embryos in July, in the western Mediterranean; Eggleston (1969) found fertile colonies in November in Manx waters.

On hard substrata offshore, to the edge of the continental shelf; widespread, though not common, off the western shores of the British Isles. Distributed from Shetland to Madeira and the western Mediterranean, extending into the English Channel but not to the North Sea.

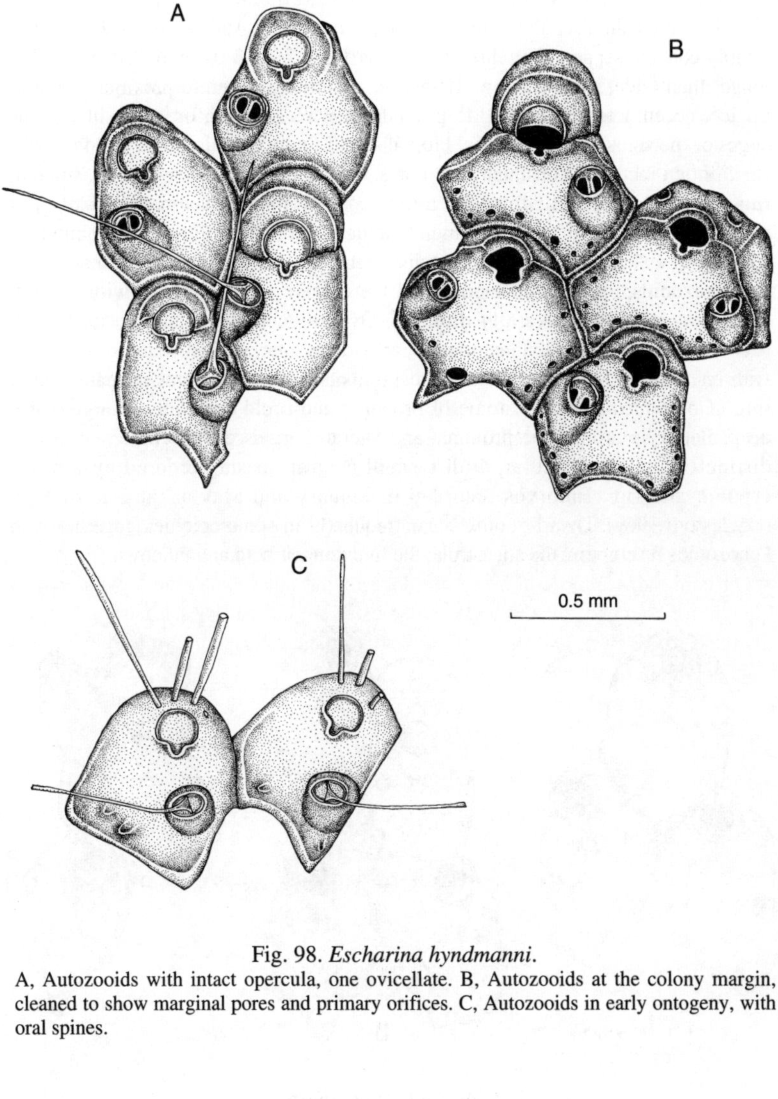

Fig. 98. *Escharina hyndmanni*.
A, Autozooids with intact opercula, one ovicellate. B, Autozooids at the colony margin, cleaned to show marginal pores and primary orifices. C, Autozooids in early ontogeny, with oral spines.

Escharina johnstoni (Quelch)
(Figs 99; 100A, B)

Lepralia simplex Johnston 1847: 305, pl. 54, fig. 4
Schizoporella simplex: Hincks 1880: 2246, pl. 35, figs 9, 10
Escharina johnstoni: Hayward & Ryland 1979: 194, figs 10B, 81

Colony a small, discoid or irregular, unilaminar patch. Autozooids oval, elongate, steeply convex, separated by distinct grooves; 0.5-0.7 x 0.3-0.4 mm. Primary orifice longer than wide; sinus shallow, U-shaped, occupying one-third proximal width of orifice, accentuated by broad, deep condyles which extend, on each side, to the edges of the sinus. No oral spines. Frontal shield finely granular, with up to a dozen small, often inconspicuous, marginal pores. Orifice encircled by a circular, smooth-rimmed, slightly flared peristome; a low, conical umbo developed medially just proximal to the lip of the peristome. Avicularia sporadic, infrequent to abundant, sometimes absent; budded from marginal pores on the proximo-lateral margins of the autozooids, and around the base of the ovicell; rostrum acute, with a short, triangular mandible, laterally directed. Ovicell recumbent on the substratum between distally succeeding autozooids, as wide as long, slightly flattened frontally, with one to three, or more, short, conical umbones; marginal pores indicate a basal ring of pore chambers; peristome of brooding autozooid especially enlarged, as a deep, flared lip around the proximal and lateral margins of its orifice. Ancestrula distinctive, almost circular, with a small central opesia bordered by a broad cryptocystal rim. Embryos recorded in January and May at the Isle of Man (Eggleston, 1969). Dwarf zooids occur frequently in some colonies, together with kenozooids resembling the ancestrula; the functions of both are unknown.

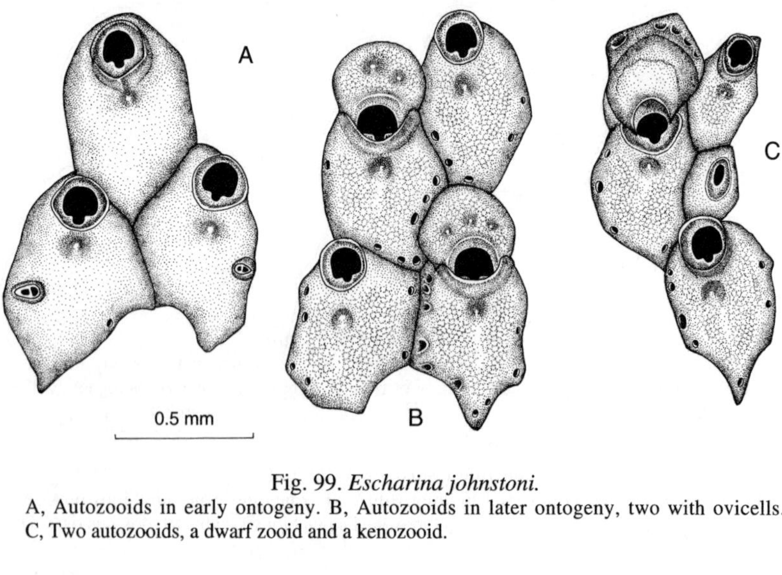

Fig. 99. *Escharina johnstoni.*
A, Autozooids in early ontogeny. B, Autozooids in later ontogeny, two with ovicells. C, Two autozooids, a dwarf zooid and a kenozooid.

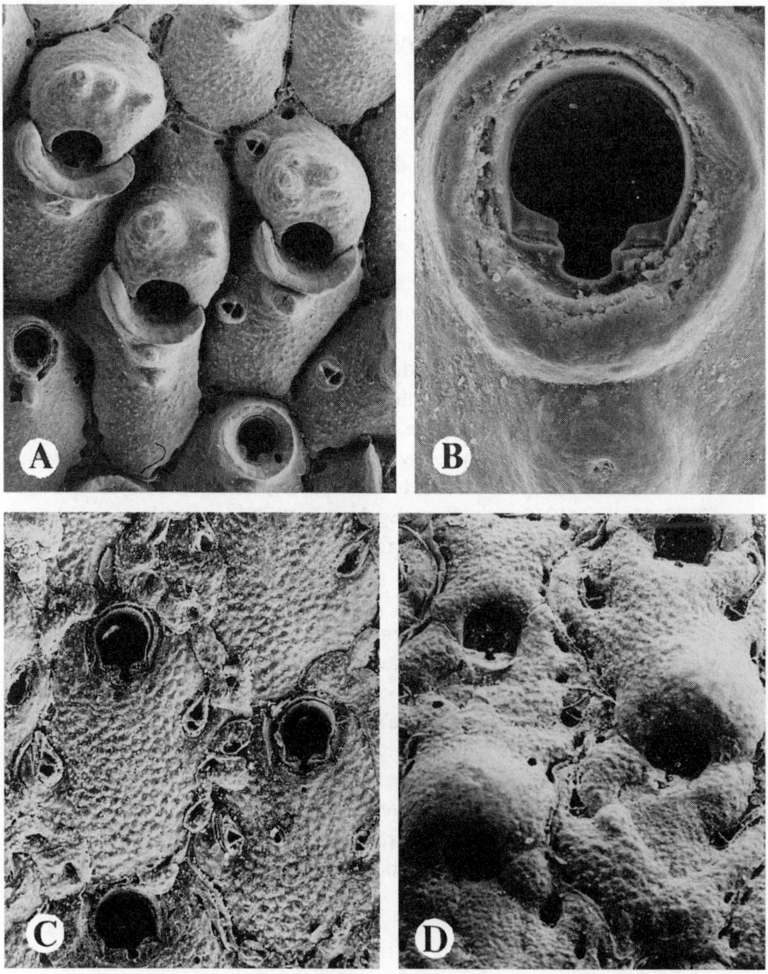

Fig. 100.
A, B, *Escharina johnstoni*.
A, Portion of a colony; x40. B, The primary orifice; x280.

C, D, *Escharina vulgaris*.
C, Autozooids with numerous avicularia; x80. D, Autozooids in later ontogeny, some with ovicells; x70.

Offshore, perhaps to the continental shelf edge, on small stones and shell. It is distributed along the western coasts of the British Isles, from Shetland to the Channel approaches, and has recently been recorded from the Atlantic coasts of Spain (Lanza Suàrez & Fernàndez Pulpeiro, 1984), and the Straits of Gibraltar (López de la Cuadra & Garcia Gómez, 1988). In the western Channel it is often abundant on small shell substrata, especially the inner surfaces of Veneracea valves.

Escharina vulgaris (Moll)
(Figs 100C, D; 101)

Eschara vulgaris var. *a* Moll 1803: 55, pl. 3, figs 10A, B
Schizoporella vulgaris: Hincks 1880: 244, pl. 15, figs 5-6; pl. 37, fig. 7
Escharina vulgaris: Hayward & Ryland 1979: 192, fig. 80

Colony encrusting, unilaminar, developing irregular, spreading patches. Autozooids oval to hexagonal, convex, separated by distinct grooves; 0.45-0.6 x 0.3-0.4 mm. Primary orifice longer than wide, proximal border almost straight, with a short, rounded, notch-like sinus; condyles short, thick, rounded; four delicate, distal oral spines, present in early ontogeny only. Frontal shield finely granular, with few, widely spaced, indistinct marginal pores. Avicularia typically paired, proximo-lateral to orifice, on the margins of the autozooids, budded from marginal pores; rostrum slender, acute, directed distally or disto-laterally, bearing an elongate, setiform mandible. Ovicell recumbent on distally succeeding autozooid, more or less spherical, finely granular, imperforate; embryos said to be orange-red (Gautier, 1962) or yellow (Eggleston, 1970), present June to February in the western Mediterranean, throughout the year around the Isle of Man. Ancestrula oval, tatiform, with smooth gymnocyst and narrow cryptocystal rim; opesia bordered by about nine spines.

Calcification thickens in later ontogeny. The orifice becomes surrounded by a low, thick, rim, obscuring the spine bases and often developed as an indistinct medio-proximal umbo. The interstices between autozooids are partly infilled with rounded masses of calcification, apparently derived from the marginal pores, which may bear additional avicularia, with varying orientation.

On hard substrata, especially shells and small stones, from the lower shore to about 100 m depth. Distributed along western coasts of Britain as far north as the Isle of Man and Antrim; not reported from Shetland, nor from the east coasts of Britain, the Low Countries or Denmark. Ranges south to Madeira, and occurs throughout the Mediterranean.

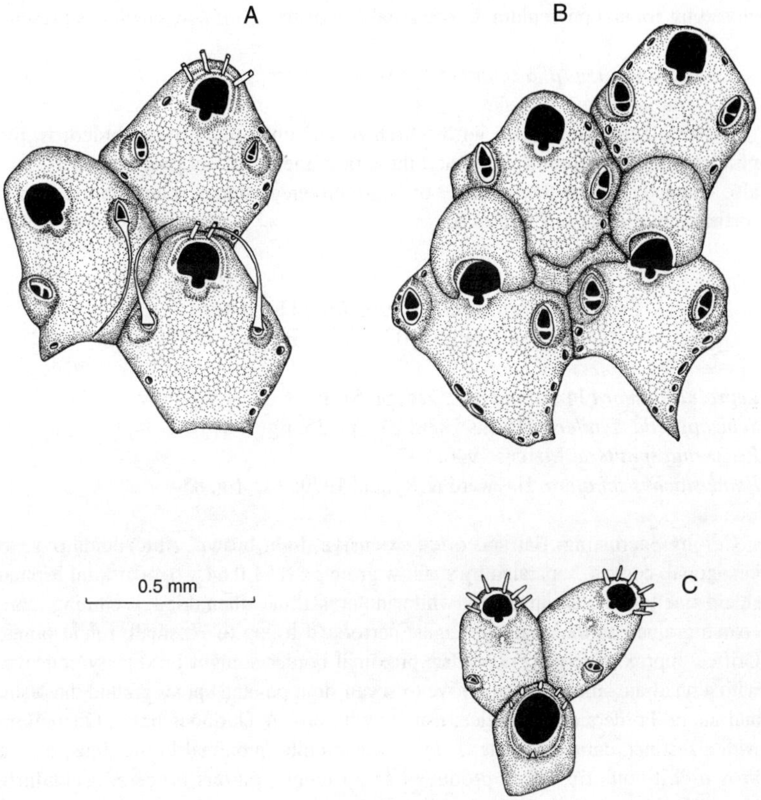

Fig. 101. *Escharina vulgaris*.
A, Three typical autozooids. B, Ovicellate zooids. C, The ancestrula and first two zooids budded from it.

Genus PHAEOSTACHYS Hayward, 1979

Frontal shield with scattered pores. Orifice sinuate, with thick oral spines. One or more frontal umbones, surmounted by thickened chitinous spines. Avicularia adventitious, frontal, often replaced by spines. Ovicell prominent, imperforate, not closed by zooidal operculum. Large distal and disto-lateral pore chambers present.

Type species: *Lepralia spinifera* Johnston, 1847

Differs from *Escharina*, with which it was until recently included, in the perforation of the frontal shield and the larger, more distinct pore chambers. It is also characterised by the presence of large, brown chitinous spines on the frontal surface, not present in *Escharina*.

Phaeostachys spinifera (Johnston)
(Figs 102; 103A, B)

Lepralia spinifera Johnston 1847: 324, pl. 57, fig. 6
Schizoporella spinifera: Hincks 1880: 241, pl. 35, figs 6-8
Escharina spinifera: Marcus 1940: 247
Phaeostachys spinifera: Hayward & Ryland 1979: 202, fig. 85

Colony encrusting, flat and often extensive, light brown. Autozooids oval to hexagonal, convex, separated by shallow grooves; 0.54-0.64 x 0.4-0.5 mm. Frontal shield finely granular, appearing white and crystalline when dried, becoming more coarse grained as the autozooid ages; perforated by up to 30 small, round pores. Orifice approximately semicircular; proximal border straight or slightly concave, with a quadrate sinus medially. Five to seven stout pointed spines around the distal and lateral borders of the orifice, usually with brown, chitinous bases. Operculum with a distinct, curved sclerite. A low conical umbo proximal to the sinus, with a brown chitinous tip, often produced as an erect, spinous process. Avicularia sporadic, situated on the frontal wall, lateral or proximal to the umbo, often absent or replaced by a chitinous spine on a conical base, similar to the suboral umbo; mandible acute, triangular, directed laterally. In older autozooids, additional spines may be developed on the frontal wall proximal and lateral to the umbo. Ovicell prominent, globular, surface coarse-grained and rugose with coarse flutings arising from the distobasal edge. Embryos orange-red. Ancestrula tatiform, with a smooth imperforate gymnocyst, a narrow cryptocyst and about twelve spines surrounding an extensive frontal membrane. Embryos recorded between December and August, with a peak in March-June, off the Isle of Man (Eggleston, 1969).

Essentially a shallow water species, encrusting stones, shells, algae such as *Furcellaria*, and *Laminaria* holdfasts; ranging from the lower shore, where it is often abundant, to about 100 m. Off all British coasts but of limited distribution elsewhere; reported from Danish waters and probably widespread in the southern North Sea, but reaching its northern limit around Faroe; not found in western Norway. Absent from the Mediterranean.

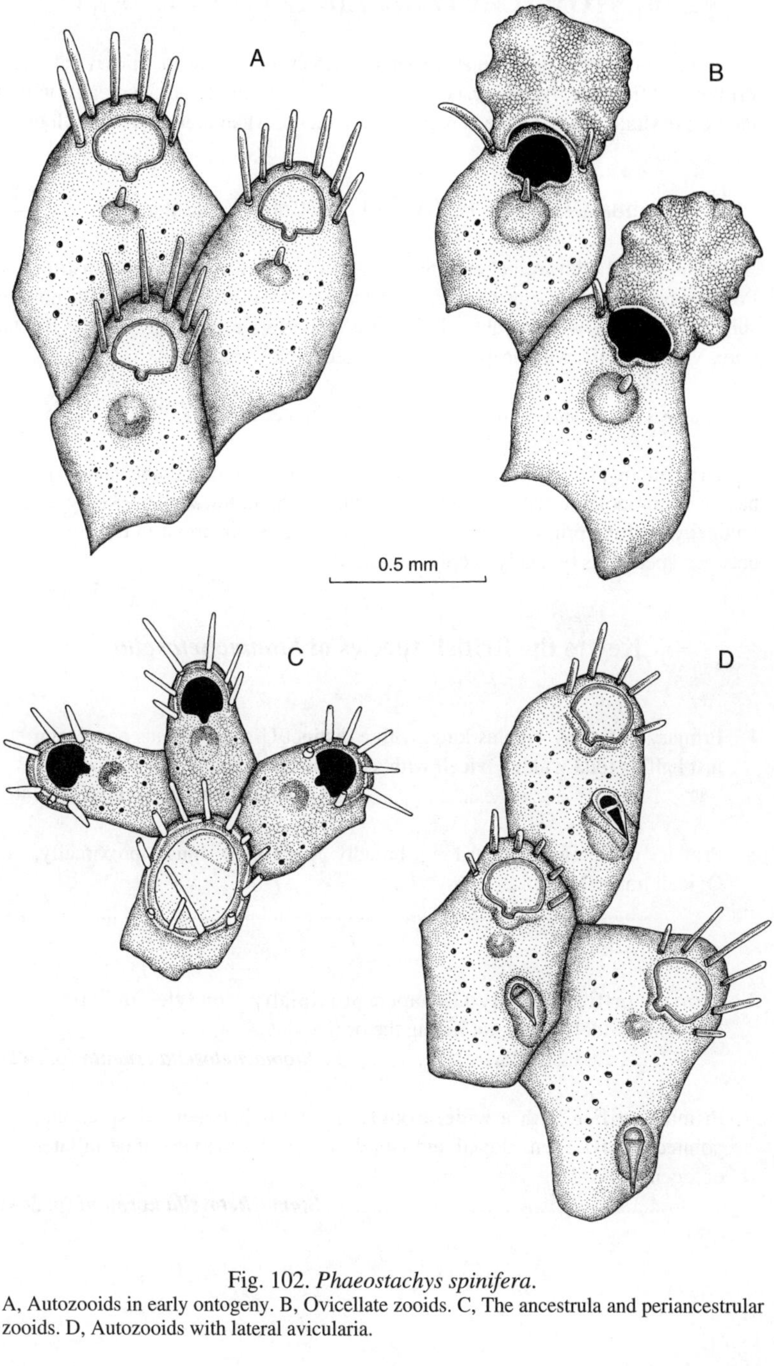

Fig. 102. *Phaeostachys spinifera*.
A, Autozooids in early ontogeny. B, Ovicellate zooids. C, The ancestrula and periancestrular zooids. D, Autozooids with lateral avicularia.

Family STOMACHETOSELLIDAE Canu & Bassler, 1917

Colony encrusting, unilaminar. Frontal shield cryptocystidean, thickly calcified. Primary orifice deeply immersed, surrounded by a peristome. Typically with a medio-proximal notch. Ovicells present. Avicularia, when present, adventitious.

Genus STOMACHETOSELLA Canu & Bassler, 1917

Frontal shield with frontal and marginal pores, typically irregularly distributed. Primary orifice concave or distinctly sinuate proximally; condyles small; no oral spines. Peristome well developed. Ovicells imperforate, or with a single frontal pore. Vertical walls with deeply recessed septula. No avicularia.

Type species: *Stomachetosella crassicollis* Canu & Bassler, 1917

The genus is predominantly Arctic or boreal in distribution, and many species have a circumpolar distribution. In all, the frontal shield thickens markedly in later ontogeny, and the primary orifice, which is especially important in discriminating between species, is typically deeply immersed.

Key to the British species of *Stomachetosella*

1. Primary orifice as wide as long, with a distinct U-shaped sinus occupying just half its total width. Ovicell with a conspicuous, single frontal pore
.... ***Stomachetosella sinuosa*** (p. 246)

Primary orifice wider than long, broadly concave or sinuate proximally. Ovicell imperforate
.... **2**

2. Primary orifice broadly U-shaped proximally, condyles indistinct. Peristome widely open, not hiding the orifice
.... ***Stomachetosella cruenta*** (p. 242)

Primary orifice with a wide, trough-like sinus between conspicuous, pointed condyles; enveloped and largely hidden by the peristome in later ontogeny
.... ***Stomachetosella normani*** (p. 244)

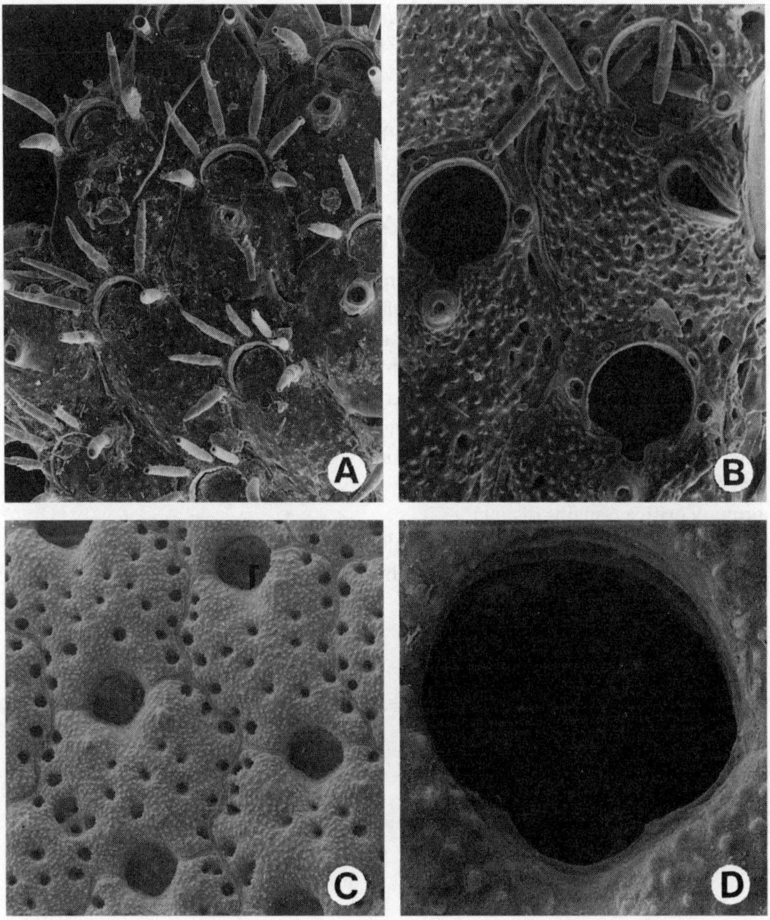

Fig. 103.
A, B, *Phaeostachys spinifera*.
A, Autozooids with cuticle and spines intact; x45. B, Cleaned autozooids; x80.

C, D, *Stomachetosella cruenta* (Faroe).
C, Group of autozooids; x50. D, The primary orifice; x250.

Stomachetosella cruenta (Busk)
(Figs 103C, D; 104)

Lepralia violacea var. *cruenta* Busk 1854: 69, pl. 110, fig. 1
Schizoporella cruenta: Hincks 1880: 270, pl. 30, fig. 5
Stomachetosella cruenta: Hayward & Ryland 1979: 210, fig. 89

Colony a thick, reddish brown crust. Autozooids oval to hexagonal, or distinctly irregular in outline, convex, separated by deep grooves; 0.5-0.7 x 0.4-0.5 mm. Primary orifice wider than long, proximal edge with a shallow, U-shaped sinus occupying about half its width; condyles tiny, indistinct. Peristome low, thick, widely open proximally and not obscuring the orifice; typically thickened proximo-laterally, forming a pair of stout knobs. Frontal shield thick, coarsely nodular; pores especially pronounced around the margins, less so medially. Ovicell smooth, imperforate; rapidly immersed and not visible in frontal view. In later ontogeny the frontal shield becomes greatly thickened and the primary orifice deeply immersed; the secondary orifice is more or less pyriform, but the primary orifice remains distinct.

Offshore, on stones and shell. Boreal-Arctic, with circumpolar distribution, in the eastern Atlantic ranging south to Shetland, Orkney and the northern coast of Scotland.

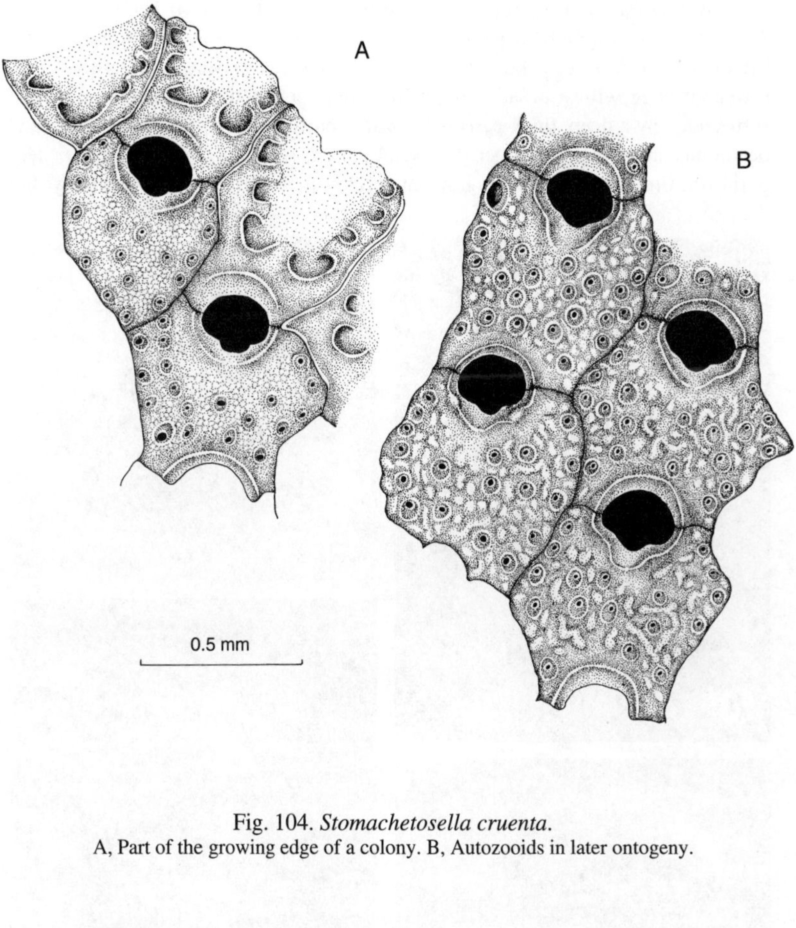

Fig. 104. *Stomachetosella cruenta*.
A, Part of the growing edge of a colony. B, Autozooids in later ontogeny.

Stomachetosella normani Hayward
(Figs 105A, B; 106)

Stomachetosella normani Hayward 1994: 198, figs 9D, E

Colony a thick, spreading sheet, yellowish-white to dull reddish brown. Autozooids hexagonal to polygonal, flat or slightly concave, separated by thickened sutures; 0.5-0.6 x 0.3-0.5 mm. Primary orifice wider than long, proximal edge with a broad, trough-like sinus between short, acute condyles; surrounded by a deep, thick peristome, with a narrow, medio-proximal notch, and becoming progressively immersed and obscured; secondary orifice more or less pyriform. Frontal shield finely granular, evenly perforated in early ontogeny, but

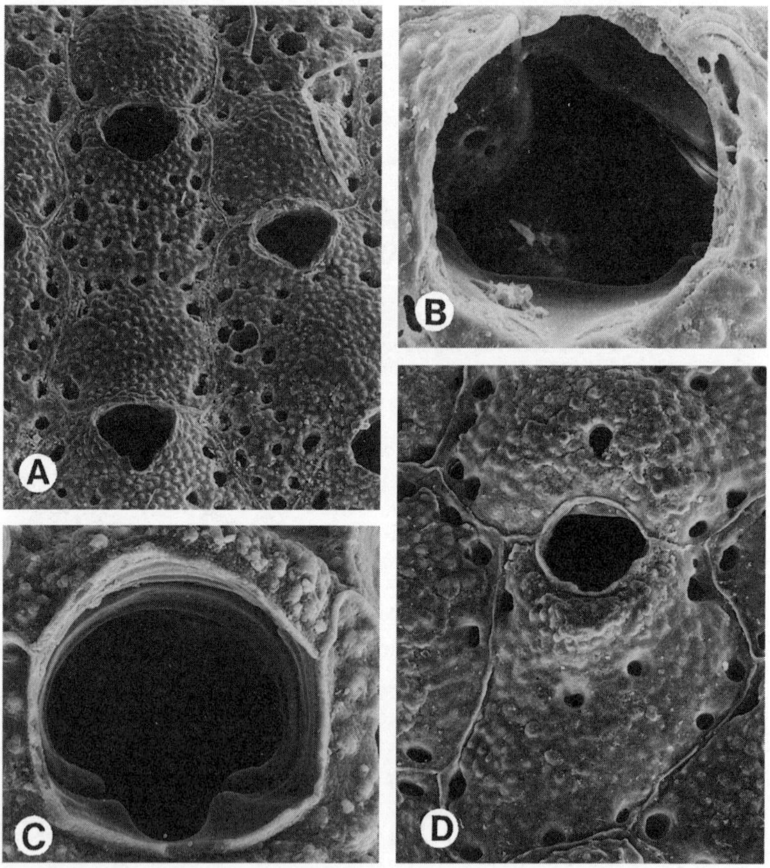

Fig. 105. A, B, *Stomachetosella normani* (The paratype specimen).
A, Ovicelled zooids; x70. B, The primary orifice; x240.
C, D, *Stomachetosella sinuosa* (Arran).
C, Primary orifice; x300. D, Ovicellate zooid; x120.

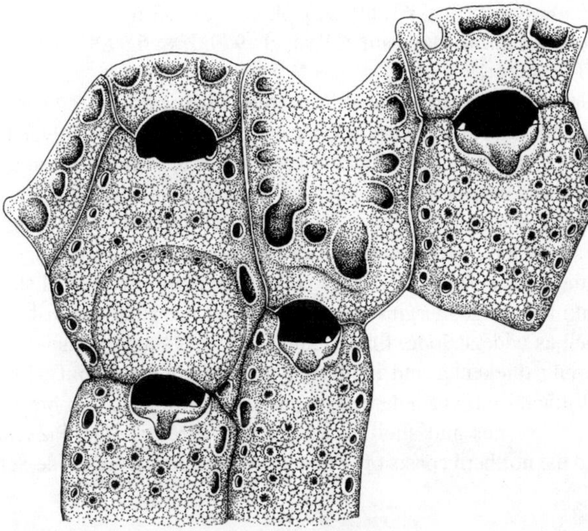

Fig. 106. *Stomachetosella normani*.
Group of autozooids at the colony margin, one with an ovicell.

with frontal wall thickening marginal pores become especially pronounced while those more medial in position become rather inconspicuous. Ovicell as wide as long, finely granular, imperforate, becoming steadily immersed and rather inconspicuous in later ontogeny.

Offshore, on stones and shell. Presently known only from Shetland and Faroe, where it occurs together with *S. cruenta*, but perhaps more generally distributed in the cold northeast Atlantic.

Stomachetosella sinuosa (Busk)
(Figs 105C, D; 107)

Lepralia sinuosa Busk 1860a: 125, pl. 24, figs 2, 3
Schizoporella sinuosa: Hincks 1880: 266, pl. 42, figs 1, 3-6
Stomachetosella sinuosa: Hayward & Ryland 1979: 208, fig. 88

Colony an extensive, spreading sheet, yellowish-white to reddish brown or purple. Autozooids oval, rectangular, hexagonal or irregularly polygonal, flat or slightly convex, separated by thick sutures; 0.4-0.6 x about 0.3 mm. Primary orifice as wide as long, proximal border with a deep, symmetrical, U-shaped sinus occupying half its total width; condyles large, rounded. Peristome low and thickened, completely surrounding orifice, deepening steadily through ontogeny and hiding the orifice; secondary orifice distinctly pyriform. Frontal shield finely granular, with few, large marginal pores, and fewer, irregularly distributed frontal pores. Ovicell as wide as long, finely granular, with a single, conspicuous frontal pore; becoming thickened and immersed in later ontogeny, but with the pore remaining distinct. Embryos red, recorded in May at Millport, Cumbrae.

Offshore, on stones and shell. Boreal-arctic, circumpolar. Present around Shetland and the northern coasts of Scotland, southwards to the Clyde Sea.

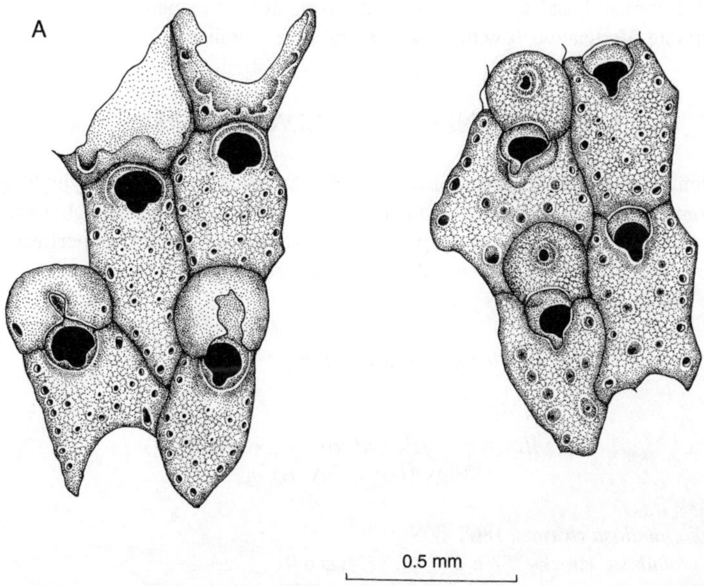

Fig. 107. *Stomachetosella sinuosa*.
A, Autozooids at the colony margin, two with developing ovicells. B, Autozooids in later ontogeny, with complete ovicells and characteristic secondary orifices.

Family GIGANTOPORIDAE Bassler, 1935

Colony encrusting or erect. Autozooids with evenly perforated frontal shield. Primary orifice shallowly sinuate, with well developed condyles. Avicularia adventitious; typically paired, lateral-oral, the rostrum of each connected by a peristomial bridge; absent in some genera. Peristome variably developed, often with a complete basal spiramen. Ovicells hyperstomial or partially immersed, imperforate. Vertical walls with basal or mural pore chambers.

Genus CYLINDROPORELLA Hincks, 1877

Colony encrusting. Primary orifice orbicular, enclosed by an erect, cylindrical peristome. Spiramen at the base of the peristome, opening immediately above the operculum. Frontal wall evenly perforated. Ovicell prominent, imperforate, opening into peristome, not closed by zooidal operculum. Avicularia and oral spines absent. Pore-chambers present.

Type species: *Lepralia tubulosa* Norman, 1868

Cylindroporella tubulosa (Norman)
(Figs 108; 109A, B)

Lepralia tubulosa Norman 1868: 308
Porina tubulosa: Hincks 1880: 230, pl. 32, figs 6-9
Cylindroporella tubulosa: Hayward & Ryland 1979: 244, fig. 105

Colony encrusting, forming small, round patches. Autozooids oval, convex, separated by shallow grooves; 0.35–0.4 x about 0.2 mm. Frontal wall smooth and vitreous, regularly perforated by large, conspicuous pores. Primary orifice orbicular, seen only in developing peripheral autozooids; elsewhere obscured by a tall (up to 0.2 mm), often curved, cylindrical peristome, perpendicular to the autozooid. At its tip the peristome has a circular orifice, the rim often produced at intervals into a series of points. Spiramen situated at the base of the peristome, opening immediately above the operculum; at first appearing as an open foramen in the developing peristome, later produced and tubiform. Ovicell recumbent on the distally succeeding autozooid, almost spherical, smooth surfaced and imperforate, opening above the orifice into the peristome. Embryos pinkish-orange to vivid orange, observed in May and October in western Norway (Ryland, 1963b).

Encrusts shells, stones and algae. Found in the Bergen fjords at about 40 m (Ryland, 1963b), but Kluge (1962, 1975) gives a range of 5– 230 m. *C. tubulosa* is an arctic-boreal species, circumpolar in distribution and rare in British waters. It has been recorded from Shetland and off the north-east coast of Scotland, these localities probably marking the southern limit of its range.

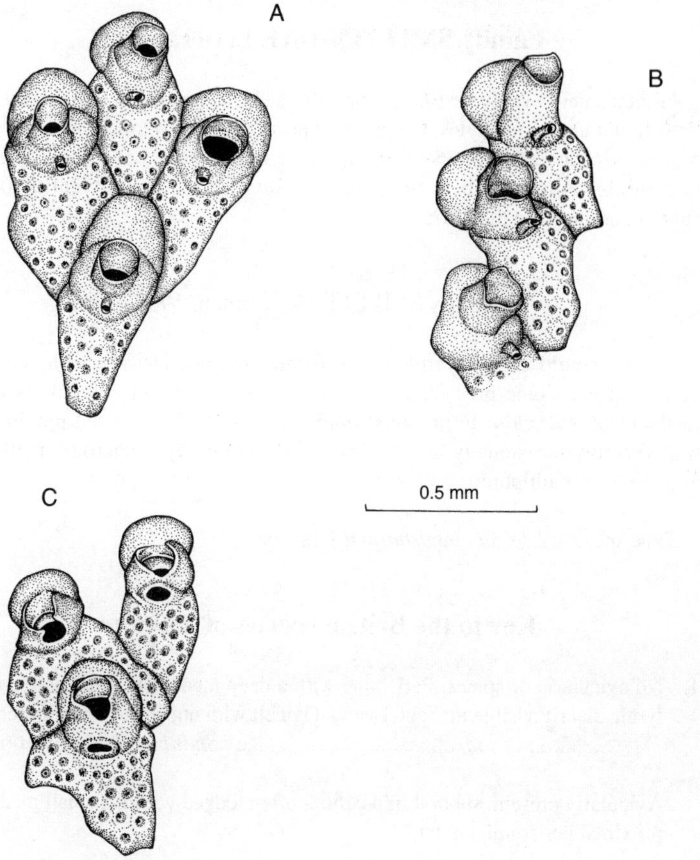

Fig. 108. *Cylindroporella tubulosa*.
A, Four typical autozooids. B, Autozooids in lateral view to show projecting spiramina. C, Autozooids in early ontogeny, showing progressive development of the spiramen.

Superfamily SMITTINOIDEA Levinsen, 1909

Family SMITTINIDAE Levinsen, 1909

Frontal shield evenly perforated, or with marginal areolae only. Primary orifice with lyrula and/or condyles; peristome typically well developed, with or without oral spines. Ovicell prominent, perforate or imperforate, not closed by zooidal operculum. Pore-chambers or septula present. Avicularia adventitious only; suboral, or lateral to the orifice.

Genus SMITTINA Norman, 1903

Frontal shield cryptocystidean, uniformly porous. Orifice with lyrula and condyles; peristome present, oral spines present or absent. Avicularia usually median, suboral, close to proximal border of orifice and directed proximally or transversely; occasionally absent. Ovicell punctured by numerous small pores. Uniporous or multiporous septula present.

Type species: *Lepralia landsborovii* Johnston, 1847

Key to the British species of *Smittina*

1. No avicularia or spines. Peristome with a deep medio-proximal notch, the lyrula clearly visible in frontal view. Ovicell with numerous small pores
.... ***Smittina cheilostoma*** (p. 258)

Avicularia present, suboral in position, often lodged within a small medio-proximal peristomial notch
.... **2**

2. Suboral avicularium orientated transversely
.... ***Smittina affinis*** (p. 254)

Suboral avicularium orientated longitudinally and directed proximally
.... **3**

3. Avicularia monomorphic, situated just proximal to peristome. Oral spines present
.... ***Smittina crystallina*** (p. 256)

Avicularia dimorphic or monomorphic, the smaller type enclosed within the peristome. No oral spines
.... **4**

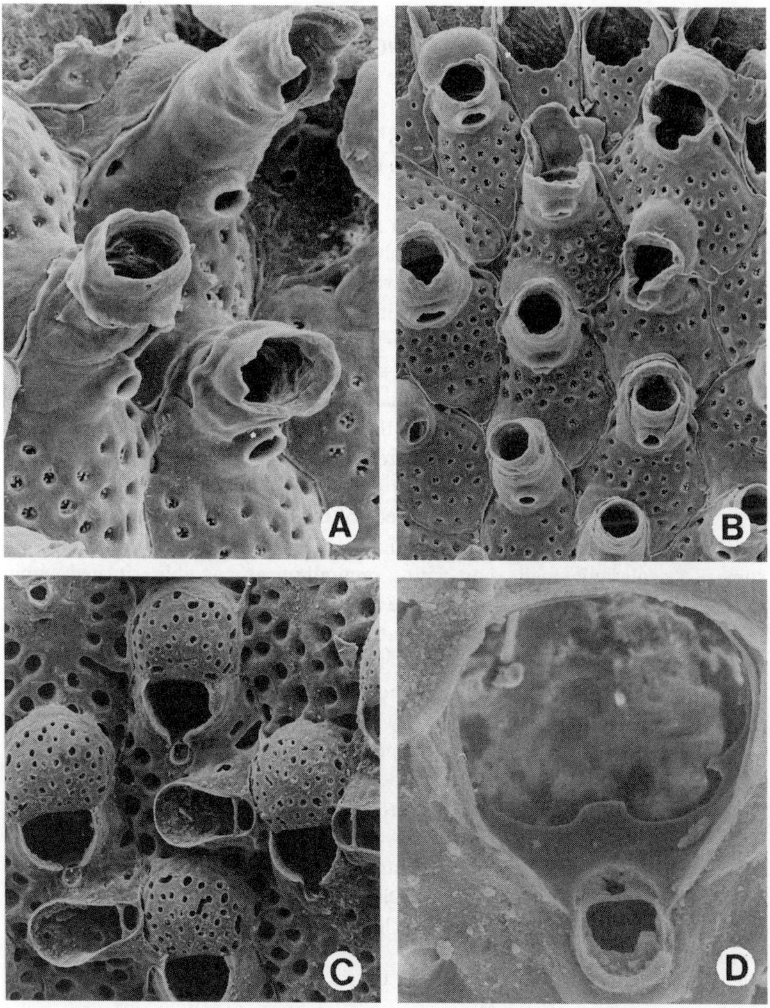

Fig. 109
A, B, *Cylindroporella tubulosa*.
A, Group of autozooids with complete peristomes; x115. B, Colony margin, with ovicells and developing peristomes; x60.

C, D, *Smittina landsborovii*.
C, Ovicellate zooids with spatulate avicularia; x60. D, The primary orifice, x250.

4. Avicularia dimorphic: small, and lodged within proximal notch of peristome; or large, spatulate, suboral or frontal. Lightly calcified. Ovicell with numerous pores .... ***Smittina landsborovii*** (p. 252)

Avicularia monomorphic, small, enclosed within peristomial rim. Thickly calcified. Ovicell sutured, with few, indistinct pores .... ***Smittina bella*** (p. 260)

Smittina landsborovii (Johnston)
(Figs 109C, D; 110)

Lepralia landsborovii Johnston 1847: 310, pl. 54, fig. 9
Smittia landsborovii: Hincks 1880: 341 (in part), pl. 48, figs 6, 9
Smittina landsborovii: Hayward & Ryland 1979: 98, fig. 34

Colony encrusting, forming broad spreading patches, or rising from the substratum as convoluted, foliaceous lobes; either unilaminar, or bilaminar, with two sheets of autozooids in contact basally. Autozooids hexagonal or quadrate, flat or slightly convex, separated by distinct sutures; 0.5-0.7 x 0.3-0.4 mm. Frontal shield of autozooid thin, smooth and fine grained, with numerous, closely-spaced, large round pores. Primary orifice semi-orbicular, broader than long, with a short, quadrate or anvil-shaped lyrula proximally, and paired lateral condyles, sharply pointed and basally deflected. Part of the distal border with a series of minute denticulations, usually seen only in carefully cleaned material. Peristome erect, thin, forming paired lateral wings, particularly prominent in fertile autozooids where they extend onto the frontal surface of the ovicell; incomplete proximally, where a deep fissure marks the position of the suboral avicularium. No oral spines. Avicularium small, short with a semi-elliptical mandible, situated within the proximal gap in the peristome. Large avicularia, up to 3 mm long, sporadically present, occasionally abundant; originating close to the peristome and directed laterally across the frontal wall of the adjoining autozooid; mandible spatulate. Rarely, the large avicularium replaces the suboral one. Ovicell prominent, broader than long and slightly flattened frontally; thin, smooth surfaced and perforated by numerous small, round or irregular pores. Embryos red (Eggleston, 1970); recorded January, March and October-November around the Isle of Man (Eggleston, 1969), and February, April and September in the Mediterranean (Gautier, 1962).

On shells, stones and other hard substrata near low water mark downwards, its lower bathymetric limit is not known. Distributed off all the western coasts of the British isles, and in the English Channel, and on the east coast from Northumberland.

Marcus (1940) recorded it from the Skagerrak and Kattegat, and it probably reaches its northern limit off western Norway (Ryland, 1963b). Found in the western Mediterranean, between 20 and 150 m (Gautier, 1962). This species has been accorded a worldwide distribution, mainly because of confusion over its identity; records from outside the north-east Atlantic area should be regarded with scepticism.

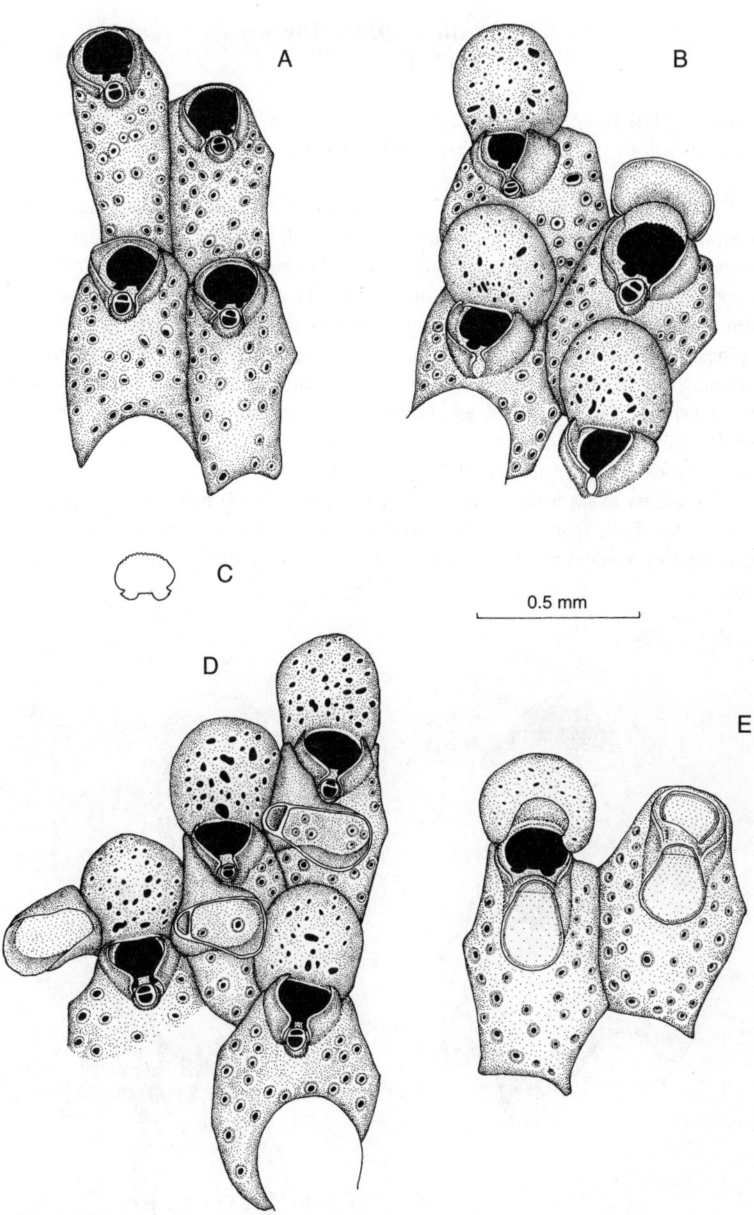

Fig. 110. *Smittina landsborovii.*
A, Autozooids in early ontogeny. B, Ovicellate zooids. C, Outline diagram of the primary orifice. D, Ovicellate zooids with spatulate adventitious avicularia. E, Autozooids with dimorphic suboral avicularia.

Smittina affinis (Hincks)
(Figs 111; 112A, B)

Smittia affinis Hincks 1862: 206, pl. 12, fig. 2; 1880: 348, pl. 49, figs 10, 11
Smittina affinis: Hayward & Ryland 1979: 102, fig. 36

Colony encrusting, reddish-orange. Autozooids oval to hexagonal, flat or slightly convex, separated by distinct sutures; 0.5-0.6 x 0.3-0.4 mm. Frontal shield smooth, fine grained, evenly punctured by round pores. Primary orifice orbicular, condyles small, often inconspicuous, lyrula quadrate, slender. Peristome erect, thin, extending onto the frontal surface of the ovicell, proximal gap broad. No oral spines. Suboral avicularium small, situated within the gap in the peristome and orientated transversely; mandible acute triangular, oblique to the frontal plane of the autozooid. Ovicell prominent, broader than long, somewhat flattened frontally and perforated by numerous, small, round or irregular pores. Embryos observed in June-September (Castric-Fey, 1971; Hayward, 1977).

This seems to be a rare species. Hincks (1862, 1880) had a single specimen, encrusting shell, from Start Bay, South Devon. It was not reported again until Castric-Fey's account of material collected from Brittany; she found it to be

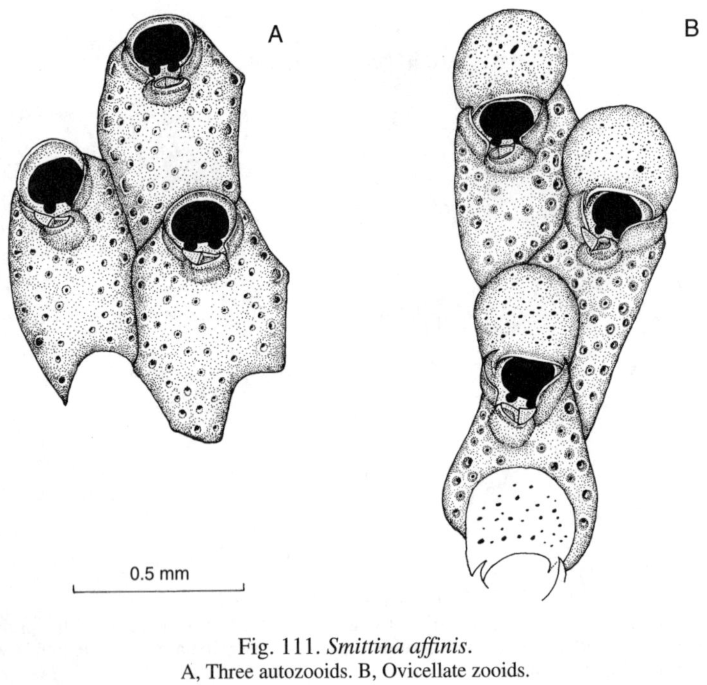

Fig. 111. *Smittina affinis*.
A, Three autozooids. B, Ovicellate zooids.

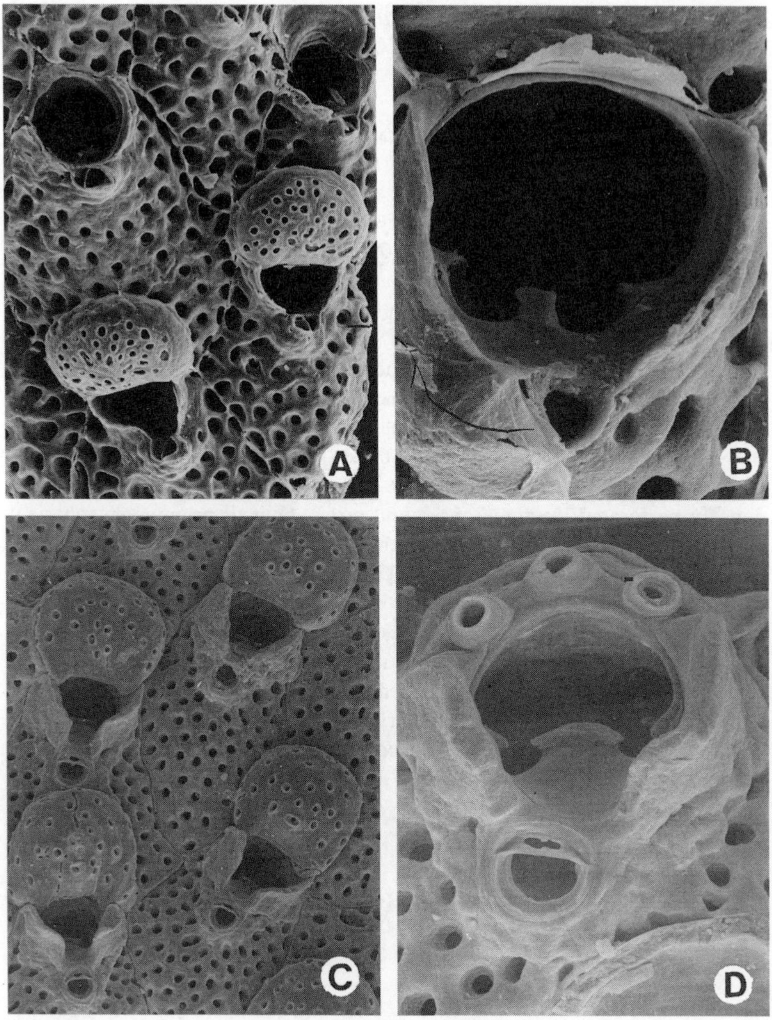

Fig. 112. A, B, *Smittina affinis* (Lundy).
A, Group of autozooids, two with ovicells; x70. B, The primary orifice; x260.

C, D, *Smittina crystallina*.
C, Ovicellate zooids; x70. D, The primary orifice; x210.

common between 2 and 18 m, and down to 30 m, encrusting hard substrata. Moore (1973) collected a single colony from a *Laminaria* holdfast off Northumberland. Subsequently (Hayward, 1977) two colonies were collected from east of Lundy, from 20 m, both encrusting stems of *Cellaria*. López de la Cuadra & Garcia-Gómez (1993) recorded it from 50 m depth in the Strait of Gibraltar.

Smittina crystallina (Norman)
(Figs 112C, D; 113)

Lepralia crystallina Norman 1867: 204
Smittia landsborovii var. *crystallina*: Hincks 1880: 341, pl. 36, fig. 2
Smittina crystallina: Hayward & Ryland 1979: 100, fig. 35

Colonies forming small, round, silvery patches. Autozooids oval to hexagonal, convex, separated by deep sutures; 0.4-0.5 x 0.25-0.3 mm. Primary orifice wider than long; lyrula broad, thick, straight-edged, with a longitudinal median ridge, occupying most of proximal border; condyles large, conspicuous, acutely pointed. Four slender, distal oral spines present in early ontogeny, later lost, their bases obscured by the distal expansion of the peristome, or the development of the ovicell. Peristome developed as tall, peaked lateral flaps, extending distally as a low, thick rim in later ontogeny, incomplete mid-proximally. Avicularium situated immediately proximal to gap in peristome; short, broadly oval, with slender crossbar, mandible almost a semicircle. Frontal shield finely granular, vitreous, densely perforated by numerous round pores. Ovicell prominent, spherical, punctured by numerous small pores. Ancestrula tatiform, oval, 0.3 x 0.25 mm; gymnocyst smooth, frontal membrane occupying about half total length, bordered by nine short spines, cryptocyst with distinctive pear-shaped opesia.

Occurs offshore, on hard substrata, especially shell. Reported from off Wick, Shetland, the Hebrides and the Antrim coast, the western approaches to the English Channel, and from the continental shelf edge of the Bay of Biscay. A number of recent records are summarised by Zabala, Maluquer & Harmelin (1993); *S. crystallina* is evidently a deep water species, occurring at 110-523 m depth in the Gulf of Cadiz, the Alboran Sea and the Strait of Gibraltar, and in the Gulf of Lions.

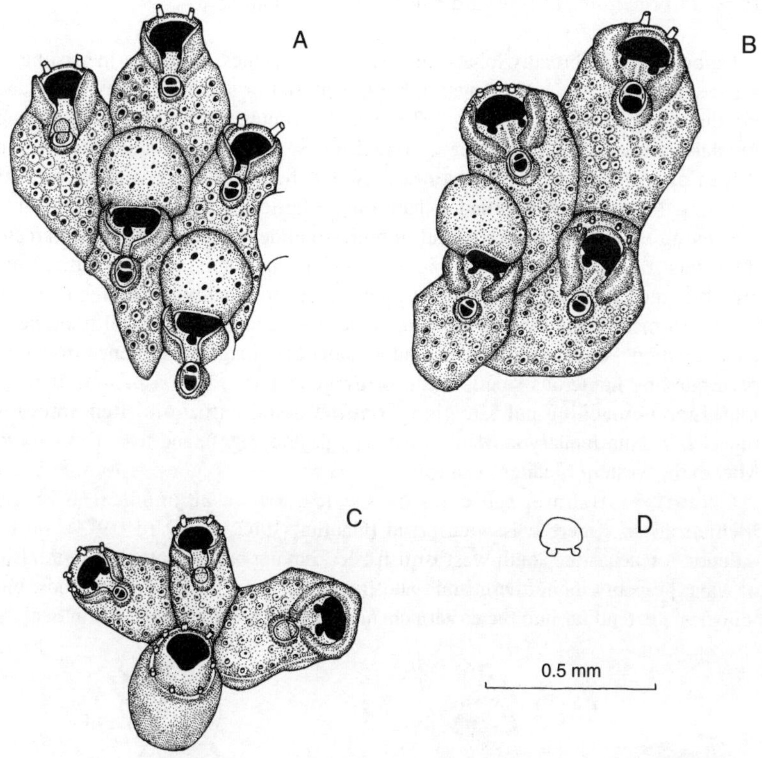

Fig. 113. *Smittina crystallina*.
A, Autozooids at the colony margin. B, Autozooids in later ontogeny, showing thickening of the peristome. C, The ancestrula and periancestrular zooids. D, Outline diagram of the primary orifice.

Smittina cheilostoma (Manzoni)
(Figs 114; 115A, B)

Lepralia cheilostoma Manzoni 1870a: 942, pl. 4, fig. 22
Smittia cheilostoma: Hincks 1880: 349, pl. 42, figs 7, 8
Prenantia cheilostoma: Hayward & Ryland 1979: 104, fig. 37

Colony forming broadly lobate, reddish-purple patches, the colour persisting in dried specimens. Autozooids oval to hexagonal, flat or slightly convex, separated by shallow grooves; 0.45-0.6 x 0.3-0.4 mm. Frontal shield thick, closely and regularly perforated with numerous round pores, each in a circular pit giving a ridged appearance to the interstices between the pores; very rugose in older colonies. Primary orifice broader than long, elliptical, with a low, broad lyrula occupying about half the proximal border; condyles short, blunt, and scarcely distinguishable. Peristome developing rapidly and enclosing the entire orifice, low and thickened with a wide U-shaped pseudosinus proximally; extending onto the frontal surface of the ovicell. No oral spines. No avicularia. Ovicell prominent, recumbent on the succeeding autozooid, broader than long, and flattened frontally; perforated by numerous small, round or irregular pores, with a peripheral ridge marking encroaching calcification from the distal autozooid. Red embryos reported in June-January in Manx waters (Eggleston, 1969), and from February to May in the western Mediterranean.

Occurring offshore, typically on shell, between 20 and 150 m in the Mediterranean, where it is widespread (Gautier, 1962; Hayward, 1974). In the Atlantic it reaches the south-west British Isles, having been reported from the Isle of Man, probably its northern limit, and frequently from the Channel Islands: but it does not extend far into the eastern channel and is absent from the North Sea.

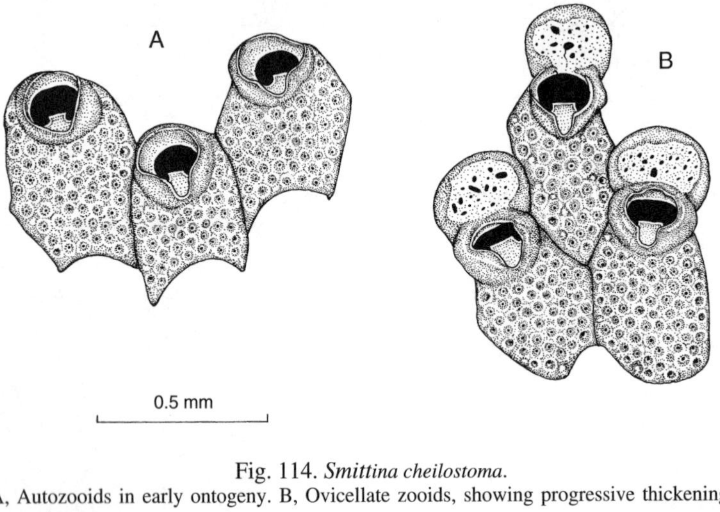

Fig. 114. *Smittina cheilostoma.*
A, Autozooids in early ontogeny. B, Ovicellate zooids, showing progressive thickening of the peristome.

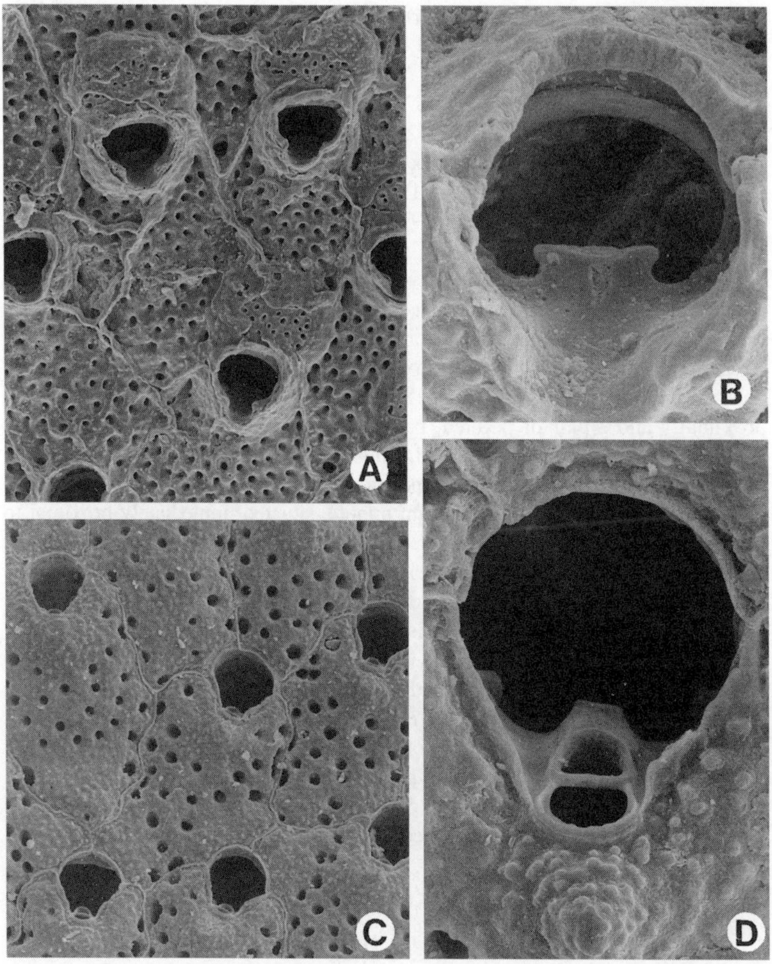

Fig. 115.
A, B, *Smittina cheilostoma*.
A, Ovicellate zooids; x70. B, The primary orifice; x260.

C, D, *Smittina bella*.
C, Group of autozooids, that on the lower left ovicellate; x60. D, The primary orifice; x240.

Smittina bella (Busk)
(Figs 115C, D; 116)

Lepralia bella Busk 1860b: 144, pl. 27, fig. 2
Smittia bella: Hincks 1880: 352, pl. 42, figs 9, 10
Porella concinna var. (*gracilis*) Hincks 1880: 324, pl. 46, fig. 9
Prenantia bella: Hayward & Ryland 1979: 106, fig. 38
Smittina bella: Hayward & Hansen 1999

Colony encrusting. Autozooids oval, hexagonal or irregular, flat or slightly convex, separated by narrow sutures; 0.6-0.7 x 0.3-0.4 mm. Frontal shield thick, granular and vitreous, closely perforated by numerous small pores; becoming thicker and more rugose with age so that the secondary orifice remains more or less level with it. Primary orifice orbicular with a small, quadrate lyrula proximally and a pair of pointed, basally deflected condyles. No oral spines. Peristome developing rapidly, encircling the orifice, which becomes deeply immersed; a proximal notch in the peristome gives a sinuate appearance to the secondary orifice. Suboral avicularium minute, situated immediately proximal to the lyrula and perpendicular to it. Ovicell indistinct, irregular in outline, subimmersed, coarsely granular, with or apparently without scattered pores; the ooecium seen only in young autozooids, being then covered by calcification extending from one or more of the distally adjacent autozooids; the cover sutured when of multiple origin.

Described originally from Shetland (Busk, 1860b), this species seems to have been confused with the subarctic *Porella concinna* var. *belli* Dawson (see Hayward, 1994) and perhaps other boreal-arctic smittinids. Its status was recently stabilised by the selection of a neotype (Hayward & Hansen, 1998); it is known with certainty from off Aberdeen, Shetland and west Norway, but its wider distribution remains to be established. All specimens presently known encrust stones from offshore habitats.

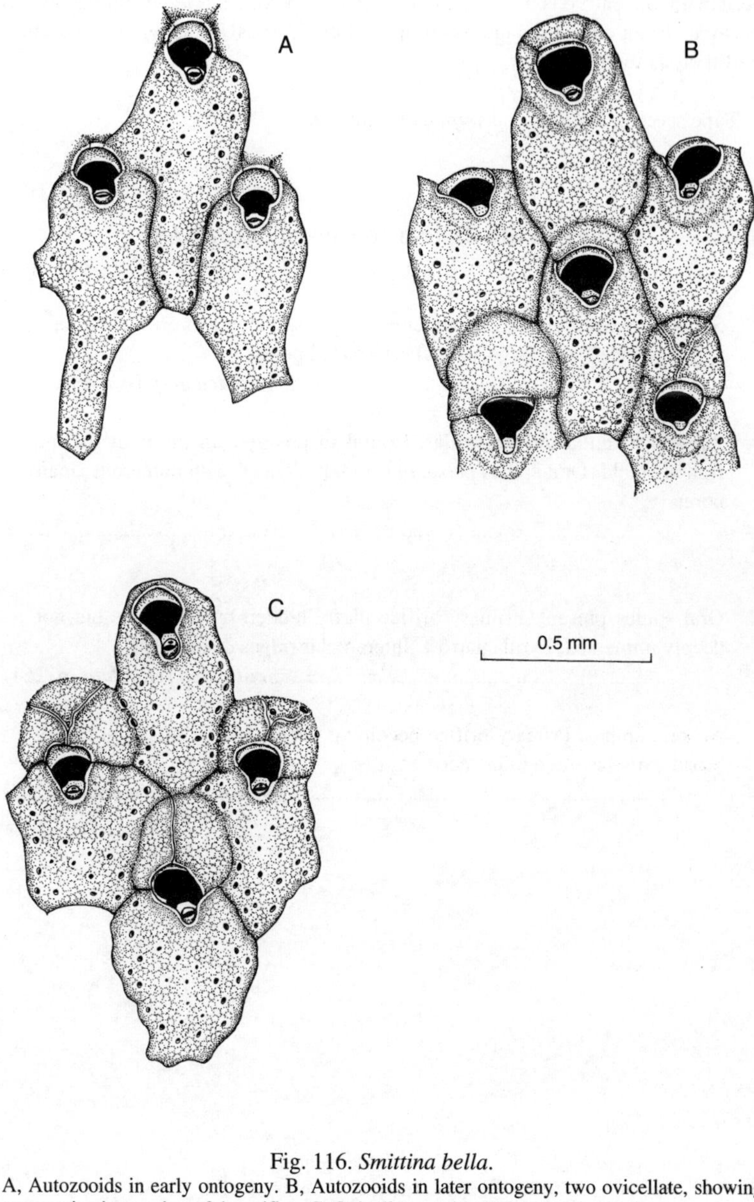

Fig. 116. *Smittina bella*.
A, Autozooids in early ontogeny. B, Autozooids in later ontogeny, two ovicellate, showing progressive immersion of the orifice. C, Ovicellate zooids with sutured ovicells.

Genus SMITTOIDEA Osburn, 1952

Colony encrusting, to erect, foliaceous. Autozooid frontal shield with marginal pores only. Primary orifice with lyrula and condyles; oral spines usually present. Avicularia adventitious, median suboral. Ovicell prominent, hyperstomial, evenly perforated; not closed by autozooid operculum. Vertical walls with uniporous or multiporous septula.

Type species: *Smittoidea prolifica* Osburn, 1952

Key to the British species of *Smittoidea*

1. Suboral avicularium short, semicircular, perpendicular to orifice. No oral spines. Ovicell with large, irregularly shaped pores

.... ***Smittoidea amplissima*** (p. 266)

Suboral avicularium triangular, frontal in position, in the plane of the frontal shield. Oral spines present or absent. Ovicell with numerous small pores

.... **2**

2. Oral spines present. Primary orifice partly hidden by peristome but not deeply immersed. Lyrula narrow. Interareolar ridges distinct

.... ***Smittoidea reticulata*** (p. 264)

No oral spines. Primary orifice becoming deeply immersed. Lyrula very broad. Areolae often indistinct

.... ***Smittoidea marmorea*** (p. 268)

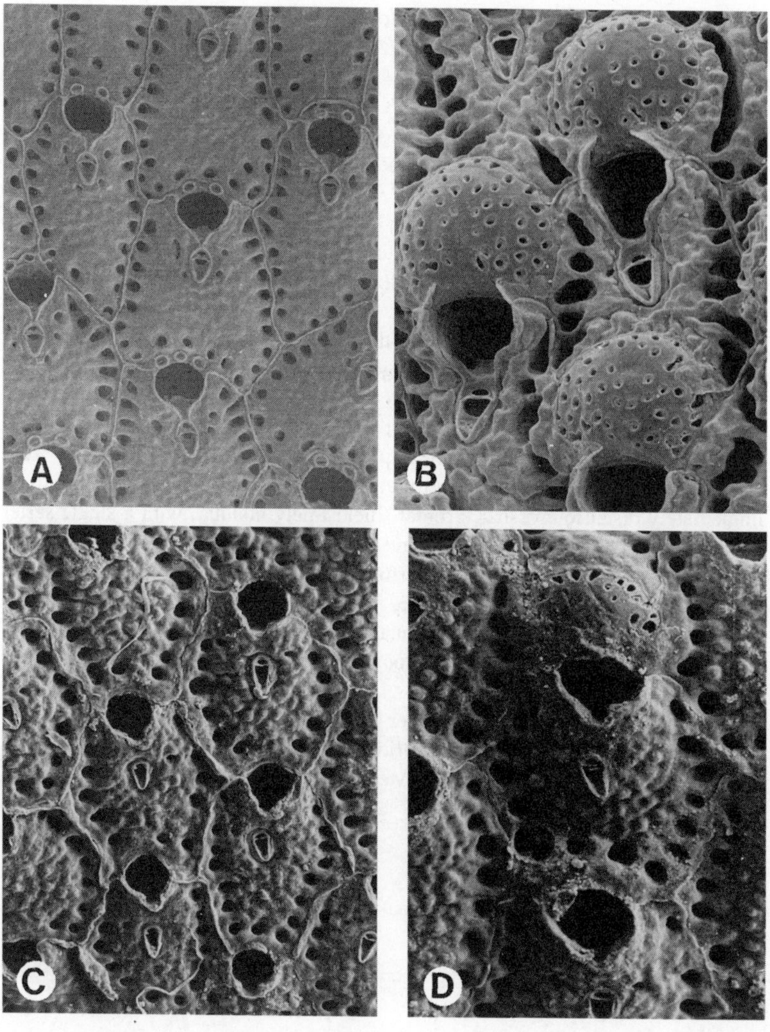

Fig. 117.
A, B, *Smittoidea reticulata*.
A, Group of autozooids; x50. B, Ovicellate zooids; x80.

C, D, *Smittoidea marmorea*.
C, Group of autozooids; x50. D, An ovicellate zooid; x80.

Smittoidea reticulata (J. Macgillivray)
(Figs 117A, B; 118)

Lepralia reticulata J. Macgillivray 1842: 467
Smittia reticulata: Hincks 1880: 346, pl. 48, figs 1-5
Smittoidea reticulata: Hayward & Ryland 1979: 108, fig. 39

Colonies developing broad, irregular sheets. Autozooids oval, rectangular to irregularly polygonal, rather flat, separated by thin, raised sutures; 0.5-0.8 x about 0.3 mm. Primary orifice wider than long; lyrula rectangular, straight-edged, occupying about half proximal border; condyles sharply pointed, downcurved; distal border finely denticulate. Two or three long, slender spines on distal border of orifice, most often lost in later ontogeny; peristome low, developed as a thin rim around the lateral borders of the orifice, deeply notched mid-proximally. Avicularium immediately proximal to peristomial notch; rostrum 0.1 mm long, rounded triangular, normal to frontal plane and proximally directed; crossbar slender, complete, without a columella, palate with a very small foramen immediately distal to crossbar. Frontal shield finely nodular, with a single series of large, closely spaced, marginal pores. Polypide with 13-15 tentacles. Ovicell globular, broader than long, closely perforated by numerous small pores. Embryos orange or yellow-orange, reported June, August-December around the Isle of Man (Eggleston, 1969), December and January in western Norway (Ryland, 1963b). Ancestrula tatiform, with narrow cryptocyst and elongate oval opesia, bordered by nine spines.

Sublittoral, on stones, shells and other bryozoans, and common as an epizoite of *Pentapora fascialis*. Common on all British coasts, in the east Atlantic distributed from the Barents Sea to the western Mediterranean and Morocco. Records from elsewhere are doubtful.

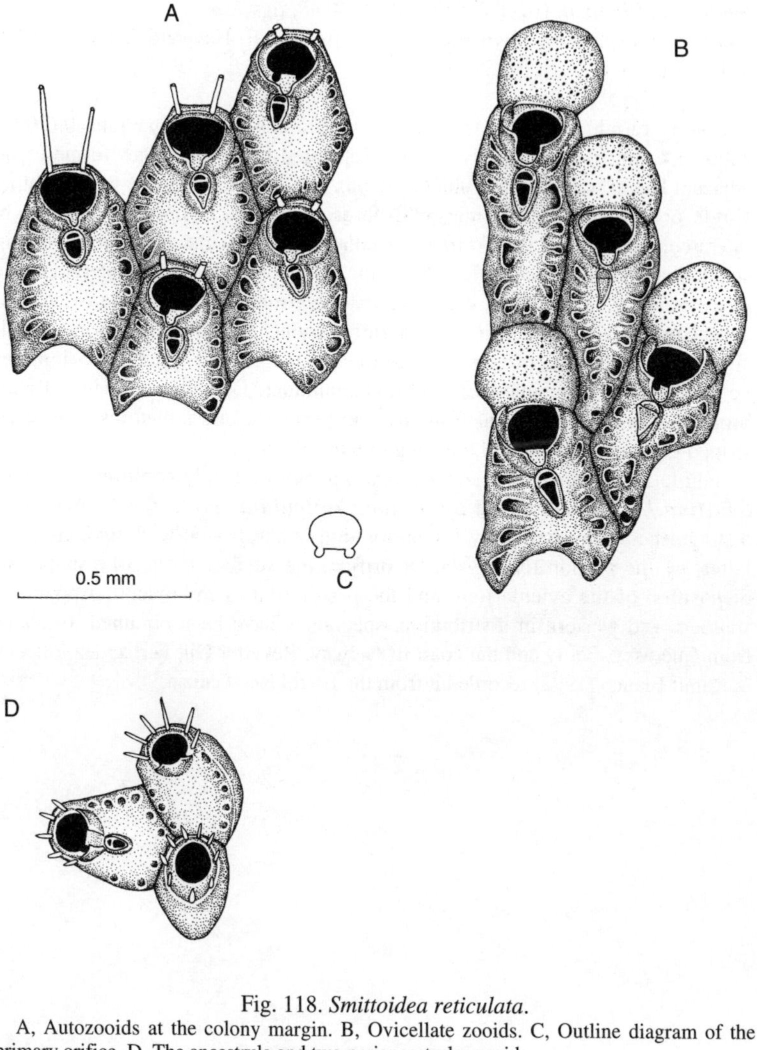

Fig. 118. *Smittoidea reticulata.*
A, Autozooids at the colony margin. B, Ovicellate zooids. C, Outline diagram of the primary orifice. D, The ancestrula and two periancestrular zooids.

Smittoidea amplissima Hayward
(Fig. 119)

Smittia landsborovii: Hincks 1880: 341 pt, pl. 48, figs 7, 8
Smittoidea amplissima Hayward 1979: 75, fig. 1D, E; Hayward & Ryland 1979: 110, fig. 40

Colony forming broad incrustations. Autozooids oval to quadrate, 0.6-0.8 x about 0.4 mm; flat or slightly convex, separated by raised ridges formed from adjacent lateral walls. Frontal shield smooth, imperforate centrally, with a distinct single or double series of marginal pores, separated in older autozooids by interareolar ridges. Primary orifice orbicular, with a slender, peg-like lyrula and small, sharply pointed condyles. Peristome developed only on the lateral borders of the orifice, erect and thin, extending onto the frontal surface of the ovicell, but absent proximally. Oral spines apparently absent. Avicularium suboral, small, situated close to the proximal edge of the orifice and typically orientated perpendicularly to it; mandible short, semicircular. Ovicell prominent, slightly broader than long, flattened frontally and perforated by numerous irregularly shaped pores. Ancestrula and breeding - no information.

A little known species, *S. amplissima* was until recently confused with both *Smittina landsborovii* and *Smittoidea reticulata*. From the former, it is distinguished immediately by the perforation of the frontal wall, and, from the latter, by the proportionately larger orifice, the smaller lyrula, the shape and disposition of the avicularium, and the perforation of the ovicell. Apparently southern and western in distribution, specimens have been obtained, offshore, from Guernsey, Scilly and the coast of Galway. Reverter Gil, Fernandez Pulpeiro & Ramil Blanco (1992) recorded it from the Ferrol Ria, Galicia.

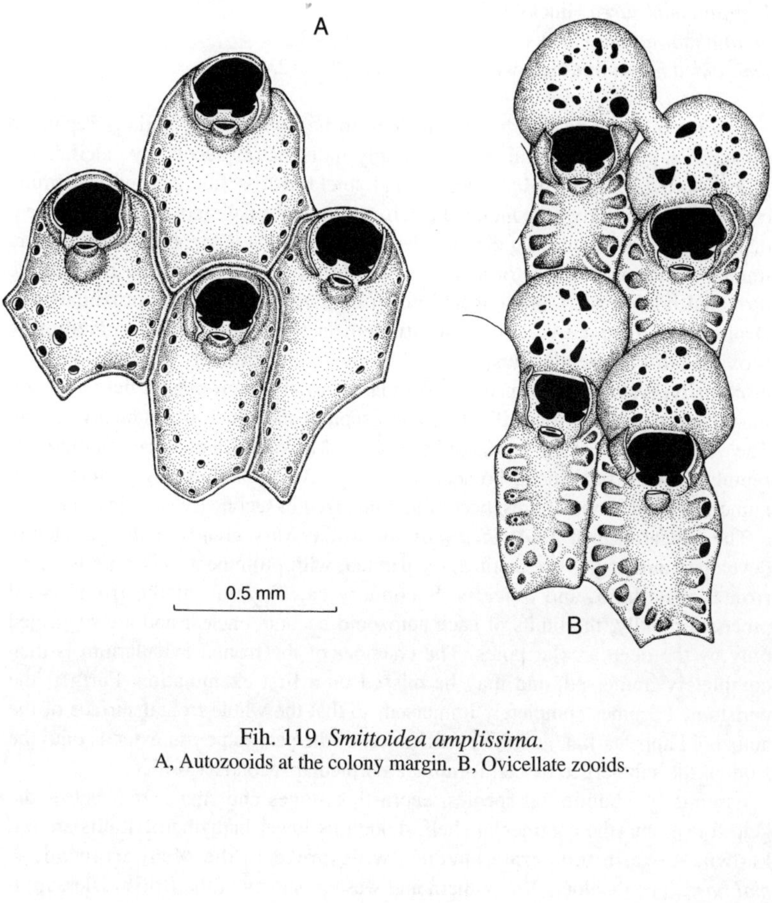

Fih. 119. *Smittoidea amplissima*.
A, Autozooids at the colony margin. B, Ovicellate zooids.

Smittoidea marmorea (Hincks)
(Figs 117C, D; 120)

Lepralia marmorea Hincks 1877: 214
Smittia marmorea: Hincks 1880: 350, pl. 36, figs 3-5
Smittoidea marmorea: Hayward & Ryland 1979: 112, fig. 41

Colony encrusting. Autozooids quadrate to hexagonal, flat, initially separated by shallow grooves, becoming increasingly indistinct as secondary calcification proceeds; 0.4-0.5 x 0.25-0.3 mm. Frontal shield at first fine grained, becoming coarse and granular with continued calcification; marginal areolae closely spaced, distinct in newly budded autozooids, less so later as the autozooids become more immersed. Primary orifice orbicular, with a broad lyrula, occupying most of the proximal border, with an elongated notch or pseudosinus on the proximal side. Oral spines absent. Avicularium situated on the frontal wall, immediately proximal to the pseudosinus, with an elongated, triangular mandible directed proximally; when first formed the avicularian cystid is prominent but is rapidly immersed in secondary calcification and becomes indistinct; occasionally absent. The vertical walls of the autozooid are pierced by a series of large, multiporous septula. Ovicell prominent, broader than long; closely and evenly perforated by numerous small, round pores, becoming immersed in secondary calcification.

The appearance of *Smittoidea marmorea* may vary greatly with age. Newly developed autozooids are distinctly separated, with prominent, raised peristomes, frontal avicularia, and ovicells. Secondary calcification of the frontal wall proceeds rapidly; the limits of each autozooid become unclear and are suggested only by the deep areolar pores. The chamber of the frontal avicularium is then completely immersed, and may be missed on a first examination. Further, the peristome becomes completely immersed, so that the whole frontal surface of the autozooid appears flat, and the calcification of the peristome rim extends onto the front of the submerged ovicell forming an orbicular secondary orifice.

Essentially a sublittoral species, encrusting stones and shells from below the kelp forest onto the continental shelf, though its lower bathymetric limits are not known. A warm temperate species, widespread in the Mediterranean, *S. marmorea* occurs along the southern and western shores of the British Isles, from the Channel Islands to Shetland. To the south, it has been reported from Madeira (Norman, 1909).

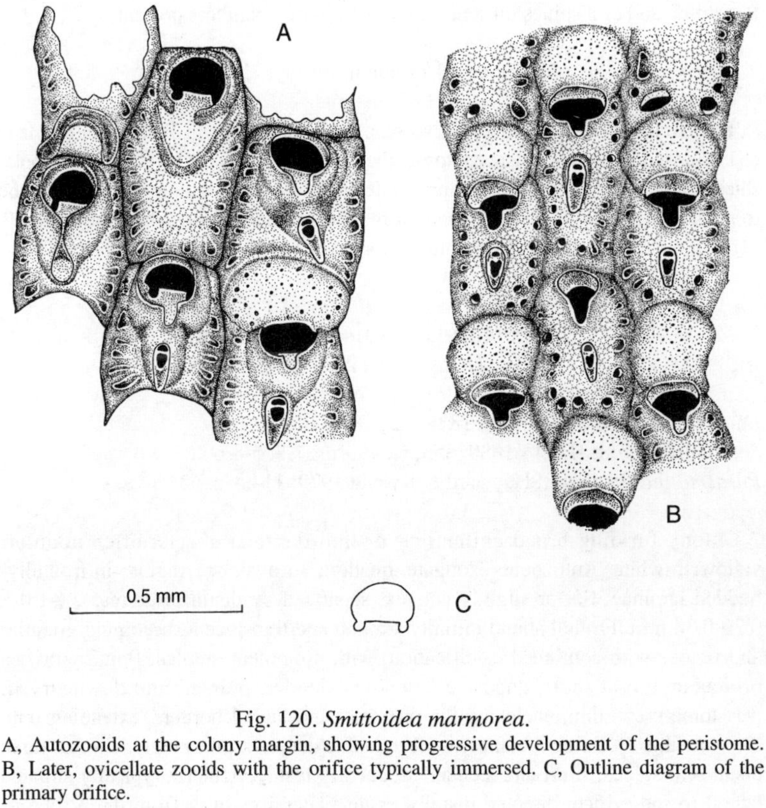

Fig. 120. *Smittoidea marmorea.*
A, Autozooids at the colony margin, showing progressive development of the peristome. B, Later, ovicellate zooids with the orifice typically immersed. C, Outline diagram of the primary orifice.

Genus PARASMITTINA Osburn, 1952

Colony encrusting. Frontal shield with marginal areolae. Primary orifice with lyrula and condyles. Ovicell prominent, with few, large perforations, not closed by zooidal operculum. Avicularia present, typically situated lateral to orifice and directed distally or disto-laterally. Additional adventitious avicularia also occur. Peristome and oral spines present. Small basal pore-chambers present.

Type species: *Smittina jeffreysi* Norman in Jeffreys, 1876

The Arctic species *P. jeffreysi* (Norman) does not occur in British waters. It is characterised by the lack of a peristome (the orifice lying almost level with the frontal shield), a very short lyrula, and a preponderance of oval, frontal avicularia over the triangular latero-oral type. Specimens from British waters previously identified as *P. jeffreysi* have all been found, on examination, to be referable to *P. trispinosa.*

Parasmittina trispinosa (Johnston)
(Figs 121; 122A, B)

Discopora trispinosa Johnston 1838: 222
Smittia trispinosa: Hincks 1880: 353, pl. 49, figs 1-3, 5-8
Parasmittina trispinosa: Hayward & Ryland 1979: 114, fig. 42

Colony forming broad unilaminar or multilaminar sheets, often nodular; yellowish-white. Autozooids elongate, quadrate to oval, or irregular in frontally-budded laminae; flat or slightly convex, separated by distinct sutures; 0.4-1.0 x 0.26-0.38 mm. Frontal shield initially smooth and translucent, becoming granular and rugose with continued calcification; with prominent areolae. Primary orifice orbicular, lyrula short, quadrate, condyles slender, pointed and downcurved; peristome erect, thin, enclosing the proximal and lateral borders, extending onto the ovicell, so forming a complete tube; with a deep slit proximally. Two or three distal oral spines, short and slender, frequently lost. Avicularia typically situated lateral to the orifice, directed distally or distolaterally, away from the orifice or towards it, encroaching on the peristome; mandible acutely triangular, usually single, occasionally paired. Other avicularia may be present, additional to or replacing the lateral oral type, usually smaller, with a triangular or elliptical mandible; closely associated with the areolae. Ovicell spherical, recumbent on the distal autozooid, thin and hyaline, but often partly covered by coarser calcification from the distal autozooid; with a few large, irregularly shaped frontal pores. Embryos red, reported throughout the year around the Isle of Man, with a peak October-January (Eggleston, 1969).

On hard substrata, shell, stone, etc., from the shallow sublittoral out onto the continental shelf. Its lower bathymetric limit is not known. Off all British coasts, often abundant on offshore shell banks. It is a cool temperate species, extending as far north as western Norway and Faroe, but probably not reaching Arctic waters. Zabala & Maluquer (1988) state that it is not present in the western Mediterranean.

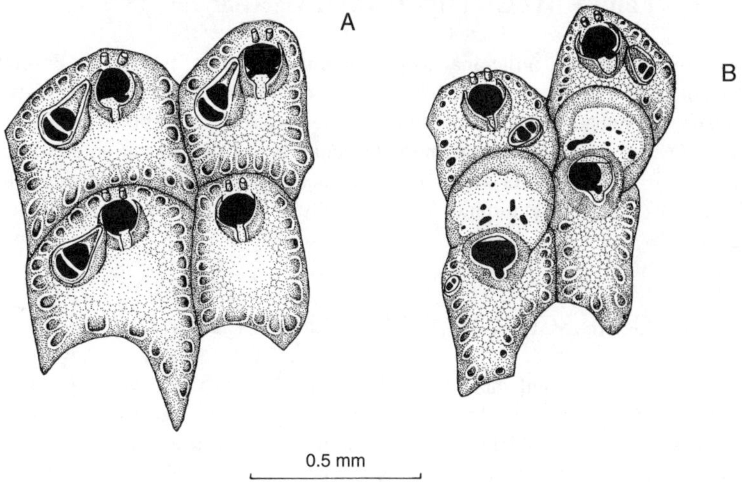

Fig. 121. *Parasmittina trispinosa*.
A, Autozooids in early ontogeny. B, Ovicellate zooids.

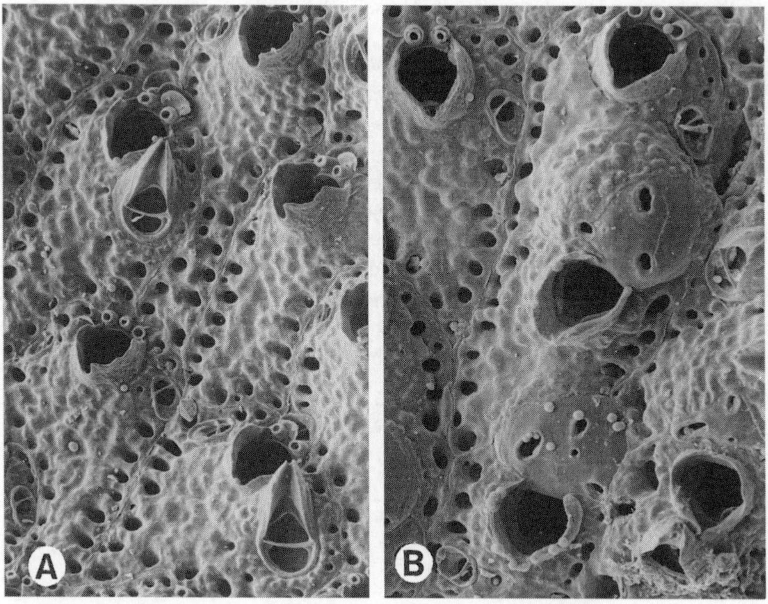

Fig. 122. *Parasmittina trispinosa* (North Sea).
A, Group of autozooids; x70. B, Ovicellate zooids; x90.

Family BITECTIPORIDAE MacGillivray, 1895

Colony encrusting, unilaminar to multilaminar; or erect, foliaceous. Frontal shield of autozooids an evenly perforated cryptocyst. Primary orifice with concave or distinctly sinuate proximal border; condyles present. Oral spines present or absent. Ovicell with ectooecium and entooecium both calcified but unfused, the ectooecium typically with pores or larger membranous fenestrae. Avicularia adventitious, often polymorphic. Vertical walls with uniporous or multiporous septula.

Genus SCHIZOMAVELLA Canu & Bassler, 1917

Colony encrusting, unilaminar or multilaminar; sheet-like, or developing partly erect, folded plates, or domed, mamillate growths. Autozooids with cryptocystidean frontal shield, evenly perforated by small pseudopores. Primary orifice sinuate; condyles present, often pronounced. Oral spines present or absent. Avicularia adventitious, monomorphic or polymorphic, usually suboral, often associated with a prominent umbo. Ovicell prominent, with conspicuous frontal perforations, not closed by autozooid operculum. Multiporous or uniporous septula present.

Type species: *Lepralia auriculata* Hassall, 1842

Ten species of *Schizomavella* are now recognised from the British sea area. Five of these were redescribed by Hayward & Thorpe (1995), who also selected neotype specimens for *S. auriculata* (Hassall) and *S. linearis* (Hassall). Two other species were newly described by Reverter Gil & Fernandez Pulpeiro (1996). *Lepralia umbonata* Busk, 1860, referred to *Schizomavella* by Ryland (1969) cannot now be recognised; there is no known type material, nor any specimen in the Busk Collection at the Natural History Museum, London, which might shed light on the identity of this taxon. At least six of the species described here occur in the Mediterranean (the geographical distributions of *S. auriculata sensu stricto* and *S. sarniensis* are still incompletely known), together with a number of other species. These are described and figured by Zabala & Maluquer (1988).

Key to the British species of *Schizomavella*

1. Primary orifice surrounded by a deep peristome, often extending on to ovicell. Five to seven slender oral spines present in young autozooids. Ovicell with scattered pores. Autozooids small, mostly <0.5 mm long
 2

 Primary orifice may be immersed and partly hidden, or obscured by a suboral umbo, but without a prominent peristome. Two to four (occasionally zero or five) oral spines in marginal autozooids. Ovicell with fewer, irregular pores frontally. Autozooids mostly >0.5 mm long
 3

2. Primary orifice almost circular, with a slender, slit-like sinus. Peristome thin, forming a complete tube but not enclosing the avicularium
 .... ***Schizomavella discoidea*** (p. 280)

 Primary orifice semicircular, with a short, quadrate sinus. Peristome substantial, enclosing the suboral avicularium and raised into a median umbo
 .... ***Schizomavella cristata*** (p. 278)

3. Primary orifice semicircular, with a broad, shallow sinus. Avicularium orientated vertically along the lateral edge of a prominent, spiked suboral umbo; mandible triangular, rostrum sharply hooked
 .... ***Schizomavella hastata*** (p. 284)

 Primary orifice U-shaped or V-shaped. Avicularia various, sometimes on a prominent suboral umbo
 4

4. Avicularia frontal in position, distant from orifice; dimorphic, either elongate triangular, or rectangular. Primary orifice with small V-shaped sinus and rounded, finely toothed condyles
 .... ***Schizomavella ochracea*** (p. 288)

 Avicularia always situated close to primary orifice, monomorphic or dimorphic
 5

5. Primary orifice drop-shaped (broadest distally) with a short U-shaped sinus. Avicularia monomorphic, oval
 6

 Primary orifice orbicular or roughly quadrangular. Avicularia monomorphic or polymorphic
 7

6. Condyles pointed. Avicularian rostrum continuous with rim of primary orifice
.... ***Schizomavella auriculata*** (p. 276)

Condyles rounded. Avicularian rostrum separate from rim of primary orifice
.... ***Schizomavella hondti*** (p. 288)

7. Avicularia polymorphic: lateral - suboral, single or paired, directed medially; or single suboral, vertically directed on a stout umbo; large frontal avicularia with narrow crescentic mandibles present on many autozooids
.... ***Schizomavella linearis*** (p. 282)

Avicularia monomorphic or polymorphic, but always median suboral
.... **8**

8. Primary orifice wider than long but almost straight-sided and appearing rectangular. Avicularia dimorphic: either short and oval or elongate-oval, parallel-sided, both with extensive palatal foramen. Frontal shield coarsely punctured with large, round pores
.... ***Schizomavella sarniensis*** (p. 290)

Primary orifice about as wide as long, lateral walls curving towards sinus, thus appearing orbicular. Avicularia monomorphic or dimorphic
.... **9**

9. Primary orifice with pronounced proximo-lateral notches in its rim; condyles flat and rounded with finely toothed edges. Avicularia dimorphic: short, elliptical, or elongate, oval to spatulate; both types with Y-shaped palatal foramen
.... ***Schizomavella cuspidata*** (p. 286)

Primary orifice with rounded proximo-lateral corners; condyles flat, triangular, with pointed corners but smooth edged. Avicularia monomorphic, short and oval with a round palatal foramen
.... ***Schizomavella teresae*** (p. 290)

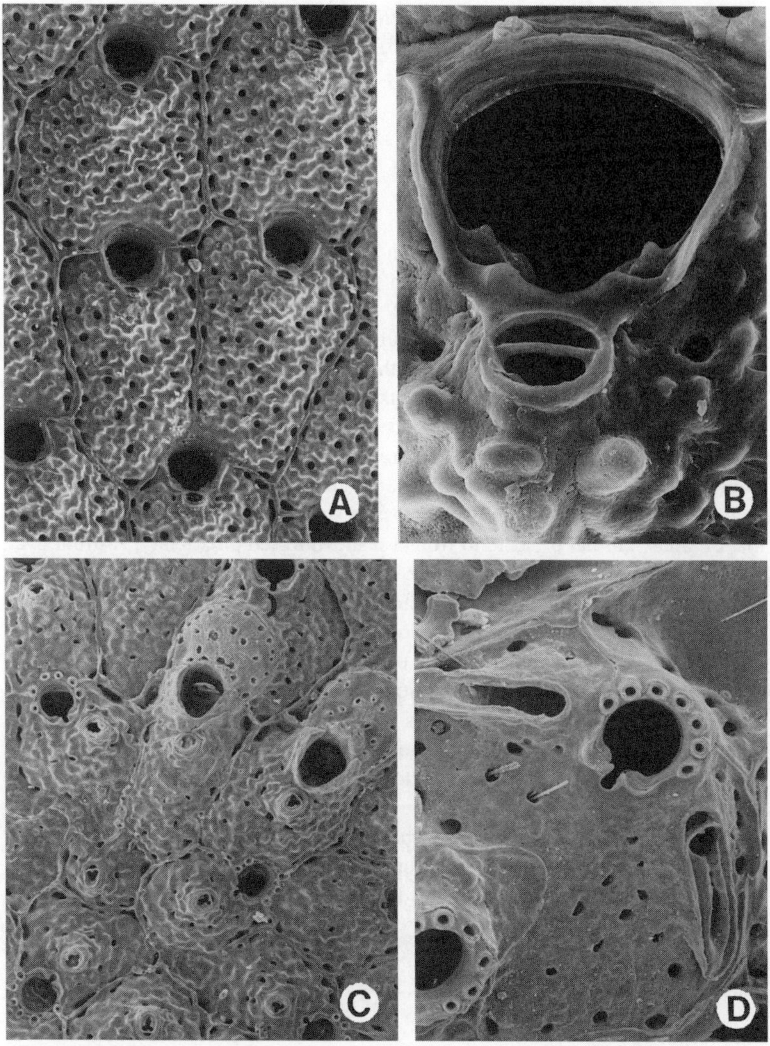

Fig. 123
A, B, *Schizomavella auriculata* (Raasay Sound).
A, Group of autozooids; x80. B, The primary orifice; x400.

C, D, *Schizomavella discoidea* (Hurd Deep).
C, Group of autozooids, two with ovicells; x70. D, An autozooid with lateral avicularia; x125.

Schizomavella auriculata (Hassall)
(Figs 123A, B; 124)

Lepralia auriculata Hassall 1842: 411
Schizoporella auriculata: Hincks 1880: 260, pl. 29, figs 3-6, 9
Schizoporella auriculata: Hayward & Ryland 1979: 176, fig. 72A-C (only); Hayward & Thorpe 1995: 663, pl. I

Colony a broad, irregular incrustation, unilaminar or multilaminar. Autozooids in radiating linear series, rectangular to polygonal, convex, separated by distinct sutures; 0.4-0.7 x 0.2-0.3 mm. Primary orifice about as wide as long, broadest distally, and appearing drop-shaped; proximal border with a shallow U-shaped sinus occupying about half its total width. Condyles distinctive: longer than wide, not extending to margins of sinus, distal edge sharply cusped. One to three delicate, impermanent oral spines may be present; inconstant, not developed by all autozooids. Frontal wall nodular, frequently with a low umbo developed immediately proximal to orifice; evenly perforated by small, round pores. Avicularium monomorphic, situated close to sinus, its rostral rim continuous with that of sinus; normal or acute to frontal plane, directed proximally; rostrum oval, about 0.04 mm long, with slender crossbar, lacking a columella. Ovicell recumbent on distally succeeding autozooid, flattened frontally, with up to 20 variably sized, irregular pores; developing a coarse peripheral ooecial cover in later ontogeny, often with a low distal umbo.

Sublittoral, on shell and other substrata. Habitat and geographical distribution imprecisely known. This species was long confused with *S. cuspidata* and has only recently been redefined by the selection of a neotype (Hayward & Thorpe, 1995). The distribution established extends from the Western Isles of Scotland to western Ireland, southwest Wales, the Channel Islands and the western English Channel, with a single record from the Moray Firth. Reverter Gil & Fernandez Pulpeiro (1996) confirmed records from Galicia and the Straits of Gibraltar, but there are at present no genuine records from the Mediterranean.

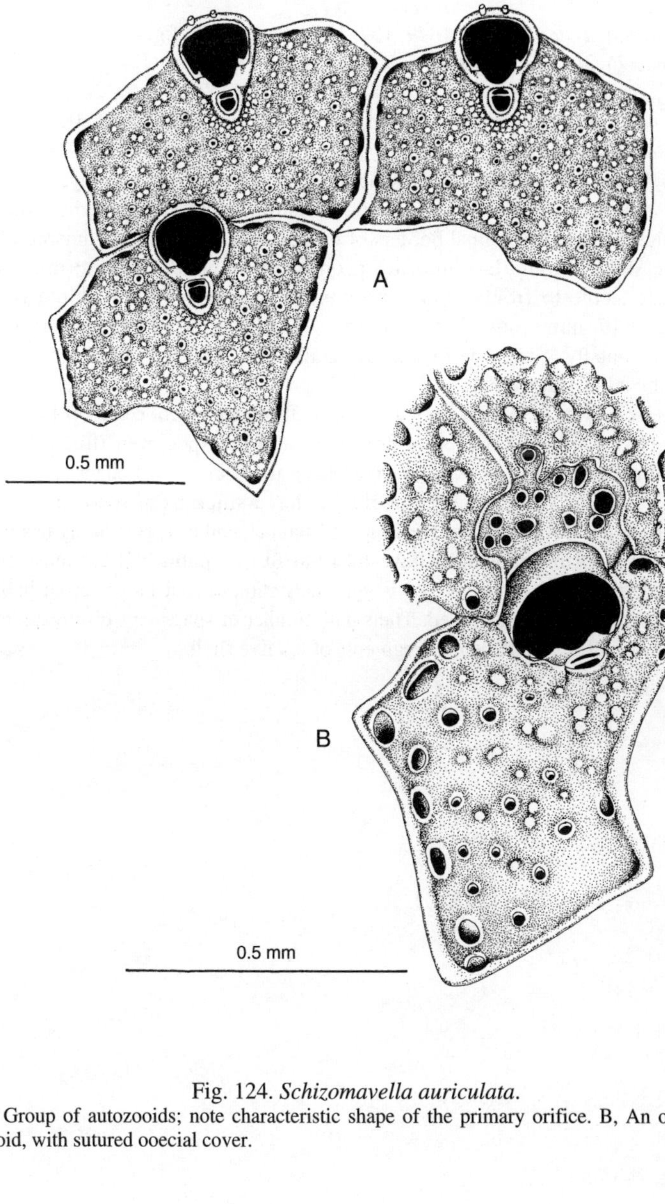

Fig. 124. *Schizomavella auriculata*.
A, Group of autozooids; note characteristic shape of the primary orifice. B, An ovicellate zooid, with sutured ooecial cover.

Schizomavella cristata (Hincks)
(Fig. 125)

Schizoporella cristata Hincks 1879: 157; 1880: 254, pl. 40, fig. 6
Schizomavella cristata: Hayward & Ryland 1979: 180, fig. 74

Colony a tiny, inconspicuous patch, unilaminar, typically on shell substrata. Autozooids oval to hexagonal, convex, separated by deep grooves; 0.34-0.4 by about 0.2 mm. Primary orifice as wide as long, with a short, rectangular sinus; four or five slender, distal oral spines. Peristome thin, erect, slightly flared, enclosing lateral and proximal borders of orifice; medio-proximally incorporating a single suboral avicularium, with prominent umbonate cystid; mandible triangular, acute to frontal plane, proximally directed. Frontal shield finely granular, with numerous, distinct pores. Ovicell slightly broader than long, flattened frontally, evenly perforated by small pores; frontal limit of ectooecium marked by a low ridge.

This species was described from a single small specimen collected off Hastings, Sussex (Hincks 1879), which is now lost, and the specimen illustrated by Hayward & Ryland (1979) was collected from the Mewstone Ledges, Plymouth by S. F. Harmer. A few more specimens have since been recovered from dredgings at the western end of the English Channel, and a single colony has been obtained from Faroe, but there have been no further published accounts of *S. cristata*. It may be that the species is genuinely rare, or that its preferred habitat has not been adequately sampled. The small number of specimens discovered this century have all been encrusting fragments of bivalve shell.

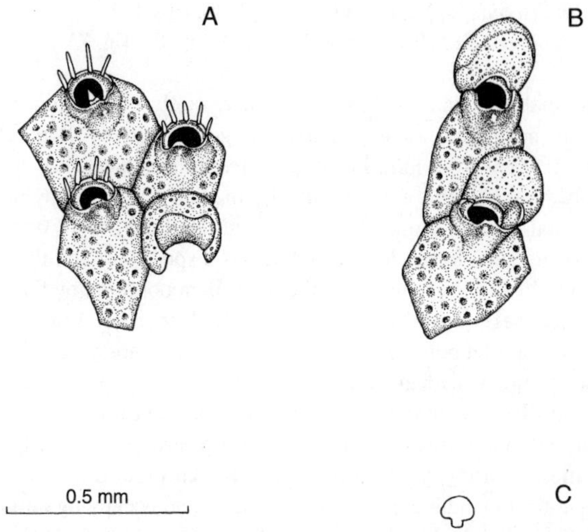

Fig. 125. *Schizomavella cristata*.
A, Autozooids in early ontogeny. B, Ovicellate zooids. C, Outline diagram of the primary orifice.

Schizomavella discoidea (Busk)
(Figs 123C, D; 126)

Lepralia discoidea Busk 1859: 66, pl. 22, figs 7, 8
Schizoporella discoidea: Hincks 1880: 265, pl. 30, figs 8, 9
Schizomavella discoidea: Hayward & Ryland 1979: 182, fig. 75

Colonies small, discoid patches, unilaminar, rarely multilaminar. Autozooids rectangular to hexagonal, or irregularly polygonal, convex. Separated by deep grooves; 0.3-0.5 x 0.2-0.3 mm. Primary orifice longer than wide; proximal edge more or less straight, with a short, slit-like median sinus occupying about one-quarter its total width; depth of sinus accentuated by broad, deep, thickened condyles; six or seven long, slender oral spines disposed around the distal half of the orifice rim, lost in later ontogeny, the bases then obscured by the development of a thickened peristomial rim. Frontal shield coarsely granular, vitreous, perforated by scattered pores. Avicularia polymorphic: most often short, elliptical, <100 μm long, either median suboral and proximally directed, or lateral suboral, proximo-laterally directed; often paired, lateral suboral, 0.2 mm long, with narrow, parallel-sided, distally rounded rostrum, proximo-laterally directed; in some colonies a minority of autozooids bear an enlarged avicularium with elongate, spatulate mandible, directed proximally and occupying almost the entire length of the frontal shield. Ovicell as wide as long, slightly flattened frontally; smooth surfaced, with about 20 relatively large, evenly distributed, tubaeform pores; developing a granular, peripheral ooecial cover and becoming partially immersed in later ontogeny; ovicelled autozooids develop a complete, tubular peristome. Embryos red, present January and September to November in Manx waters (Eggleston 1969).

Offshore, on shell and stone; reported from 20-80 m in the western Mediterranean (Gautier 1962). The type locality is Madeira, from where it ranges north, along the western coasts of Britain as far as Shetland, and throughout the Mediterranean. It occurs widely in the western English Channel and the Gulf of St. Malo, but is absent from the North Sea.

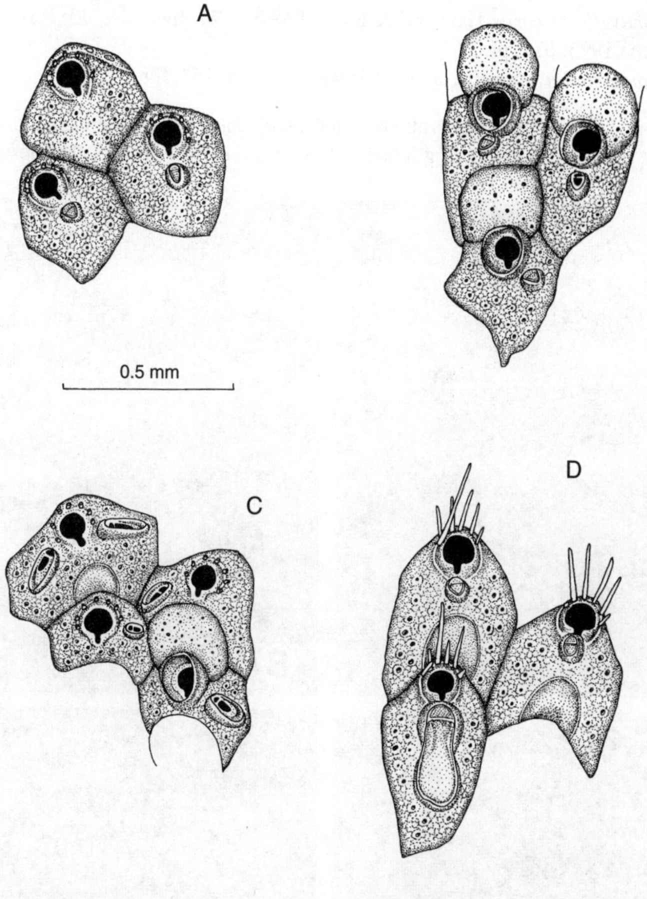

Fig. 126. *Schizomavella discoidea*.
A, Three typical autozooids. B, Ovicellate zooids, with characteristic tubular development of the peristome. C, Autozooids with lateral oral avicularia. D, Autozooids with oral spines intact, one with a dimorphic suboral avicularium.

Schizomavella linearis (Hassall)
(Figs 127; 128)

Lepralia linearis Hassall 1841: 368, pl. 9, fig. 8
Schizoporella linearis: Hincks 1880: 247, pl. 24, fig. 1; pl. 38, figs 5-9
Schizomavella linearis: Hayward & Ryland 1979: 178, figs 10A, 73; Hayward & Thorpe 1995: 671, pl. IV
Schizomavella hastiformis Hayward & Ryland 1978: 152, fig. 5C, D

Colony forming broad incrustations, initially unilaminar but developing multilaminar growth, occasionally producing lobed or folded sheets partly free of the substratum;

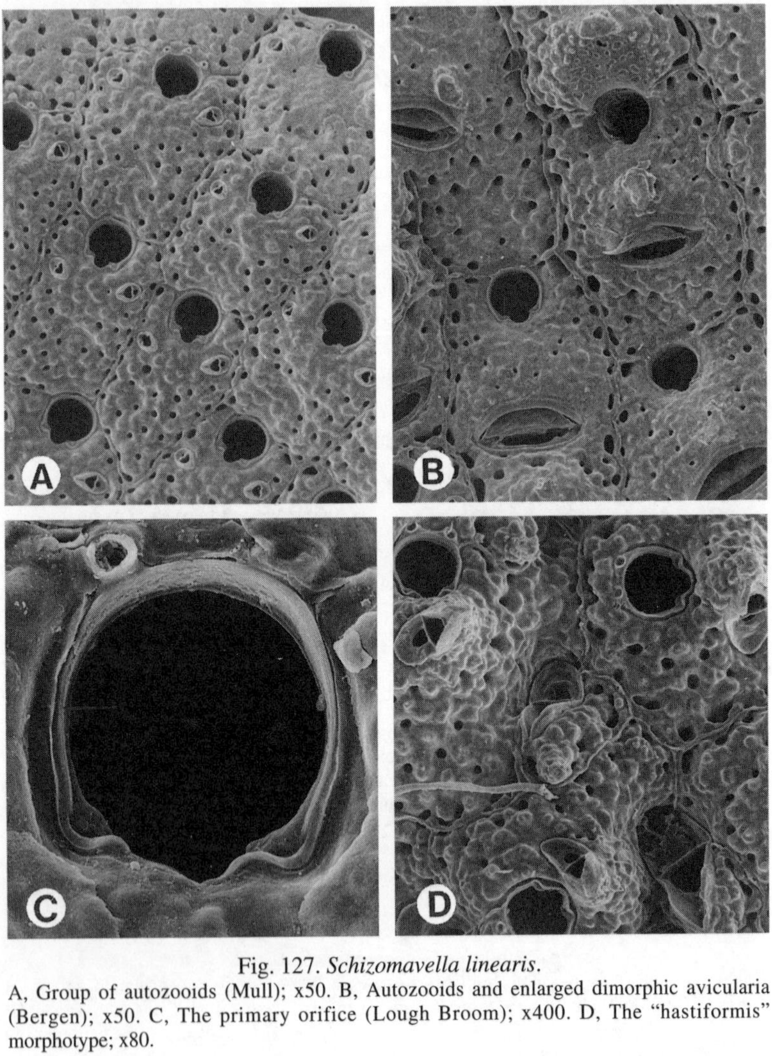

Fig. 127. *Schizomavella linearis*.
A, Group of autozooids (Mull); x50. B, Autozooids and enlarged dimorphic avicularia (Bergen); x50. C, The primary orifice (Lough Broom); x400. D, The "hastiformis" morphotype; x80.

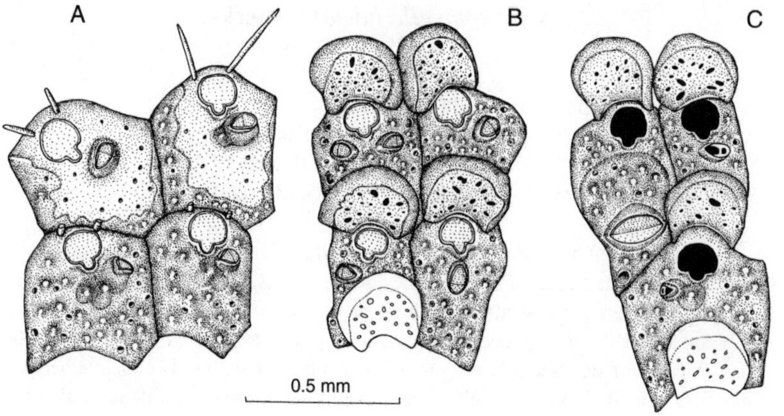

Fig. 128. *Schizomavella linearis*.
A, Autozooids in early ontogeny, with opercula and spines intact. B, Ovicellate zooids, with typical variation in position of the avicularium. C, Three ovicellate autozooids and a vicarious avicularium.

pale pink or red in colour. Autozooids in radiating linear series, rectangular or polygonal, convex, with boundaries marked by distinct grooves; 0.4-0.7 x 0.3-0.4 mm. Primary orifice about as wide as long, proximal border with a short, U-shaped median sinus occupying about half its width; proximo-lateral corners of orifice rim sharply concave, forming a shoulder on each side of the sinus. Condyles distinct on each side of the sinus, rounded, broader than long. Two slender distal oral spines present in early ontogeny. Frontal shield convex, nodular, with about 15-25 small round pores, randomly distributed. Avicularia adventitious, polymorphic; most typically proximo-lateral to orifice, single or paired, *ca.* 0.07 mm long, with short triangular rostrum, acute to frontal plane and directed obliquely medially. Often grossly enlarged, unpaired, with elongated triangular rostrum almost perpendicular to frontal plane, adjacent to or partly fused with, a stout nodular umbo situated immediately proximal to sinus. The small avicularia may occur elsewhere on the frontal wall of the autozooids; less frequent, but characteristic for the species is an enlarged, globular avicularium, covering almost the whole frontal wall of the autozooid, bearing a semi-elliptical mandible 0.15 mm wide. Ovicell recumbent on succeeding zooid, prominent, broader than long, flattened frontally, with numerous irregular frontal pores; developing a coarsely nodular ooecial cover which often bears a prominent umbo. Embryos orange-red, present in summer months. Ancestrula oval, with nine spines surrounding a broad cryptocyst; opesia sinuate.

The morphotype with the single, large suboral avicularium is confusingly similar to *S. hastata*, the taxonomic identity of which was consequently doubted for long. However, *S. hastata* is distinguished immediately by its orifice, which has a broad, shallow sinus, occupying its entire proximal border.

On rock, shell and *Laminaria* holdfasts. Often common beneath large stones on sheltered rocky shores. Widespread and common on all British coasts, extending offshore, on hard substrata, to the edge of the continental shelf. Distributed throughout the north-east Atlantic region, from west Norway and Faroe to the Mediterranean.

Schizomavella hastata (Hincks)
(Figs 129; 130A, B)

Lepralia hastata Hincks 1862: 206, pl. 12, fig. 4
Schizoporella linearis var. *hastata*: Hincks 1880: 248, pl. 33, fig. 10
Schizomavella linearis var. *hastata*: Hayward & Ryland 1979: 180, fig. 74
Schizomavella hastata: Hayward & Thorpe 1995: 668, pl. III

Colony forming broad, irregular patches, unilaminar or multilaminar; frequently only loosely encrusting substratum, developing erect, folded sheets or cylinders, especially on erect substrata and on large hydroids. Autozooids rectangular to polygonal, flat or slightly convex, separated by distinct sutures; 0.4-0.7 x 0.3-0.5 mm. Primary orifice wider than long, proximal border shallowly concave between indistinct, rounded, proximo-lateral condyles. Two or three stout distal oral spines present in early ontogeny. Frontal shield initially smooth, becoming progressively nodular in later ontogeny, perforated by large round pores. Polypide with 13 tentacles (Calvet 1900). Avicularium monomorphic, large, situated immediately proximal to sinus, rostrum perpendicular to frontal plane, facing laterally; rostrum elongate, lanceolate, 0.2 mm long, its distal rim finely toothed; cystid broad-based, produced apically, above the tip of the rostrum, as a stout conical umbo. Ovicell prominent, longer than wide, flattened frontally, with about 20 irregular pores; developing a coarsely nodular ooecial cover with a tall conical umbo.

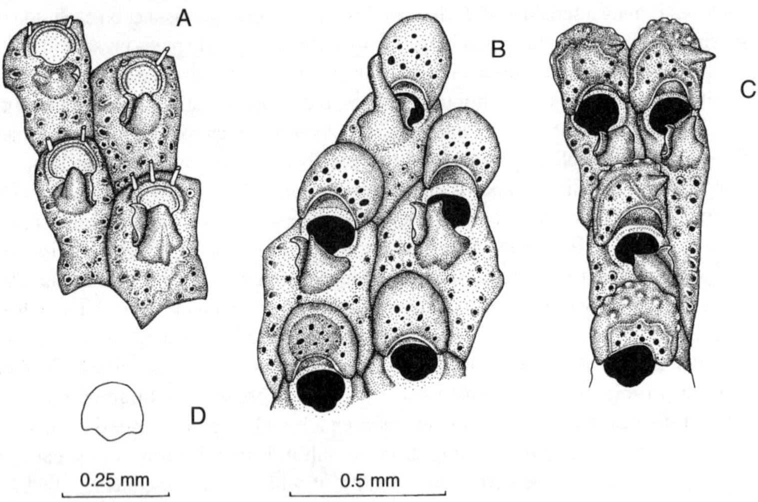

Fig. 129. *Schizomavella hastata*.
A, Autozooids in early ontogeny, with spines and opercula intact. B, Ovicellate zooids. C, Autozooids in late ontogeny, the ovicells covered with coarse secondary calcification. D, Outline diagram of the primary orifice.

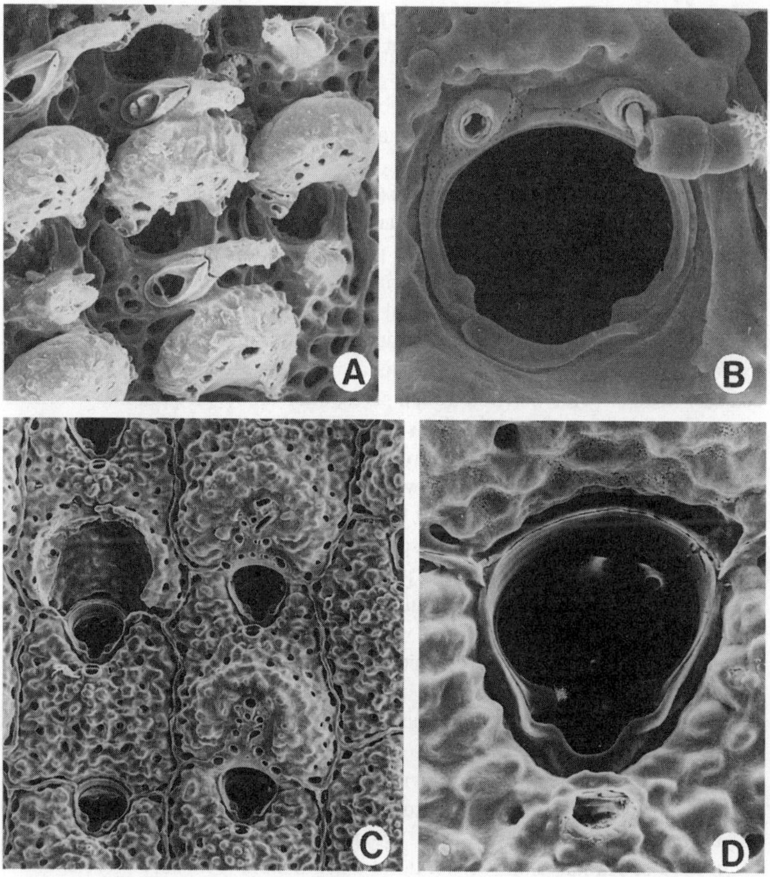

Fig. 130.
A, B, *Schizomavella hastata* (specimen from Guernsey).
A, Ovicelled zooids with umbonate avicularia; x70. B, The primary orifice; x280.

C, D, *Schizomavella hondti*.
C, Group of autozooids, two with fully developed ovicells; x70. D, The primary orifice; x270.

Embryos present in July and August in south-west Britain. Gautier (1962) reported yellow embryos in March, April and September in the western Mediterranean, and larval settlement in October.

Sublittoral, on shells, stones and large hydroids. Geographical distribution imprecisely known but present on western British coasts from Shetland to Guernsey, also recorded from Galicia, Catalonia and the Balearic Islands (Reverter Gil & Fernandez Pulpeiro, 1996).

Schizomavella cuspidata (Hincks)
(Fig. 131)

Schizoporella auriculata var. *cuspidata* Hincks 1880: 261, pl. 29, fig. 8
Schizomavella cuspidata Hayward & Thorpe 1995: 665, pl. II

Colony developing broad unilaminar or multilaminar sheets, often extensive, nodular or distinctly mamillate in the largest specimens. Autozooids in radiating linear series, rectangular to irregularly polygonal, convex, separated by distinct grooves; 0.39-0.56 x 0.23-0.35 mm. Primary orifice about as wide as long, with distinct, rather angular proximo-lateral corners; sinus a short, symmetrical U-shape, occupying about one-third width of proximal border; condyles broad, extending to edges of sinus. Three distal oral spines present. Frontal shield coarsely nodular, with scattered frontal pores, often umbonate proximal to orifice. Avicularia polymorphic, always situated immediately proximal to sinus, moderately or steeply inclined to frontal plane and proximally directed; most often short oval, 0.05 mm long, with slender crossbar and inconspicuous columella; frequently enlarged, 0.1-0.15 mm, with parallel-sided, distally rounded rostrum, or longer, with a narrowly spatulate rostrum. In all larger avicularia the crossbar has a knob-like columella, and the palate has a conspicuous Y-shaped foramen. Ovicell recumbent on distally succeeding autozooid, broader than long, slightly flattened frontally, with a variable number of large, irregular pores; developing a coarse ooecial cover which often bears a prominent nodular umbo.

This species often seems confusingly protean, with the incidence of the different sizes of avicularia varying in frequency, and the degree of development of the suboral umbo and ooecial cover similarly variable. In the most thickly calcified specimens the avicularian rostrum is often orientated almost perpendicular to the orifice plane. However, in all cases the shape of the orifice and the position and orientation of the avicularia are constant.

On hard substrata, shallow sublittoral and offshore. Probably present on all British coasts but former confusion with *S. auriculata* results in distribution of both species being unclear. *S. cuspidata* is probably the more common; it ranges from Shetland southwards to the Channel Islands, and is present throughout the Mediterranean, but precise distributional limits are not known.

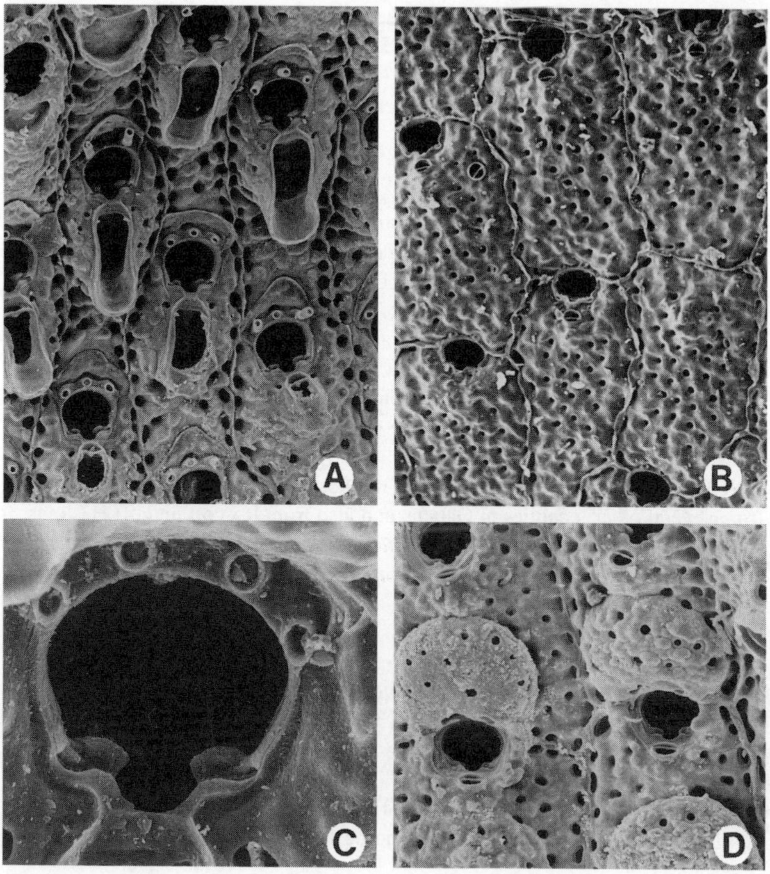

Fig. 131. *Schizomavella cuspidata*.
A, Autozooids with dimorphic avicularia; x65. B, Autozooids in later ontogeny, with small avicularia; x60. C, The primary orifice; x400. D, Ovicellate zooids; x90.

Schizomavella hondti Reverter Gil & Fernandez Pulpeiro (Fig. 130C, D)

Schizomavella hondti Reverter Gil & Fernandez Pulpeiro 1996: fig. 2; pl. 4, figs A-E

Colony a small, unilaminar patch. Autozooids rectangular to polygonal, in regular radiating rows, separated by distinct grooves; 0.41-0.51 x 0.26-0.35 mm. Primary orifice wider than long, broadest distally and distinctly drop-shaped; sinus a short, symmetrical U-shape, occupying about half width of proximal border; condyles large, rounded, rhomboidal, with faceted surfaces. Two indistinct spines present in early ontogeny. Frontal shield granular, convex and densely perforated. Avicularium monomorphic, rather small (0.04 mm long) slightly distant from sinus, almost normal to frontal plane. Ovicell about as wide as long, slightly flattened frontally, with a few irregular pores, becoming partially obscured by a nodular ooecial cover.

Sublittoral, on shell. Described from the Ferrol Ria, this species was recorded from several other localities in Galicia, and also from Roscoff and off Hastings. It seems particularly similar to *S. auriculata* but is distinguished by its orifice, which has a very much smaller sinus and proportionately larger condyles, and by its very small avicularium.

Schizomavella ochracea (Hincks) (Fig. 132A-C)

Lepralia ochracea Hincks 1862: 206, pl. 12, fig. 3
Schizoporella auriculata var. *ochracea*: Hincks 1880: 260, pl. 29, fig. 7
Schizomavella ochracea: Hayward & Thorpe 1995: 674, pl. V, figs a-c

Colony a broad, spreading unilaminar sheet, exceeding 100 mm^2. Autozooids flat, hexagonal to irregularly polygonal, separated by distinct raised sutures; 0.45-0.7 x 0.3-0.45 mm. Primary orifice as wide as long, sinus a broad, symmetrical V occupying most of proximal edge; condyles prominent, elongate oval, with coarsely toothed edge. No oral spines. Frontal shield distinctly nodular, evenly perforated by numerous closely spaced pores. Avicularia sporadic, missing from many autozooids, dimorphic: small, oval, 0.2 mm long, with complete, slender crossbar and large oval palatal foramen; gigantic, oval, triangular or rectangular, 0.3-0.35 mm long, with complete crossbar and variably sized palatal foramen; all types of avicularia are normal to frontal plane and proximally directed.

This species was first described from specimens collected off Cornwall. It was for long confused with *S. auriculata* and has only recently been redescribed (Hayward & Thorpe, 1995). Further specimens are known from the Mediterranean coast of Spain and from Malta but its distribution is otherwise unreported.

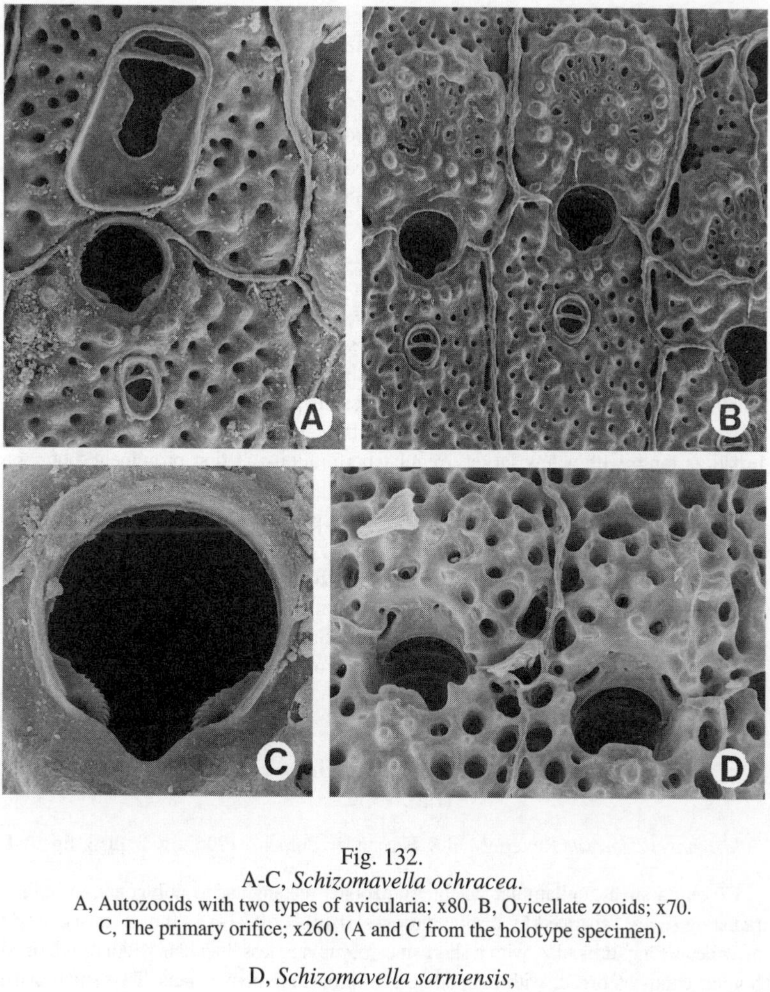

Fig. 132.
A-C, *Schizomavella ochracea*.
A, Autozooids with two types of avicularia; x80. B, Ovicellate zooids; x70.
C, The primary orifice; x260. (A and C from the holotype specimen).

D, *Schizomavella sarniensis*,
ovicellate zooids; x90.

Schizomavella sarniensis Hayward & Thorpe (Figs 132D; 133A, B)

Schizomavella sarniensis Hayward & Thorpe 1995: 675, pl. V, figs d-f

Colony unilaminar to multilaminar, developing extensive nodular sheets. Autozooids quadrangular to polygonal, convex, separated by distinct sutures; 0.45-0.60 x 0.25-0.45 mm. Primary orifice wider than long, proximal border with a short, U-shaped sinus occupying about half its width, the rim developed as distinct shoulders on each side. Condyles large, conspicuous, transversely oval, their free edges finely toothed; two or three indistinct oral spines, present in early ontogeny only. Frontal shield thickly calcified, pierced by numerous large pores; becoming very rugose in later ontogeny, typically developing a short suboral umbo. Avicularium dimorphic: most often immediately proximal to sinus, almost perpendicular to plane of orifice, small, with short semi-oval mandible; occasionally enlarged, normal to frontal plane, proximally directed and occupying most of frontal surface of autozooid. Ovicell as broad as long, with a few large, irregular perforations; at first prominent, but soon enveloped in a thickening, rugose ooecial cover and immersed. In later ontogeny the primary orifice becomes surrounded by a low, spiky peristomial rim which extends distally onto the frontal surface of the ovicell.

Offshore, on hard substrata. Only recently distinguished from other species of *Schizomavella, S. sarniensis* is presently known from Salcombe, Devon, from the western end of the English Channel, and from Santander, northern Spain. Reverter Gil & Fernandez Pulpeiro (1996) have also recorded it from Galicia, and from the coast of Algeria.

Schizomavella teresae Reverter Gil & Fernandez Pulpeiro (Fig. 133C-E)

Schizomavella teresae Reverter Gil & Fernandez Pulpeiro 1996: fig. 2; pl. 3, figs A-F

Colony a small, unilaminar patch. Autozooids rectangular to polygonal, in regular radiating series, separated by distinct grooves; 0.4-0.5 x 0.22-0.34 mm. Primary orifice as wide as long, orbicular, with a short sinus occupying less than half width of proximal border; condyles broad, with pointed corners projecting above sinus. Two small distal oral spines, present in early ontogeny only. Peristome developed as a lobed or spiky rim completely encircling the orifice. Frontal shield rugose, with few, large pores; becoming increasingly nodular in later ontogeny and developing pronounced spikes. Avicularium monomorphic, situated immediately proximal to sinus, perpendicular to plane of orifice, and enclosed within peristome. Ovicell wider than long, with a few large ectooecial pores immediately above its aperture; eventually partly obscured by rugose, spiky calcification, and the porous area covered by the distal expansion of the peristome. Pink embryos reported in April, and larval settlement in late summer (Reverter Gil & Fernandez Pulpeiro, 1996). Ancestrula oval, with nine spines bordering the opesia.

Sublittoral, on shell. This species was described from the Ferrol Ria, Galicia, and reported also from the Bay of Biscay, off Roscoff and from several localities in the English Channel. It is most similar to *S. cuspidata*, but is distinguished from that species by its smaller, more rounded orifice, small sinus and smooth, pointed condyles.

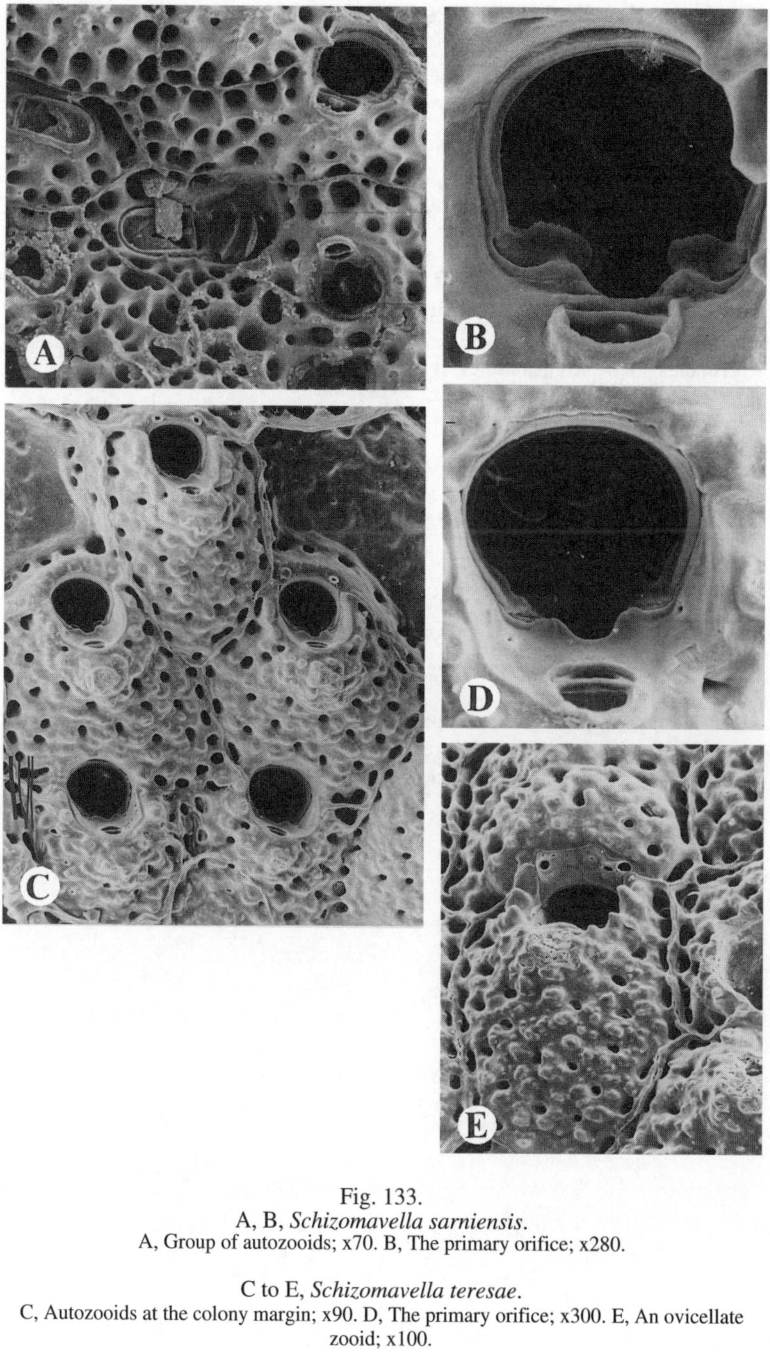

Fig. 133.
A, B, *Schizomavella sarniensis*.
A, Group of autozooids; x70. B, The primary orifice; x280.

C to E, *Schizomavella teresae*.
C, Autozooids at the colony margin; x90. D, The primary orifice; x300. E, An ovicellate zooid; x100.

Family MICROPORELLIDAE Hincks, 1880

Primary orifice semicircular: ascus opening via an ascopore, proximal to the orifice. Frontal shield with scattered pores. Ovicell present. Oral spines and avicularia present or absent.

Genus MICROPORELLA Hincks, 1877

Colony encrusting. Primary orifice semicircular, ascopore distinct, on a raised base. Oral spines present. Ovicell prominent, hyperstomial, imperforate, closed by zooidal operculum. Frontal shield with scattered pores. Avicularia adventitious, with acute or setiform mandibles. Basal pore-chambers large and distinct.

Type species: *Eschara ciliata* Pallas, 1766

Key to British species of *Microporella*

1. Oral spines slender, small. Ovicell uniformly granular. Avicularia latero-frontal with acute or short setiform mandible. Widespread and common
.... ***Microporella ciliata*** (p. 296)

Oral spines thick, proximal pair often forked. Ovicell with a marked frontal ridge. Avicularia lateral to orifice, with elongate setiform mandible. South-western only; rare
.... ***Microporella appendiculata*** (p. 294)

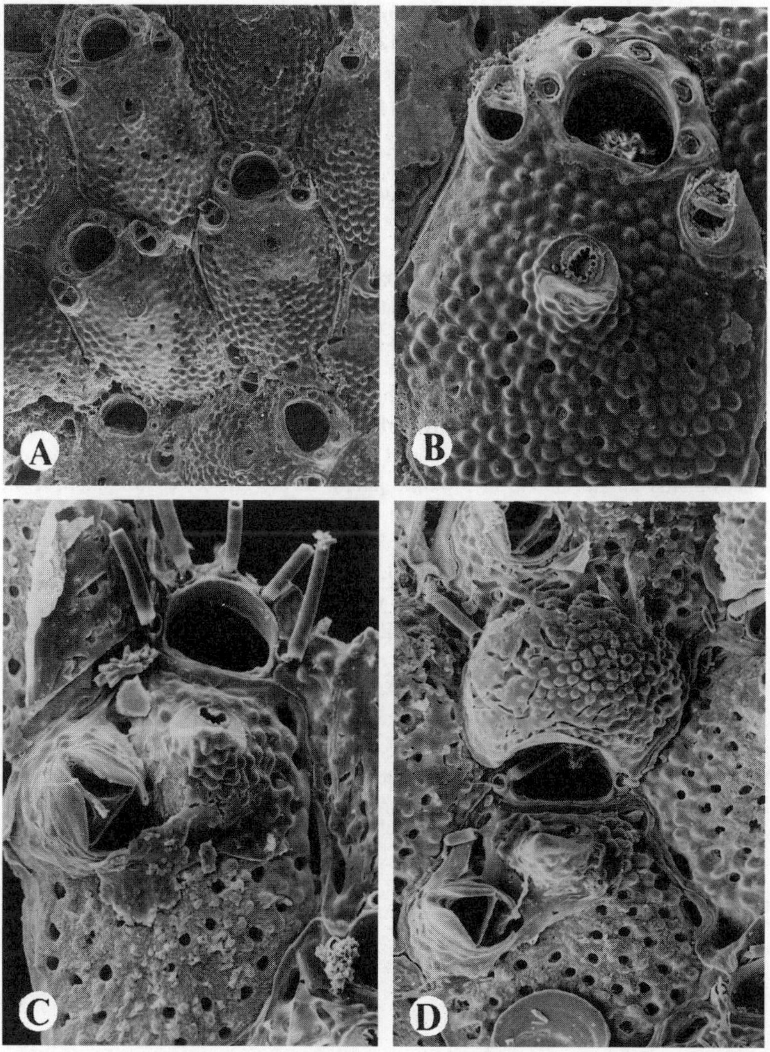

Fig. 134.
A, B, *Microporella appendiculata*.
A, Group of autozooids; x50. B, Detail of a single autozooid; x110.

C, D, *Microporella ciliata*.
C, An autozooid at the colony margin; x140. D, An ovicellate zooid; x110.

Microporella appendiculata (Heller)
(Figs 134A, B; 135)

Lepralia appendiculata Heller 1867: 31, pl. 2, fig. 8
Microporella marsupiata: Gautier 1962: 175; Hayward & Ryland 1979: 220, fig. 94
Microporella pseudomarsupiata Aristegui 1984: 184, fig. 46; Zabala & Maluquer 1988: 141, pl. 19, fig. C

Colony encrusting, light brown. Autozooids oval to hexagonal, flat or slightly convex, separated by shallow grooves; 0.5-0.65 x 0.35-0.45 mm. Frontal shield finely granular, with small scattered pores. Primary orifice semicircular, closed by a dark brown operculum, with distinct marginal sclerite; five short, thick oral spines, each with a brown chitinous base, only the proximal pair persisting in ovicellate autozooids and frequently forked distally. Ascopore crescentic, with a denticulate edge, raised on a cup-shaped prominence. Avicularia typically paired, occasionally single, lateral or proximo-lateral to the orifice; cystid small, oval, supporting a long setiform mandible. Ovicell spherical, prominent, with a transverse frontal rib. Ancestrula undescribed.

The specimen illustrated here was collected from the western end of the English Channel. It was originally attributed to the Madeiran species *Microporella marsupiata* (Busk), which Gautier (1952) showed to be widely distributed in the Mediterranean, and represented the most northerly record of that species (Hayward & Ryland, 1979). However, Aristegui (1984) showed that the Madeiran species was quite different from that described by Gautier (1962), for which he introduced the name *M. pseudomarsupiata* Aristegui. Both species occur around the Canaries. It is now apparent that the correct name should be *M. appendiculata* (Heller), which was first described from the Adriatic and synonymised with *M. marsupiata* by Norman (1909).

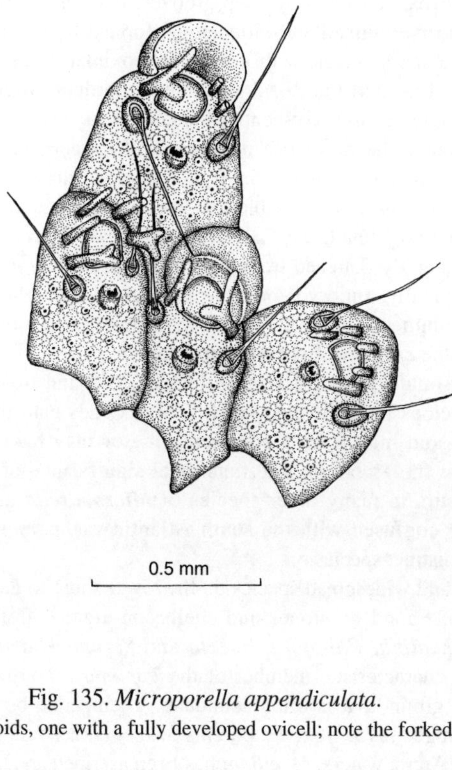

Fig. 135. *Microporella appendiculata*.
Group of autozooids, one with a fully developed ovicell; note the forked oral spines.

Microporella ciliata (Pallas)
(Figs 134C, D; 136)

Eschara ciliata Pallas 1766: 38
Microporella ciliata: Hincks 1880: 206, pl. 28, figs 1-5, 7-8; Hayward & Ryland 1979: 222, fig. 95

Colony encrusting, often extensive; silvery or vitreous. Autozooids oval to hexagonal, convex, separated by deep grooves; 0.46-0.54 x 0.3-0.4 mm. Frontal shield thick, coarse grained and vitreous, perforated by numerous minute pores. Primary orifice semicircular, with a straight proximal border; six erect, pointed spines disposed around the distal and lateral borders, often missing in older autozooids. Ascopore large, crescentic; situated on the distal face of a small umbo. Avicularium single, lateral or just proximal to the ascopore, on the right or left; rostrum short, triangular, acute to the frontal plane of the autozooid, supporting a short setiform mandible (of variable length), directed laterally or disto-laterally. Polypide with 13-14 tentacles (Calvet, 1900). Ovicell prominent, more or less spherical, but slightly flattened frontally; surface uniformly granular, often with a small umbo apically. Embryos orange-red, present throughout the year, with a peak in late summer. Ancestrula tatiform, gymnocyst smooth, with about nine spines around the extensive frontal membrane.

In the "personate" form of Hincks (1880: 207) and other authors, a thick peristome develops in fertile autozooids. This extends onto the ovicell, encloses the ascopore, and may form a complete tube or may have a small proximal foramen above the ascopore. The functional significance of this feature is not known; it occurs in many other species of *Microporella* and, taxonomically, should not be confused with the south Atlantic var. *personata* of Busk, now regarded as a distinct species.

A common and widespread species on the lower shore to at least 100 m, off all British coasts. Found on stones and shells, on algae (often abundantly, with *Celleporella hyalina, Callopora lineata* and *Scrupocellaria reptans*, on kelp fronds) and a characteristic member of the *Laminaria* holdfast community. On offshore shell grounds, it is often abundant. A temperate-boreal species; it was reported by Kluge (1975) from the Barents Sea, but otherwise does not seem to be widespread in Arctic waters. *M. ciliata* has been ascribed an almost cosmopolitan, but probably spurious, distribution.

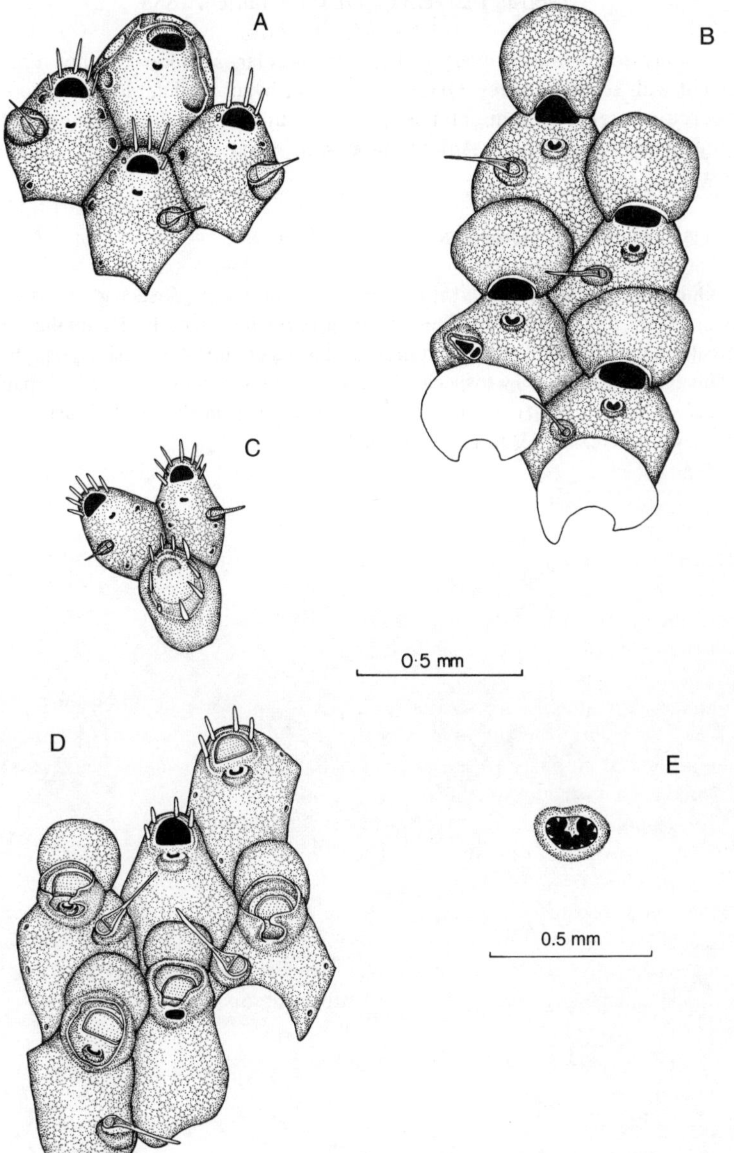

Fig. 136. *Microporella ciliata*.
A, Autozooids at the colony margin. B, Ovicellate zooids. C, The ancestrula, with the first two zooids budded from it. D, The "personate" form, showing the peristome in different stages of development. E, The ascopore in enlarged view.

Genus FENESTRULINA Jullien, 1888

Colony encrusting. Primary orifice semicircular, ascopore distinct. Frontal shield with scattered pores. Ovicell prominent, hyperstomial, closed by zooidal operculum; entooecium calcified, ectooecium largely membranous, with conspicuous basal pores. Avicularia absent. Basal pore-chambers large and distinct.

Type species: *Cellepora malusii* Audouin, 1826

Only one species, *F. malusii*, occurs in British waters. A second, *F. joannae* (Calvet), is found in the Mediterranean. It is distinguished by its smaller size, greater number of oral spines, prominent umbo and almost exclusively epiphytic habit (Gautier, 1962). Two species have been newly described from the Spanish coast: *F. asturiasensis* Alvarez (1992) is known from the north coast and *F. barrosoi* Alvarez (1993) from the Sea of Alboran.

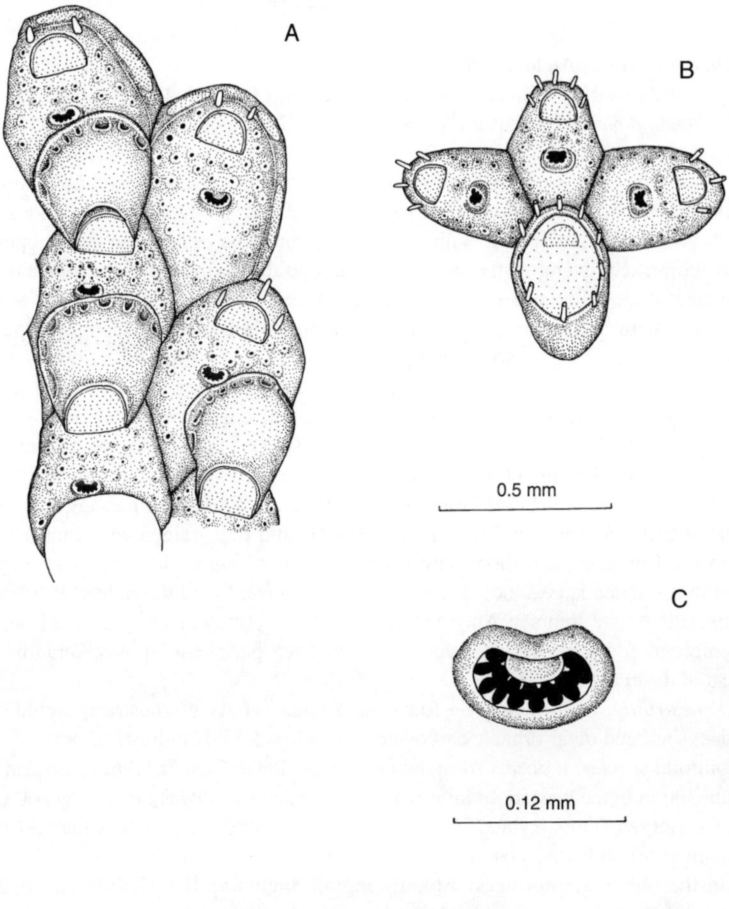

Fig. 137. *Fenestrulina malusii*.
A, Autozooids at the colony margin, three with ovicells.
B, The ancestrula and periancestrular zooids. C, The ascopore in enlarged view.

Fenestrulina malusii (Audouin)
(Figs 137; 138A, B)

Cellepora malusii Audouin 1826: 239, pl. 8, fig. 8
Microporella malusii: Hincks 1880: 211, pl. 28, figs 9-10, pl. 29, fig. 12
Fenestrulina malusii: Hayward & Ryland 1979: 224, fig. 96

Colony forming brilliant white patches. Autozooids oval, separated by distinct grooves; 0.4-0.6 x 0.3-0.4 mm. Frontal shield convex, smooth and porcellanous at first, later white and chalky; with a number of round pores between the ascopore and the orifice, lateral to the orifice, and around the periphery of the autozooid, the central proximal area being imperforate. Orifice semicircular, proximal edge straight, with a thin rim and a pair of minute, pointed teeth. Operculum pale brown with a strong, distinct marginal sclerite. Two or three short distal oral spines present in young autozooids. Ascopore broad, crescentic, inner rim with small, but conspicuous, denticulations; proximal edge raised and thickened. Polypide with 14-15 tentacles (Calvet, 1900). Ovicell prominent, recumbent on the distally succeeding autozooid, distinctive in appearance: ectooecium calcified only along its distal basal edge, appearing as a ridge encompassing but not touching the endooecium. The calcified endooecium fine grained, less translucent than the frontal calcification, with a single series of pores along its distal edge giving it a fluted appearance. Embryos yellow-orange; found throughout the year, especially in late summer. Ancestrula elliptical, tatiform, with an extensive frontal membrane and a negligible cryptocyst; about ten peripheral spines. Budding a triad of distal autozooids.

Fenestrulina malusii may be found on a wide variety of substrata, including stone, shell and other organic carbonates, algal fronds and holdfasts. Essentially a sublittoral species, it occurs from the kelp forest down. Gautier (1962) recorded it from 300 m in the western Mediterranean, but stated that it was rare below 100 m; in the Bergen fjords Ryland (1963a) recorded it only in depths less than 50 m. Common off all British coasts.

In the temperate northeast Atlantic region, including the Mediterranean, *F. malusii* seems to be a well defined species. It should be noted, however, that its unknown type locality probably lies on the Egyptian Red Sea coast and that Audouin's species may not be the same as that recognised as *F. malusii* in European marine faunas. *Fenestrulina malusii* has been reported from all the world's seas but re-examination of records from beyond European coasts will probably show these reports to be the result of incorrect identification. The genus is in need of careful taxonomic review.

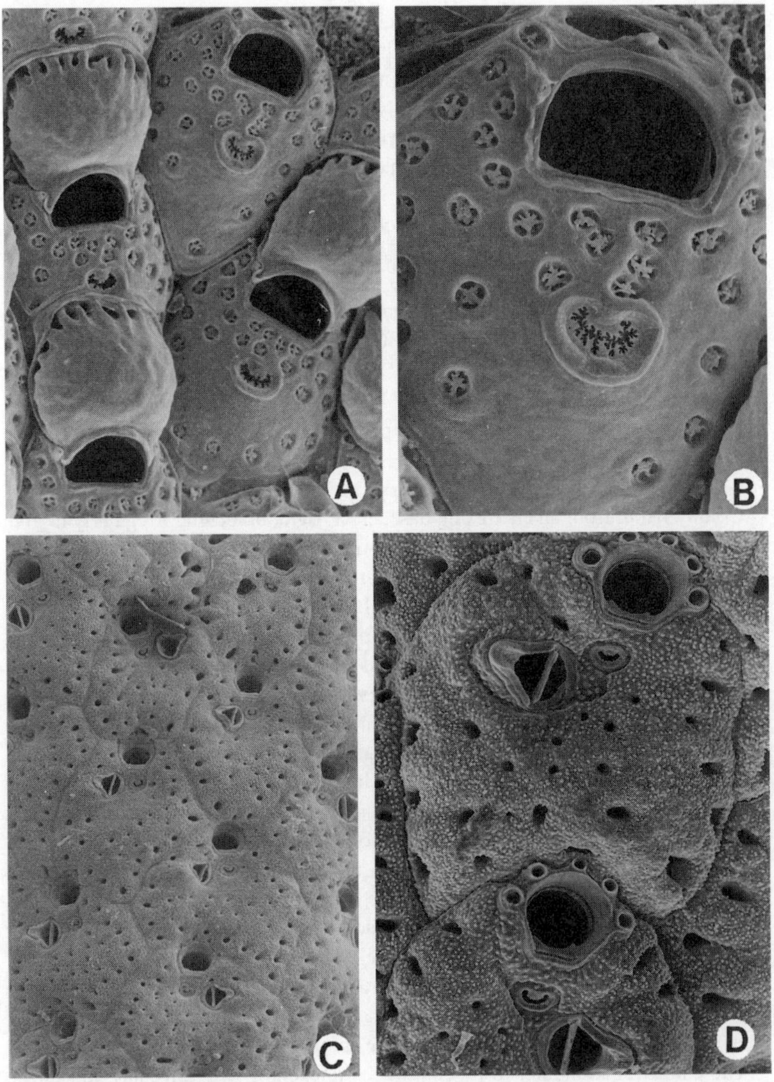

Fig. 138.
A, B, *Fenestrulina malusii.*
A, Autozooids at the colony margin, with ovicells; x70. B, A single autozooid; x150.

C, D, *Diporula verrucosa.*
C, Portion of a colony; x33. D, Detail of two autozooids in earlier ontogeny; x80.

Genus DIPORULA Hincks, 1879

Colony erect. Primary orifice semicircular, ascopore proximal to lower rim. Frontal shield with scattered pores and very large marginal areolae. Avicularium adventitious, proximo-lateral to the orifice, developing from a marginal pore. Ovicell present, perforated. Small multiporous septula present.

Type species: *Eschara verrucosa* Peach, 1868

Diporula verrucosa (Peach)
(Figs 138C, D; 139)

Eschara verrucosa Peach 1868: 116
Diporula verrucosa: Hincks 1880: 220, pl. 31, figs 1-2; Hayward & Ryland 1979: 226, fig. 97

Colony rising from an encrusting base, erect, rigid, branching irregularly to give a flat, fan shape; branches cylindrical. Autozooids in six to nine longitudinal series, opening all around the branch; broad, hexagonal to oval, convex and separated by distinct grooves, becoming less distinct as secondary calcification proceeds; 0.6-0.7 x 0.5-0.6 mm. Primary orifice approximately semicircular, distal rim finely denticulate, proximal edge shallowly concave between two prominent condyles, immersed in a low thickened peristome, three to five short, delicate spines distally, impermanent and usually missing in older autozooids. Operculum dark brown. Frontal shield finely granular, becoming increasingly smooth and vitreous with increasing calcification; large, conspicuous marginal areolae alternating with thickened ridges; numerous frontal pores in the central area, each in a shallow pit. In older autozooids both types of pore become obscured. Ascopore crescentic, proximal to the orifice, with a finely denticulate rim. Avicularium single, proximo-lateral to the orifice, on the right or the left; cystid arising from one of the areolae, large and blister-like with a short, triangular rostrum directed disto-laterally; mandible elongate, setiform. Ovicell described by Hincks (1880) as globular, with scattered pores. Small septula, with two or three pores, recessed in the lateral walls; and also present between the main cavity of each autozooid and the chambers below the areolae.

The colony is strengthened by continuous thickening of the frontal wall of the autozooids. The orifice becomes deeply immersed and the marginal and frontal pores are occluded, only the ascopore and the avicularium remaining distinct.

A warm temperate species, *D. verrucosa* is found from the coasts of West Africa, throughout the Mediterranean and northwards to the extreme southwest of the British Isles. It is here rare and has been reported on a few occasions from off the south Cornish coast. Gautier (1962) obtained it from 70-120 m in the western Mediterranean.

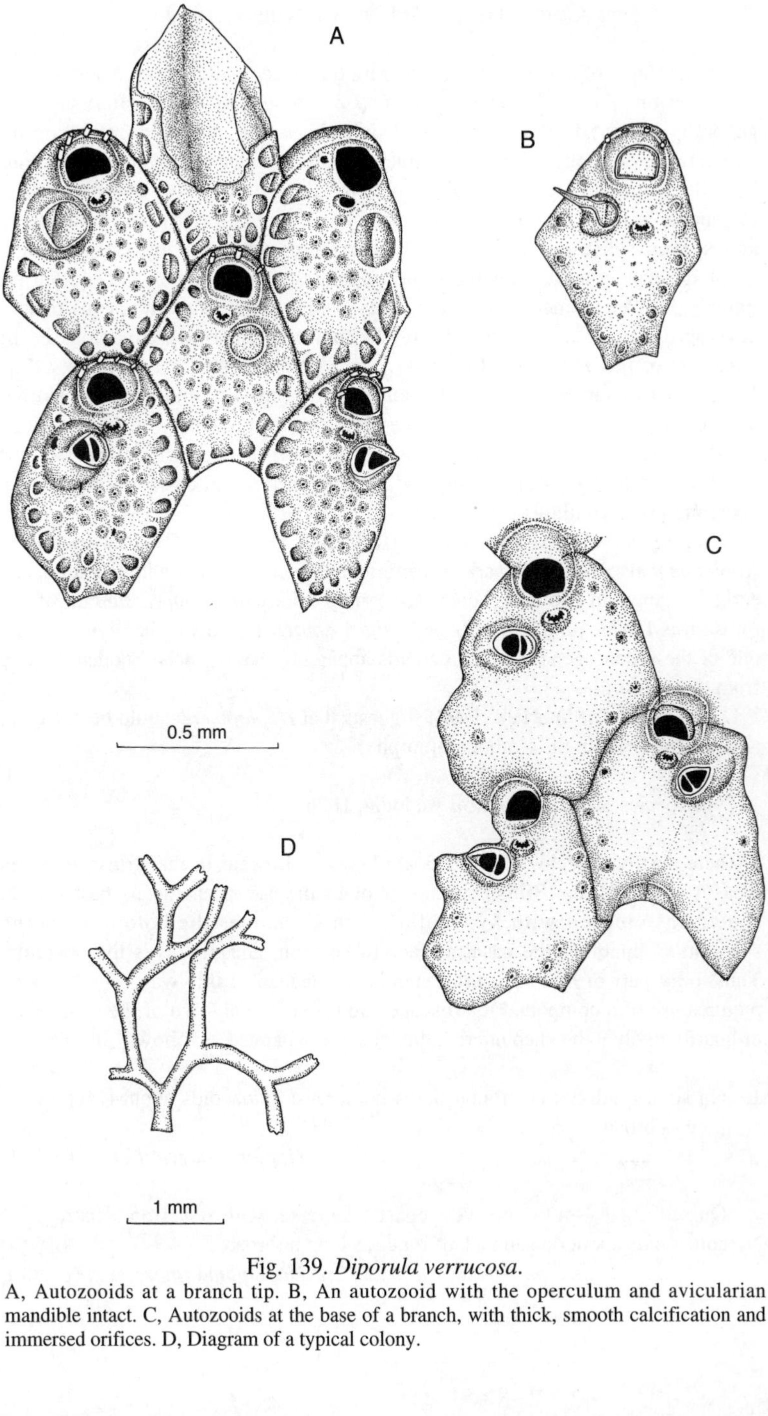

Fig. 139. *Diporula verrucosa.*
A, Autozooids at a branch tip. B, An autozooid with the operculum and avicularian mandible intact. C, Autozooids at the base of a branch, with thick, smooth calcification and immersed orifices. D, Diagram of a typical colony.

Genus HAPLOPOMA Levinsen, 1909

Frontal shield of autozooids convex, with translucent, often transversely striated calcification. Pores always present, few or numerous, in marginal series or generally scattered; often extended distally by a thin suture; simple, with the lumen open, or "stellate", with the lumen almost occluded by inwardly projecting, broadly triangular teeth (detail which is visible only with magnifications obtainable with a compound microscope, see Fig. 145G). Orifice of ancestrula almost always sinusoid, with rounded anter and poster of variable size and form; in all subsequent autozooids the poster separates as an ascopore linked with the roughly D-shaped anter by a suture. Interzooidal connections by means of well developed pore chambers, which may be tubular (Fig. 145I), cavernous in the thickness of the wall base (Fig. 144E), or forming blister-like chambers (Fig. 142C). No avicularia. Ovicellate autozooids indistinguishable from other autozooids, separable only by orifice size; or conspicuously shorter. Ovicell subglobular, porous, surrounded by a ring of pore chambers, recumbent on the distal autozooid; its opening combined with the orifice and closed by the zooidal operculum (i.e., cleithral).

As in *Hippothoa* and *Celleporella* (Ryland & Gordon, 1977), the species of *Haplopoma* also display remarkable differences in ancestrular budding pattern and early astogeny, no two being alike. The young colony of *H. impressum* illustrated by Barrois (1877, pl. 16, fig. 2, as *Mollia granifera*) seems to be atypical, with one of the second generation autozooids apparently having been budded directly from the ancestrula.

D. P. Gordon (*in litt.,* Feb. 1998) suggests that *Haplopoma* should be placed in a new family within the Umbonulomorpha.

Type species: *Flustra impressa* Audouin, 1826

There are five species of *Haplopoma* known, all occur in the British Isles sea area. A key to all the British species involves the use of characters best seen in specimens cleaned with hypochlorite; this removes the colony from the substratum, enabling the basal surface to be seen, and dissolves the opercula. Thus, only part of one's material should be cleaned in this way. The key also requires use of a compound microscope. However, as only two of the species are ordinarily likely to be encountered, they may be separated as follows:

1. On stones, all coasts; frontal pores numerous; autozooids about twice as long as broad
.... ***Haplopoma graniferum*** (p. 314)

On red algae, south and west coasts; delicate, with few frontal pores; autozooids about one and a half times as long as broad
.... ***Haplopoma impressum*** (p. 308)

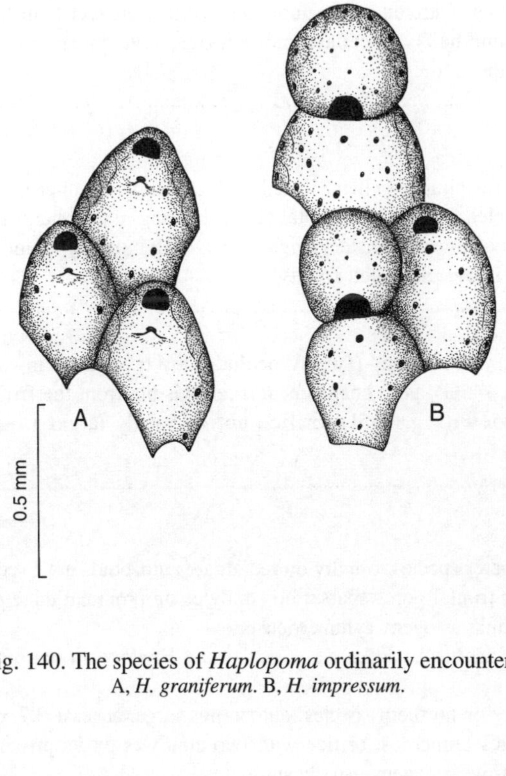

Fig. 140. The species of *Haplopoma* ordinarily encountered
A, *H. graniferum*. B, *H. impressum*.

Key to all the British species of *Haplopoma*

1. Frontal pores in a single series near the margin of the autozooid; orifice as long as broad; ovicellate autozooids much shorter than non-ovicellate autozooids

.... ***Haplopoma sciaphilum*** (p. 310)

Frontal pores scattered, often numerous; orifice broader than long (often by about one-half); ovicellate and non-ovicellate autozooids about the same length

.... **2**

2. Frontal pores simple (when viewed with a compound microscope); pore chambers clearly visible in frontal view, showing in peripheral autozooids as a series of internal blisters (Figs 142C, 144B); orifice larger (by about one-third) in ovicellate autozooids

.... **3**

Frontal pores "stellate" (largely occluded by triangular, inward-facing teeth (Fig. 145G); pore chambers scarcely visible from the front, tubular in basal view (Fig. 145I); orifice not or hardly larger in ovicellate autozooids

.... **4**

3. Shallow-water species, usually on red algae; autozooids not exceeding 0.6 x 0.4 mm; frontal pores sparse; no condyles on proximal edge of orifice; embryos pink; astogeny symmetrical

.... ***Haplopoma impressum*** (p. 308)

Deep water or northern species; autozooids large, at least 0.7 x 0.5 mm; frontal pores numerous; orifice with two condyles on its proximal edge; embryos white; astogeny usually spiral

.... ***Haplopoma planum*** (p. 312)

4. Autozooids very small, not exceeding 0.5 x 0.3 mm; prominent paired projections beside the orifice; basal surface with a central uncalcified window (Fig. 145E); on red algae, Channel Islands southwards

.... ***Haplopoma bimucronatum*** (p. 316)

Autozooids larger, at least 0.6 x 0.3 mm; no projections beside the orifice (or, at most, a pair of minute "ears"); basal surface not with a central uncalcified window; a common intertidal species, especially on stones

.... ***Haplopoma graniferum*** (p. 314)

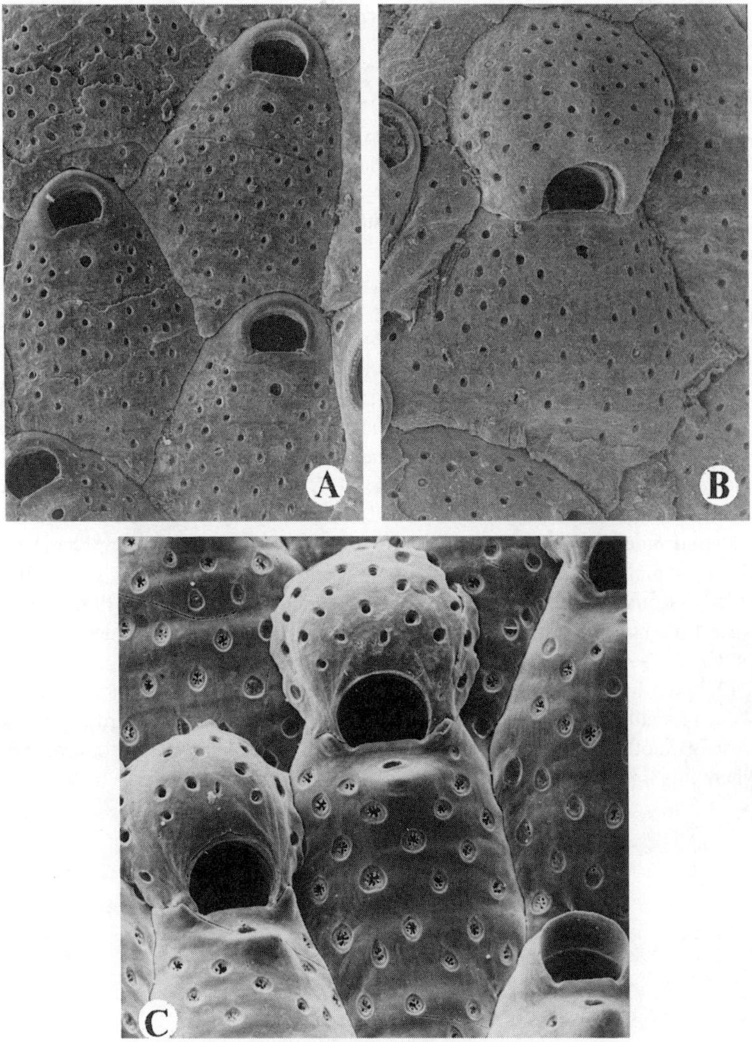

Fig. 141.
A, B, *Haplopoma planum*.
A, Group of autozooids; x60. B, An ovicellate zooid; x70.

C, *Haplopoma graniferum*; x80.

Haplopoma impressum (Audouin)
(Fig. 142)

Flustra impressa Audouin 1826: 240, pl. 10, fig. 6
Microporella impressa var. *glabra* Hincks 1880: 214, pl. 26, fig. 10
Haplopoma impressum : Hayward & Ryland 1979: 230, fig. 98

Colony encrusting, typically forming small vitreous patches. Autozooids of one type, hexagonal-oval, often tapering proximally, separated by sulci; 0.5-0.55 x 0.3-0.35 mm. Frontal calcification thin, translucent (opaque when dry), convex; with or without a small umbo proximal to the ascopore. Ascopore round, about 30 µm diameter, not joined to the orifice by a clearly visible suture. Frontal pores few, scattered, without a distally directed suture, the lumen simple (i.e., not "stellate" as in Fig. 145G). Orifice D-shaped; proximal margin straight, without condyles; about 70 x 100 µm. Pore-chambers blister-like, visible from the front; a pair or triad distally and three, sometimes four, on each side; the external pore oval, situated distally in its chamber (Fig. 142C). Ovicellate autozooids similar to others but with orifice larger, about 80 x 120 µm. Ovicells subglobular, with a few scattered pores, sometimes keeled; with about seven peripheral pore chambers. Embryos pink. Ancestrula oval, non-porous; orifice sinusoid; anter subcircular, poster (sinus) wide, the two parts separated by lateral cusps (Fig. 142D). Ancestrula budding a distal triad of autozooids; early astogeny symmetrical (Fig. 142E).

Embryos have been found throughout the year. Found most typically on red algae in shallow water just offshore. Recorded in Britain southwards from the west of Scotland; not known in the north-east. Also from the Mediterranean, where it is found on the sea-grass, *Posidonia*, and from the Red Sea.

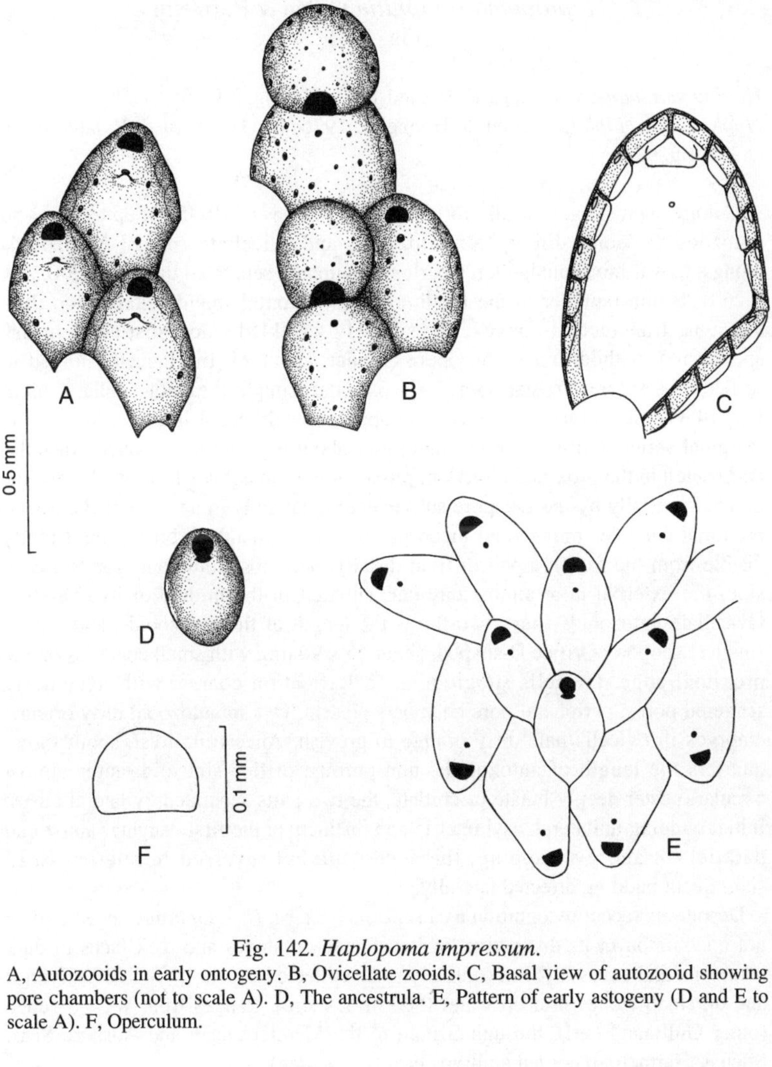

Fig. 142. *Haplopoma impressum.*
A, Autozooids in early ontogeny. B, Ovicellate zooids. C, Basal view of autozooid showing pore chambers (not to scale A). D, The ancestrula. E, Pattern of early astogeny (D and E to scale A). F, Operculum.

Haplopoma sciaphilum Silén & Harmelin
(Fig. 143)

Haplopoma impressum [in part]: Ryland 1963c: 11, fig. 2 J only
Haplopoma sciaphilum Silén & Harmelin: 1976: 61; Hayward & Ryland 1979: 232, fig. 99

Colony encrusting, typically forming small patches of 10-25 but up to about 50 autozooids. Zooids dimorphic: autozooids and ovicellate zooids. Autozooids elongate, oval-hexagonal, often tapering proximally, separated by sulci, 0.5-0.55 x 0.26-0.29 mm (smaller in the Mediterranean). Frontal shield calcification thin, vitreous, translucent; convex, transversely wrinkled (most distinctly in dry specimens); without umbo. Ascopore circular, about 25 μm diameter, joined to orifice by a suture. Frontal pores large, circular, simple (i.e., not "stellate" as in Fig. 145G), of uniform size, placed at approximately equal intervals in a single marginal series. Orifice shortly linguiform, about 90 x 90 μm; rounded distally, constricted in the proximal direction; proximal margin convex towards the orifice, divided centrally by the ascopore suture, ending laterally in acute-angled corners; proximal condyles minute and inconspicuous. Pore chambers blister-like (faintly visible from the front); a pair or triad distally, and most commonly four on each side; the external pore almost circular, situated in the middle of its chamber. Ovicellate autozooids short, two-thirds the length of the autozooids, and with a smaller ascopore. Orifice D-shaped, about 55 x 90 μm, with small condyles on the proximal edge. Ovicells subglobular, calcification coarse, with irregularly scattered pores; peripheral pore chambers present. Ova in autozooid dirty orange; embryos in ovicells pale dirty orange to greyish. Ancestrula oval, about three-quarters the length of autozooids, non-porous; orifice sinusoid: anter almost circular, poster deeply lunate or cordate, the two parts separated by lateral cusps. Initial budding unilateral, asymmetric and distinctive: the first daughter autozooid parallel to, and overlapping, the ancestrula but reversed in relation to it; subsequent budding directed laterally.

Despite its recent recognition as a separate species, *H. sciaphilum* appears to be not uncommon in its distinctive habitat. Found on stones and rock faces in dark situations and caves to 40 m (unlikely to be collected except by SCUBA-diving); and on small stones in deep water (500 m or more). Ranges from West Sweden (outer Gullmar Fjord), through Britain to the Mediterranean and Adriatic Seas. Silén & Harmelin recorded embryos in July (Sweden).

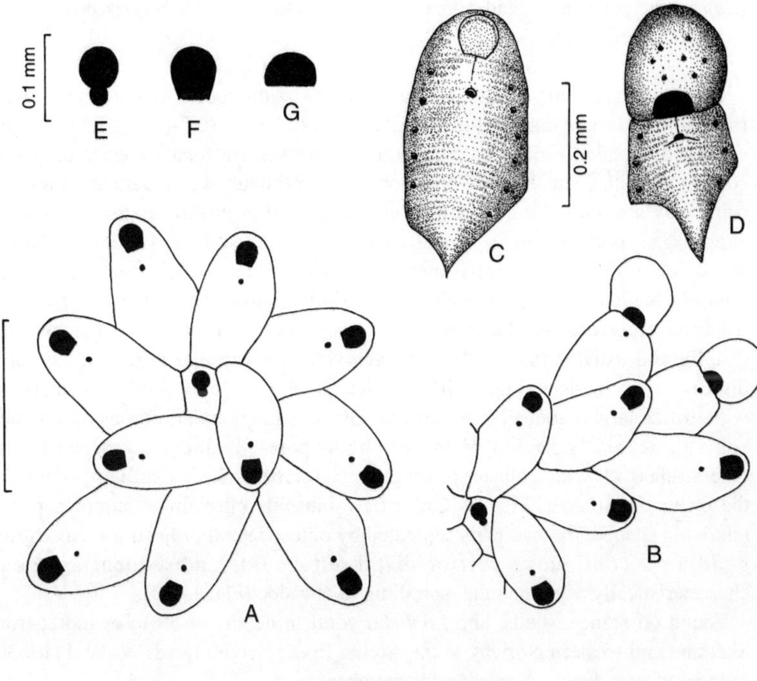

Fig. 143. *Haplopoma sciaphilum.*
A, Small colony to show special early astogeny and partial overgrowth of ancestrula. B, Part of colony including ancestrula and ovicellate zooids. C and D, Autozooid and ovicellate female zooid (drawn at twice standard scale). E, F and G, Orifices of ancestrula, autozooid and ovicellate female zooid. (Specimens from Abereiddy, Pembrokeshire, Wales).

Haplopoma planum Ryland
(Figs 141A, B; 144)

Haplopoma impressum: Marcus 1940: 216, fig. 111; Kluge 1962: 520, fig. 363; 1975: 632

Haplopoma planum Ryland 1963c: 14, figs 1 D-F, G, 2 G; Hayward & Ryland 1979: 234, fig. 100

Colony encrusting, forming small patches. Autozooids of one type, oval-rhomboidal, larger than in other species of the genus (0.7-0.9 x 0.5-0.6 mm), separated by sulci. Frontal calcification thin, translucent (opaque when dry), flat, sometimes with a small umbo. Ascopore circular, about 28 μm diameter, joined to orifice by a suture. Frontal pores numerous, with a distally extending suture, simple (i.e., not "stellate"), spread over most of the surface. Orifice D-shaped, about 100 x 135 μm; proximal margin straight, with a condyle near each of the obtusely angled corners. Pore chambers faintly visible and blister-like from the front; cavernous in the thickness of the wall when seen basally; a pair or triad distally and usually four to five on each side; the external pore oval, situated distally in its chamber (Fig. 144B, E). Ovicellate autozooids similar to others but with orifice larger, about 120 x 160 μm. Ovicells subglobular, porous throughout, sometimes faintly keeled; with peripheral pore-chambers. Embryos white. Ancestrula oval, with a cluster of small pores (perhaps closing with age) joined to the orifice by sutures (Fig. 144C); orifice sinusoid: anter almost circular, poster (sinus) U-shaped, the two parts separated by obtuse lateral projections. Ancestrula budding a contiguous pair of distal autozooids; subsequent astogeny characteristically asymmetrical, spiralling to one side (Fig. 144D).

Found on stones, shells, and *Lophelia* coral in depths of 50 m or more; from Shetland and western Norway to the Arctic. In the Bergen fjords, Ryland (1963b) found embryos during August and September.

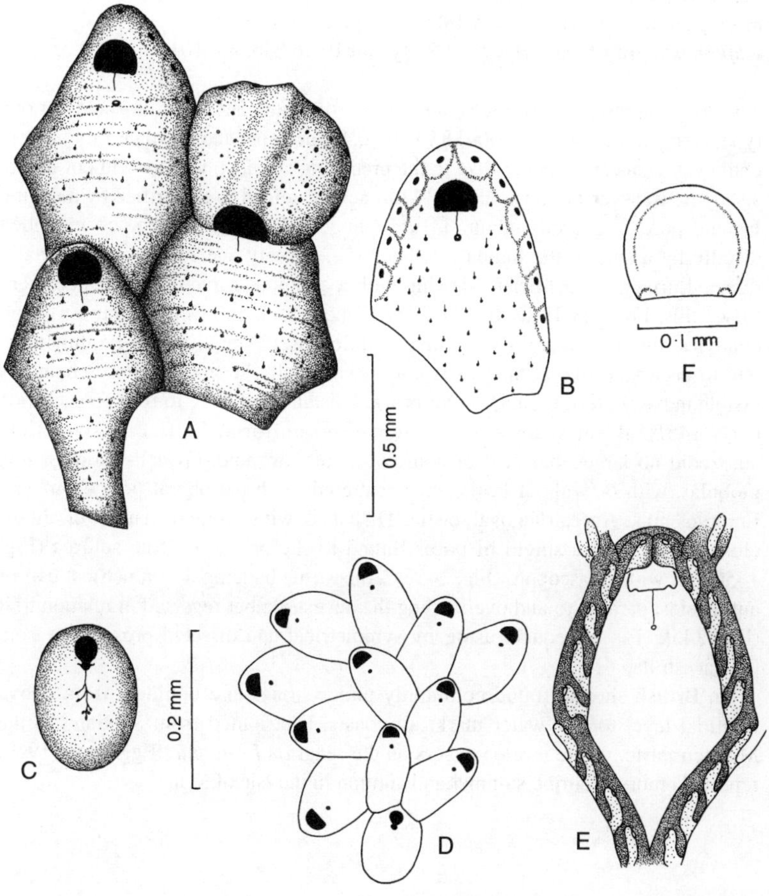

Fig. 144. *Haplopoma planum*.
A, Group of zooids. B, Autozooid in frontal view showing appearance of pore chambers. C, The ancestrula. D, Pattern of early astogeny. E, Autozooid in basal view, showing detail of pore chambers (not to scale A). F, Operculum. (Specimens from Bergen Fjord).

Haplopoma graniferum (Johnston)
(Figs 141C; 145)

Lepralia granifera Johnston 1847: 309, pl. 54, fig. 7
Microporella impressa: Hincks 1880: 214, pl. 26, figs 9, 11
Haplopoma graniferum: Hayward & Ryland 1979: 236, fig. 101

Colony encrusting, forming small vitreous or white patches. Autozooids of one type, elongate-hexagonal, 0.60-0.85 x 0.2-0.4 mm, separated by sulci. Frontal wall convex, translucent (thicker than in the preceding species), with or without umbo. Ascopore transversely oval, about 30 μm across, joined to the orifice by a suture. Frontal pores numerous, about 14 μm (range 9-19 μm) in diameter, extended distally by a suture; the lumen "stellate" (i.e., largely occluded by centripetally directed, triangular denticles: viewing with a compound microscope is necessary to see this, Fig. 145G). Orifice D-shaped, about 95 x 140 μm, without proximal condyles, its distal margin sometimes a little raised; small, lateral prominences, one to each side like minute ears, may occasionally be present. Pore chambers visible in basal view, tubular; a pair or triad distally and three to four on each side (Fig. 145I). Basal surface of the autozooid uniform. Orifice of ovicellate autozooid no larger than that of non-ovicellate autozooid. Ovicells more or less globular, with or without keel; pores scattered; with peripheral pore chambers. Embryos pink. Ancestrula oval; orifice D-shaped, with a minute, narrow or almost closed sinus, and a shield of pores linked to the orifice by fine sutures (Fig. 145C); or with an ascopore (Fig. 145F). Ancestrula budding a symmetrical pair of autozooids, parallel to and overlapping the ancestrula but reversed in relation to it (Fig. 145E, F); subsequent astogeny symmetrical and directed proximally from the ancestrula.

On British shores, found commonly under stones and boulders from above midtidal level to low water mark; all coasts. Distributed from Shetland to the Mediterranean, where it often occurs on the seagrass *Posidonia*. Eggleston (1969) reported embryos during summer and autumn in the Isle of Man.

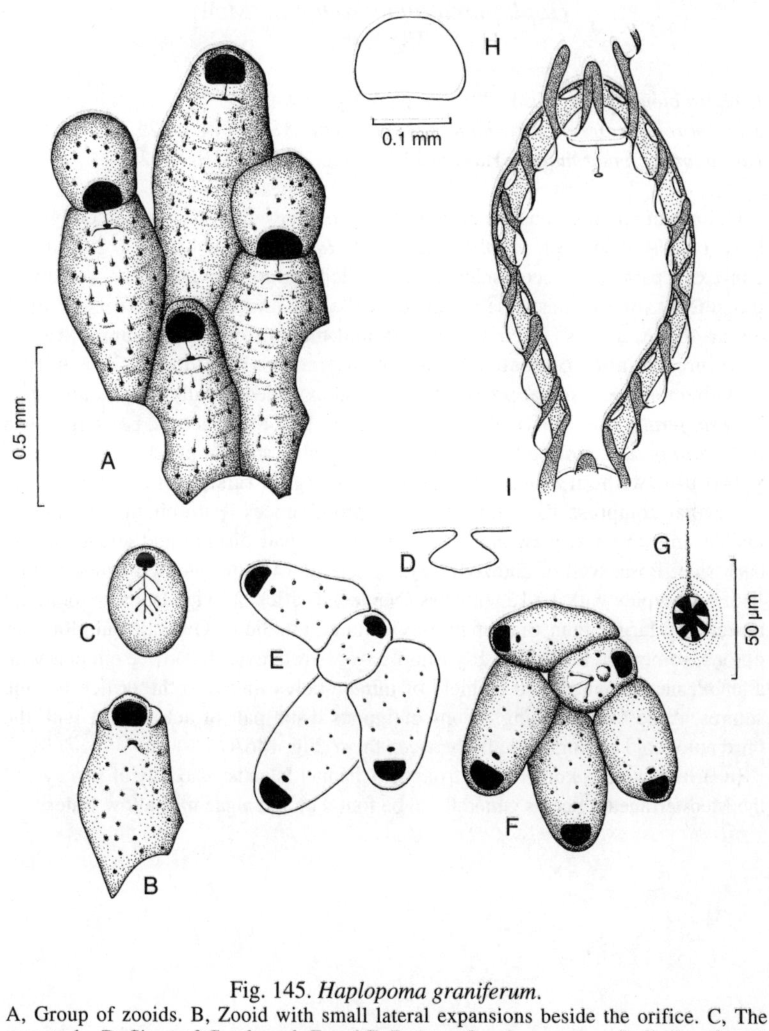

Fig. 145. *Haplopoma graniferum*.
A, Group of zooids. B, Zooid with small lateral expansions beside the orifice. C, The ancestrula. D, Sinus of C enlarged. E and F, Pattern of early astogeny. G, Stellate frontal pore. H, Operculum. I, Autozooid in basal view to show detail of pore chambers (E and F drawn from South Pembrokeshire specimens).

Haplopoma bimucronatum (Moll)
(Fig. 146)

Eschara bimucronata Moll 1803: 65, pl. 4, fig. 18A-C
Microporella impressa var. ?*bimucronata*: Hincks 1880: 214, pl. 29, figs 10-11
Haplopoma bimucronatum: Hayward & Ryland 1979: 238, fig. 102

Colony small, often uneven or nodular. Autozooids all of one type, more or less hexagonal, smaller than in other species of *Haplopoma* (0.4-0.5 x 0.2-0.3 mm); convex, separated by deep sulci. Frontal calcification thick, vitreous, somewhat translucent, often transversely ridged, umbonate. Ascopore transversely oval, about 40 µm across; often situated on and hidden by a lunate umbo, facing proximally; not connected with the orifice by a suture visible with a stereomicroscope. Frontal pores "stellate" and extended distally by a suture (as in *H. graniferum*, Fig. 145G); about 19 µm (range 14-24 µm) across, i.e., larger than in *H. graniferum*. Orifice D-shaped, curved distally, straight proximally, about 85 x 140 µm; without condyles; its lateral margins raised into conspicuous, somewhat compressed, distally-directed prominences. Pore chambers tubular, visible only in basal view and then indistinct; a pair distally and about four on each side. Basal wall of autozooid with a central, oval, uncalcified window (Fig. 146E). Polypide with 10-12 tentacles (Naples). Orifice of ovicellate autozooid not noticeably larger than that of non-ovicellate autozooid. Ovicells subglobular, porous, somewhat keeled. Embryos pink. Ancestrula oval; its orifice circular with a small, narrow sinus, and a shield of minute pores linked to the orifice by fine sutures. Ancestrula budding a non-contiguous distal pair of autozooids, with the third autozooid arising medially between them (Fig. 146A).

In British waters, known only from the Channel Islands; also Bay of Biscay and the Mediterranean. Seems generally to be found on red algae in shallow water.

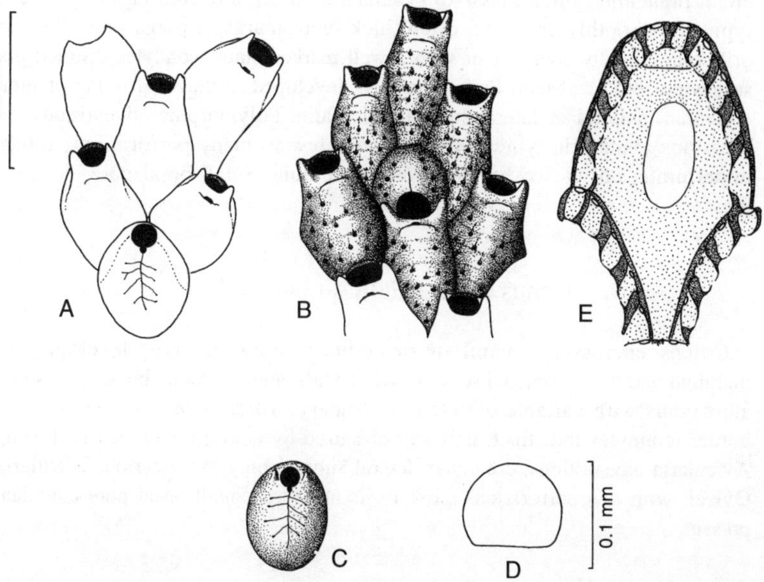

Fig. 146. *Haplopoma bimucronatum*.
A, The ancestrula and periancestrular zooids. B, A group of autozooids, one ovicellate. C, The ancestrula. D, Operculum. E, Autozooid in basal view showing non-calcified central area and detail of pore chambers. (Specimen A from Jersey, Channel Islands).

Superfamily CELLEPOROIDEA Johnston, 1838

Family CELLEPORIDAE Johnston, 1838

Colony encrusting, multilaminar, typically domed, mamillate or nodular; or erect, branching, often massive. Frontal shield of autozooid cryptocystidean, typically smoothly calcified, often thick, with marginal pores only. Primary orifice proximally concave, or with a well marked sinus; condyles distinct; oral spines present or absent. Peristome well developed, usually with one or more columnar, suboral or lateral suboral avicularia. Polymorphic adventitious and vicarious avicularia typical. Ovicell with few to many perforations; rarely, imperforate; not closed by autozooid operculum. Small basal pore chambers present.

Genus CELLEPORA Linnaeus, 1767

Colony encrusting, mamillate or nodular, often massive, developing by marginal budding of ordered series of autozooids, and by frontal budding of single autozooids with variable orientation. Primary orifice with concave proximal border, condyles indistinct, partially obscured by a well developed peristome. Avicularia adventitious, columnar, lateral suboral only; no vicarious avicularia. Ovicell with few perforations close to its aperture. Small basal pore chambers present.

Type species: *Millepora pumicosa* Pallas, 1766

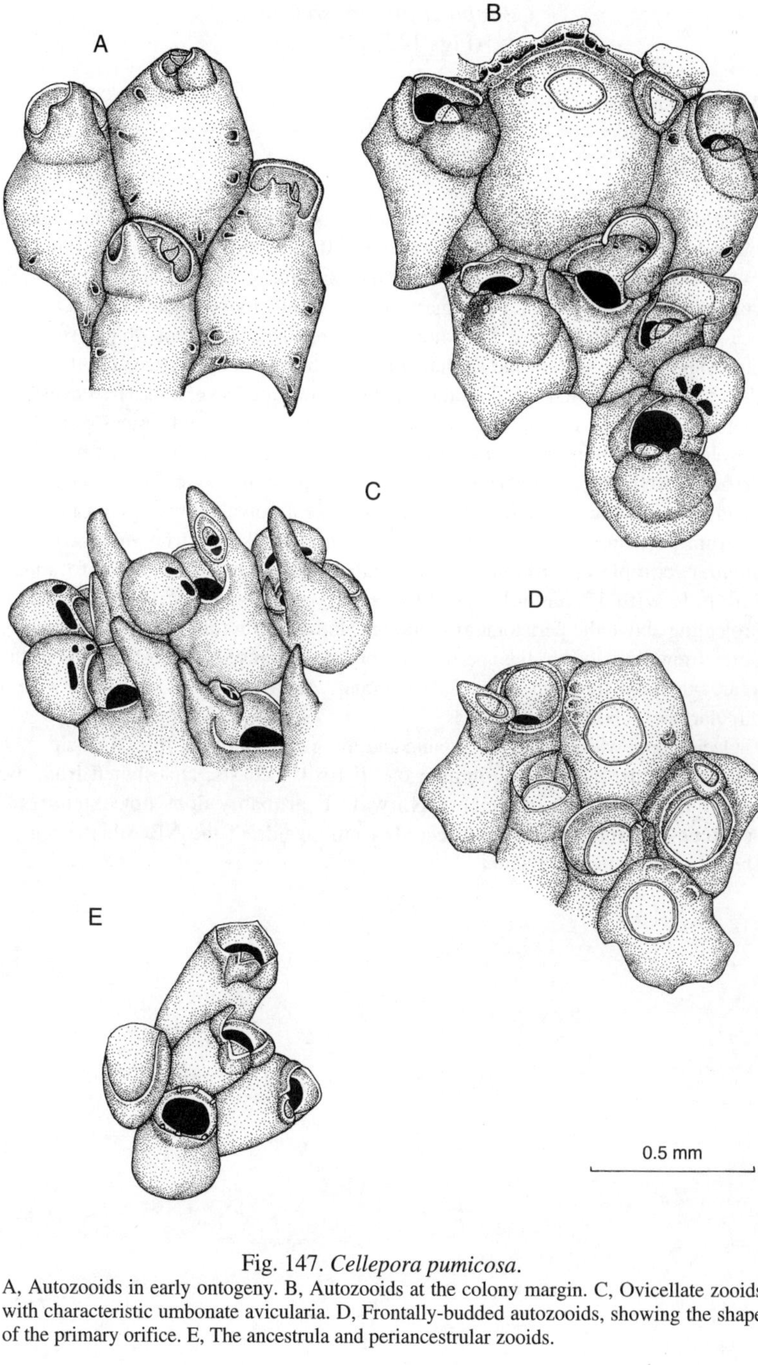

Fig. 147. *Cellepora pumicosa.*
A, Autozooids in early ontogeny. B, Autozooids at the colony margin. C, Ovicellate zooids with characteristic umbonate avicularia. D, Frontally-budded autozooids, showing the shape of the primary orifice. E, The ancestrula and periancestrular zooids.

Cellepora pumicosa (Pallas)
(Figs 147; 148A, B)

Millepora pumicosa Pallas 1766: 254
Cellepora pumicosa: Hincks 1880: 398, pl. 54, figs 1-3; Hayward & Ryland 1979: 274, fig. 119

Colonies developing as lenticular patches, becoming domed or nodular, often massive, may exceed 20 mm diameter; often forming stout, multilaminar cylinders around erect substrata. Decumbent autozooids oval to hexagonal, convex, with distinct boundaries marked by shallow grooves; commonly 0.5-0.7 x 0.3-0.5 mm; superposed laminae consist entirely of erect or partly erect autozooids, with little of the frontal shield visible. Primary orifice slightly wider than long, the proximal edge broadly and shallowly concave; condyles developed merely as proximo-lateral thickenings of the orifice rim. No oral spines. Peristome developed as a thin, raised rim around the lateral and proximal borders of the orifice; proximally it incorporates a tapered, columnar avicularian cystid, the distal tip of which is frequently developed as a conspicuous spiked process; rostrum perpendicular to orifice, facing laterally, bluntly triangular, crossbar slender, complete, without a columella; palate with an extensive foramen. Polypide with 17 tentacles (Calvet, 1900). Ovicell prominent, spherical, projecting above the peristome rim, thinly calcified, with four to six variably sized pores frontally, close to the aperture. Embryos red, present in most months of the year, but most frequent in the late summer. Ancestrula oval, tatiform, with a circular opesia bordered by spines.

Occurs on a wide variety of organic and inorganic substrata, from the kelp zone offshore to about 100 m. Common on all British coasts. Distributed from the Mediterranean north to western Norway. It probably does not extend into subarctic or Arctic waters, and records from outside of the NE Atlantic region should be regarded with doubt.

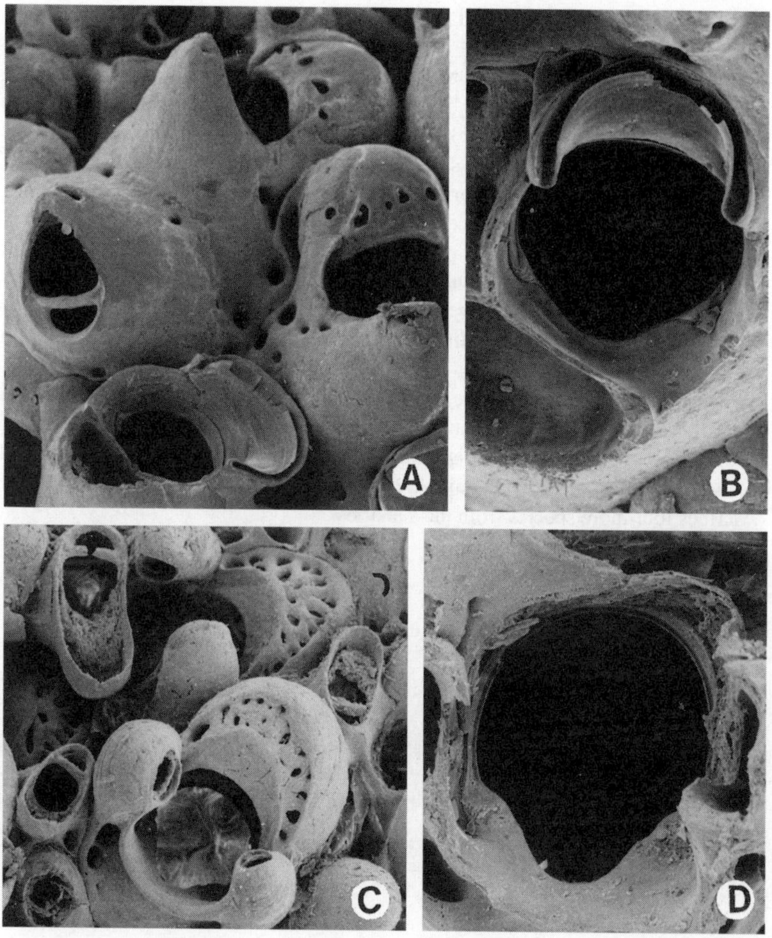

Fig. 148.
A, B, *Cellepora pumicosa* (Lewis).
A, Ovicellate zooids and well-developed avicularia; x90. B, The primary orifice; x160.

C, D, *Celleporina hassallii* (Guernsey).
C, Ovicellate zooids and vicarious avicularia; x100. D, The primary orifice; x200.

Genus CELLEPORINA Gray, 1848

Colony encrusting, lenticular to mound-like, pisiform, nodular or cylindrical; multilaminar; autozooids recumbent at colony margins, suberect, with varying orientation, elsewhere. Frontal shield cryptocystidean, with few, conspicuous marginal pores. Primary orifice sinuate, typically flanked by paired (rarely, 1-3) columnar, adventitious avicularia; peristome variably developed, oral spines absent. Vicarious avicularia present. Ovicell hyperstomial, prominent, not closed by autozooid operculum; with a perforated area of entooecium (tabula) frontally, bordered by a thickened rim. Small basal pore chambers present.

Type species: *Lepralia hassallii* Johnston, 1847

Four species are present in the British fauna. *C. hassallii* var. *tubulosa* is here regarded as a distinct species, and *Lagenipore pygmaea* (Norman) is also recognised as belonging to *Celleporina.* Additional species are described for the western Mediterranean by Zabala & Maluquer (1988), while Aristegui (1989) describes four more species from the Canaries.

Key to the British species of *Celleporina*

1. Peristome developed as a concave rim between the prominent lateral oral avicularia, the orifice remaining distinct. Vicarious avicularia frequent
   .... ***Celleporina hassallii*** (p. 324)

   Peristome a tall tube, hiding the orifice, or with a stout, prominent, medio-proximal mucro
   .... **2**

2. Peristome with a thick, medio-proximal mucro. Lateral oral avicularia relatively large, constant. Sinus a narrow U-shape, occupying only half proximal width of orifice
   .... ***Celleporina decipiens*** (p. 326)

   Peristome tall, cylindrical, without a thickened mucro. Sinus broad and shallow, occupying most of proximal edge of orifice
   .... **3**

3. With a pair of conspicuous, slender lateral oral avicularia projecting above peristome rim. Vicarious avicularia frequent. Tabula with large, conspicuous pores
   .... ***Celleporina tubulosa*** (p. 330)

   One (rarely two) tiny lateral oral avicularium on peristome rim. No vicarious avicularia. Tabula with few indistinct pores
   .... ***Celleporina pygmaea*** (p. 328)

Celleporina hassallii (Johnston)
(Figs 148C, D; 149)

Lepralia hassallii Johnston 1847: 304, pl. 14, fig. 3
Cellepora costazii: Hincks 1880: 411, pl. 55, figs 11-13
Celleporina hassallii: Hayward & Ryland 1979: 276, fig. 120

Colony a rounded to irregular mound on flat substrata, or a substantial pisiform, cylindrical or nodular growth on erect substrata; often exceeding 5 mm diameter. Primary orifice slightly longer than wide, with a deep, U-shaped sinus occupying whole width of proximal edge, and constituting almost one-third total orifice length; condyles scarcely developed. Lateral oral avicularia paired; rostrum oval, almost perpendicular to orifice plane, its long axis directed proximo-laterally; crossbar complete, without a columella. Peristomial rim present proximally between the avicularia but deeply concave, so that the orifice remains visible. Frontal shield smooth, with few, large marginal pores. Vicarious avicularia frequent, often abundant; rostrum 0.2-0.35 mm long, narrowly spatulate; crossbar complete, without a columella, palate with a large foramen. Ovicell prominent, hemispherical; tabula closely perforated by two rows of large, irregular pores; proximal rim of tabula typically thickened and peaked medially. Embryos red, present in most months of the year, with a peak in late summer. Ancestrula tatiform; opesia oval, bordered by about seven spines.

Lower shore and sublittoral, on rocks, shells, hydroids and erect bryozoans, and often common on kelp holdfasts. Widespread, occurring commonly on all British coasts. Distributed in the temperate northeast Atlantic from western Norway to southwest Europe, and throughout the Mediterranean.

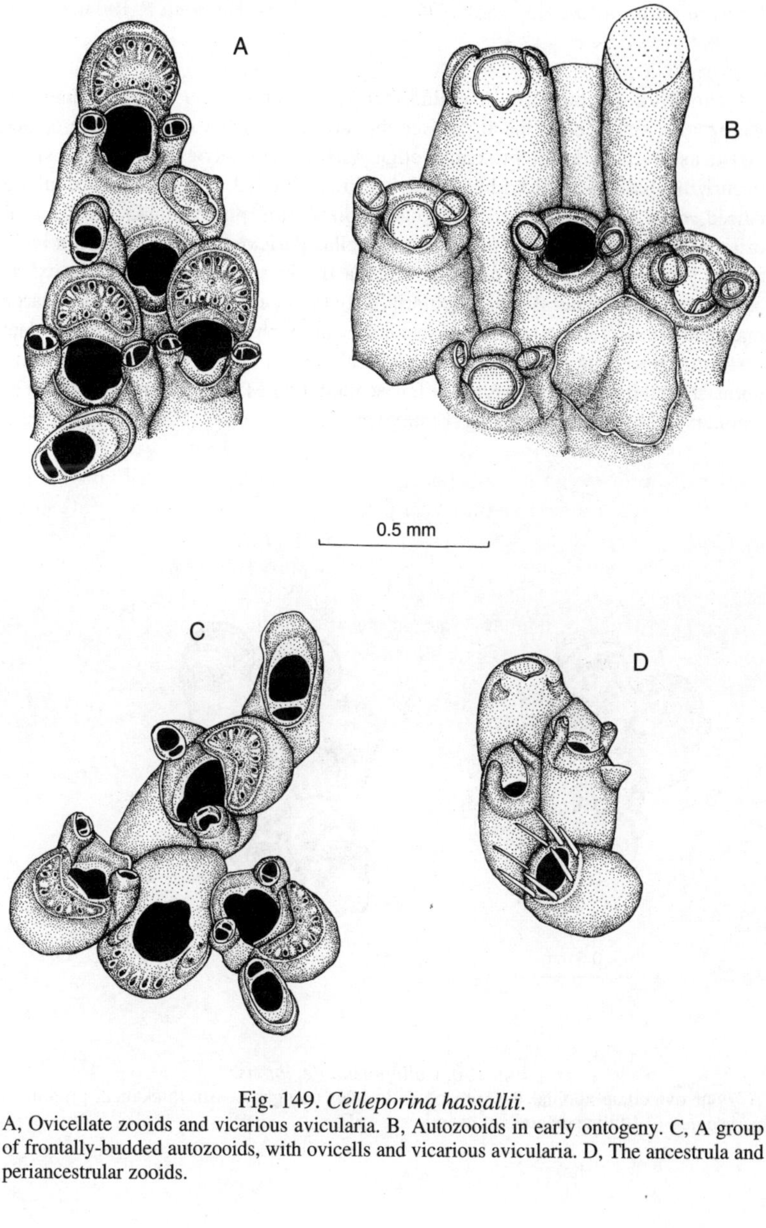

Fig. 149. *Celleporina hassallii.*
A, Ovicellate zooids and vicarious avicularia. B, Autozooids in early ontogeny. C, A group of frontally-budded autozooids, with ovicells and vicarious avicularia. D, The ancestrula and periancestrular zooids.

Celleporina decipiens Hayward
(Figs 150; 151A, B)

Celleporina decipiens Hayward 1976: 325, fig. 4A, B; Hayward & Ryland 1979: 278, fig. 121

Colony domed, pisiform or cylindrical, rarely larger than 2-3 mm diameter. Autozooids small, erect distal portion commonly 0.3 x 0.2 mm. Primary orifice about as wide as long, proximal edge with a symmetrical U-shaped sinus occupying half its total width; condyles low, rounded. Lateral oral avicularia paired; rostrum oval, with finely toothed distal rim, perpendicular to plane of orifice; crossbar slender, without a columella. Peristome entire, comparatively thick, produced medio-proximally into a thick, rounded mucro, almost as prominent as the lateral oral avicularia. Frontal shield smooth, with few, large marginal pores. Vicarious avicularia sporadic, often absent; small, with a spatulate rostrum. Ovicell more or less hemispherical; tabula with a single row of small pores around its distal border, and a few scattered pores medially; proximal rim of tabula typically peaked medially. Embryos red.

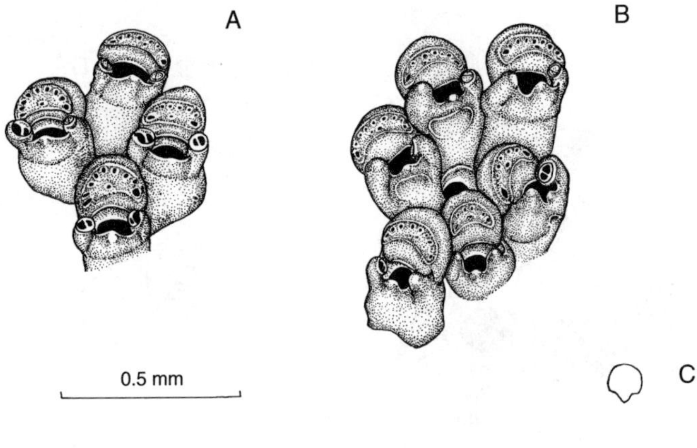

Fig. 150. *Celleporina decipiens*.
A, Four ovicellate zooids. B, Autozooids in later ontogeny, with thickened, prominent peristomes. C, Outline diagram of the primary orifice.

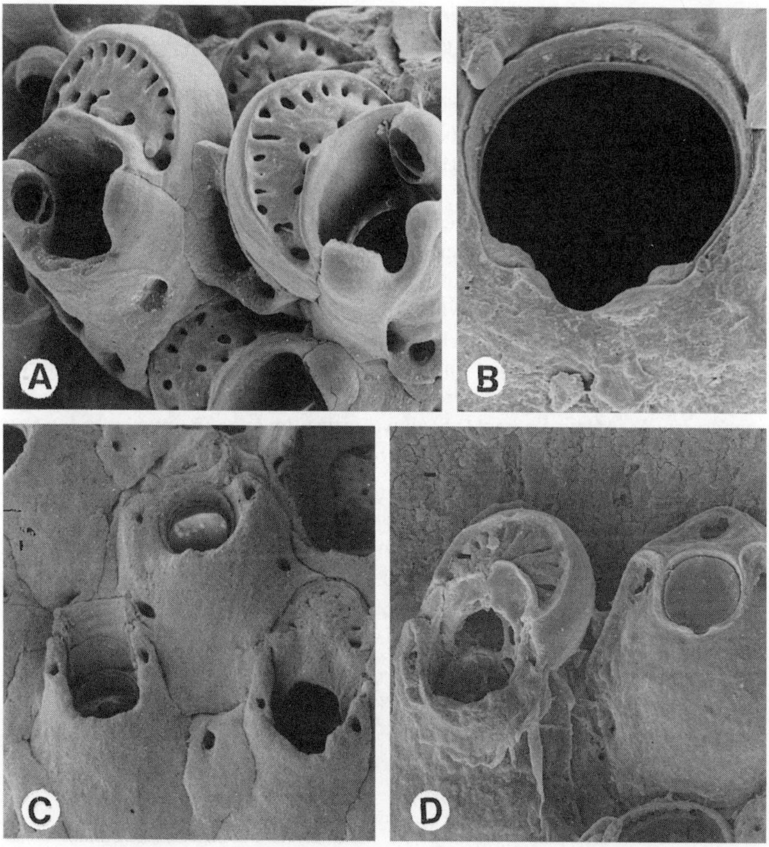

Fig. 151.
A, B, *Celleporina decipiens* (Lundy).
A, Ovicellate zooids; x150. B, The primary orifice; x450.

C, D, *Celleporina pygmaea*.
C, Autozooids in late ontogeny; x120.
D, Autozooids at the colony margin, showing the ovicell and the primary orifice; x120.

This inconspicuous species is found typically on hydroid stems or on erect bryozoans. It occurs sublittorally, from 10 to at least 50 m, and has been collected from around the Channel Islands, Scilly, Lundy, South Uist and St. Kilda. Zabala & Maluquer (1988) figured specimens from an unrecorded western Mediterranean locality.

Celleporina pygmaea (Norman)
(Figs 151C, D; 152)

Celleporella pygmaea Norman 1868: 308; Hincks 1880: 415
Lagenipora pygmaea: Ryland 1963b: 30; Hayward & Ryland 1979: 282, fig. 123
Celleporina pygmaea: Hayward 1994: 204, fig. 13c

Colony small, white, thickly calcified, developing a rounded or lobed patch rarely exceeding 3-4 mm diameter. Autozooids small, oval, convex, separated by shallow grooves; 0.3-0.4 x about 0.2 mm. Primary orifice orbicular, 0.08 mm long; condyles minute, visible with SEM; tapered proximo-medially. Peristome tall, cylindrical, its rim slightly flared, produced into a number of short, blunt processes. One, or rarely two, tiny avicularia laterally on the peristome rim, the oval rostrum laterally directed. Frontal shield smoothly calcified, with few (*ca* 6) inconspicuous marginal pores. Ovicell wider than long, its aperture arched; tabula narrowly crescentic, with fine ridges between distinct distal pores. Embryos colourless.

Offshore, on stones, shells, and the coral *Lophelia*. Rarely reported. Recorded from below 50 m depth in western Norway (Ryland 1963b), and from Faroe and the Shetlands. This tiny species is easily confused with *Lagenipora lepralioides*, which is distinguished by its noticeably nodular calcification. Both occur with greatest frequency on small pebbles, and are invariably damaged by collection.

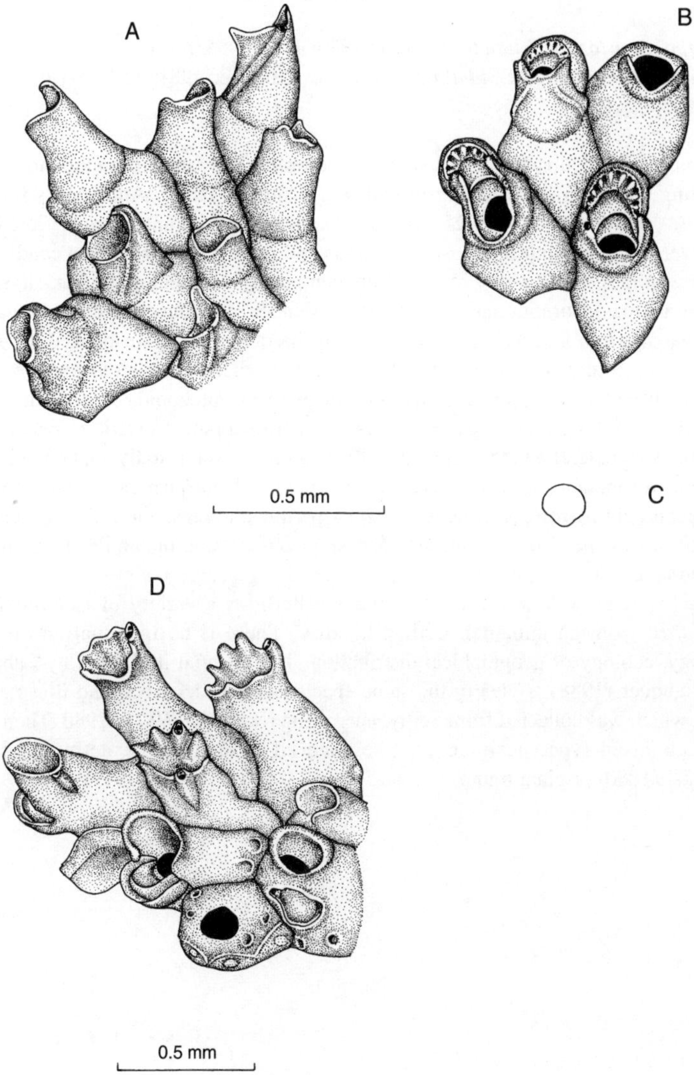

Fig. 152. *Celleporina pygmaea*.
A, A group of autozooids, with the peristomes typically of varying length. B, Ovicellate zooids. C, Outline diagram of the primary orifice. D, Autozooids at the colony margin, with largely undamaged peristomes.

Celleporina tubulosa (Hincks)
(Fig. 153)

Cellepora costazii var. *tubulosa* Hincks 1880: 412, pl. 55, fig. 13
Celleporina hassalli var. *tubulosa*: Zabala & Maluquer 1988: 159, text-fig. 441, pl. 26, fig. C

Colony a lobed, unilaminar sheet, developing low, multilaminar mounds; forming an irregular spreading growth 8 x 8 mm or larger. Autozooids flask-shaped, small, 0.4-0.5 x 0.25-0.3 mm, excluding peristome. Primary orifice transversely oval, with broad, shallow sinus and minute, inconspicuous condyles. No oral spines. Peristome 0.25 mm deep, completely surrounding orifice, slightly flared at rim; incorporating on each side a slender, columnar avicularium, each originating from a septulum disto-lateral to the orifice; rostrum oval, 0.06 mm long, with complete crossbar and finely toothed distal rim; acute to plane of orifice, directed obliquely distally. Frontal wall of autozooids finely nodular, vitreous, with few, widely spaced but distinct marginal pores. Vicarious avicularia broadly spatulate, 0.2-0.25 mm long, 0.12-0.15 mm broad; distally cupped, with a large oval foramen and a complete crossbar. Ovicell hemispherical, frontal tabula comparatively narrow, with large, irregular peripheral pores, and a few scattered frontal pores; peristome continues across proximal side of tabula, forming a closed tube.

This species was described by Hincks (1880), as a variety of *Celleporina hassallii*, from an unnamed British locality. There is no information on its biology, ecology or geographical distribution. The material illustrated by Zabala & Maluquer (1988) is clearly the same species as that described and illustrated here, which was collected from Scilly, encrusting rock at ELWS, in 1980. There is no recognised type specimen, but the material figured here exactly matches Hincks' (1880) original figure.

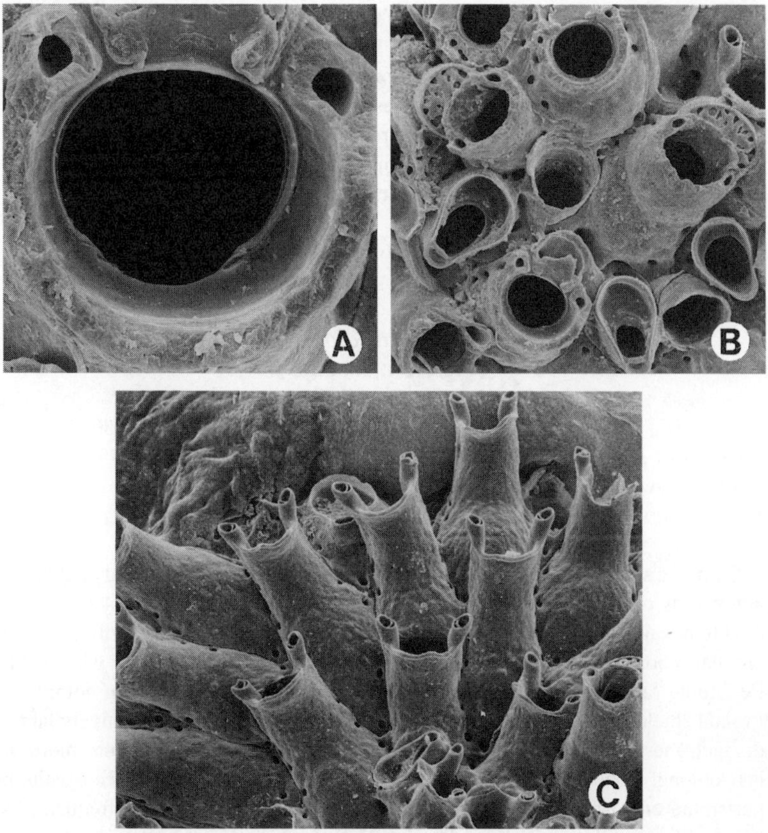

Fig. 153. *Celleporina tubulosa*.
A, The primary orifice; x250. B, Ovicellate zooids and vicarious avicularia; x60. C, Autozooids at the colony margin, with undamaged peristomes; x50.

Genus LAGENIPORA Hincks, 1877

Colony encrusting, unilaminar. Autozooid frontal shield with few marginal pores. Primary orifice orbicular; condyles indistinct, sinus poorly defined; no oral spines. Peristome well developed, entire, hiding orifice when fully developed. Ovicell hyperstomial, ectooecium incomplete frontally, entooecium imperforate. No avicularia. Large kenozooids sporadically present.

Type species: *Celleporella lepralioides* Norman, 1868

Lagenipora lepralioides (Norman)
(Figs 154; 155A, B)

Celleporella lepralioides Norman 1868: 222, pl. 7, figs 4,5; Hincks 1880: 414
Lagenipora lepralioides: Ryland 1963a: 32; Hayward & Ryland 1979: 280, fig. 122; Hayward 1994: 204, fig. 13A, B
Lagenipora socialis Hincks 1877: 215; 1880: 235, pl. 34, figs 7, 8

Colony a small, thick, white patch, 2-3 mm diameter; rounded, lobed or stellate, autozooids often in disorderly series. Autozooids very small, 0.3-0.4 mm long, oval to hexagonal, convex, separated by indistinct grooves. Primary orifice almost circular, about 0.08 mm diameter, with minute condyles visible only with SEM. Peristome tall, thick, obscuring orifice, its rim slightly flared when complete. Frontal shield thick, vitreous, nodular, bordered by three to seven relatively large, distinct, marginal pores. Ovicell slightly broader than long, its aperture more or less oblong; frontal entooecium smooth, bordered by a thick rim of ectooecium; peristome encloses ovicell laterally in later ontogeny. Kenozooids frequent, as large as autozooids, or much smaller, infilling spaces between autozooids. Earliest autozooids (? ancestrula) with proportionately longer peristome, its rim produced as slender, pointed processes

Offshore, on shell fragments and especially, small pebbles. Most specimens damaged on collection and thus not readily recognised. Geographical and bathymetric distribution incompletely known. It was described originally from Shetland; it has been recorded with certainty from the English Channel, the Channel Islands and the western Mediterranean, and from Faroe. Nothing is known of its biology or ecology.

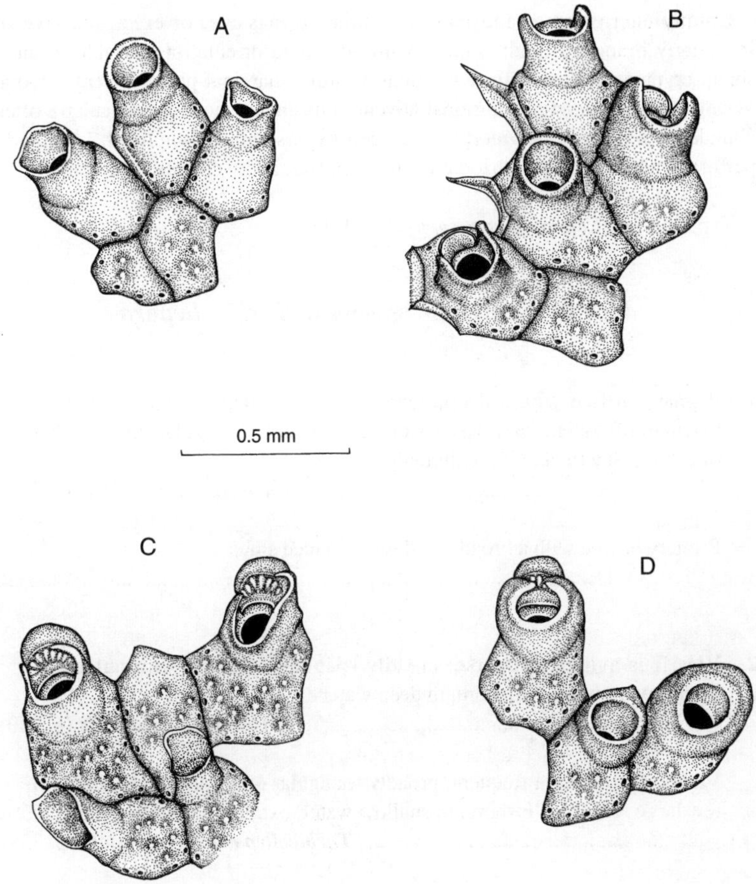

Fig. 154. *Lagenipora lepralioides*.
A, Three autozooids and two kenozooids. B, Autozooids at the colony margin, with developing ovicells. C, Ovicellate zooids. D, Autozooids in later ontogeny, with thickened peristomes.

Genus TURBICELLEPORA Ryland, 1963

Colony encrusting, mamillate to nodular or massive; or erect, massive or irregularly branched; multilaminar with autozooid orientation variable. Frontal shield cryptocystidean, with few, often indistinct marginal pores. Lateral suboral avicularia characteristic; additional adventitious and/or vicarious avicularia often abundant, frequently polymorphic. Ovicell hyperstomial, prominent, irregularly perforate, not closed by autozooid operculum. Basal pore chambers present.

Type species: *Cellepora coronopus* Wood, 1844

Key to the British species of *Turbicellepora*

1. Primary orifice with a deep, narrow U- or V-shaped sinus. Vicarious avicularia frequent, narrowly spatulate or broadly triangular, the crossbar of each with a thick, bifid columella
.... ***Turbicellepora avicularis*** (p. 336)

Primary orifice with a broad, shallow U-shaped sinus
.... **2**

2. Vicarious avicularia sparse, broadly spatulate. Ovicell with numerous small, round pores. Offshore, in deep water
.... ***Turbicellepora boreale*** (p. 340)

Vicarious avicularia frequent, broadly triangular or elliptical. Ovicell with few large foramina. Inshore, in shallow water, extreme southwest only
.... ***Turbicellepora magnicostata*** (p. 338)

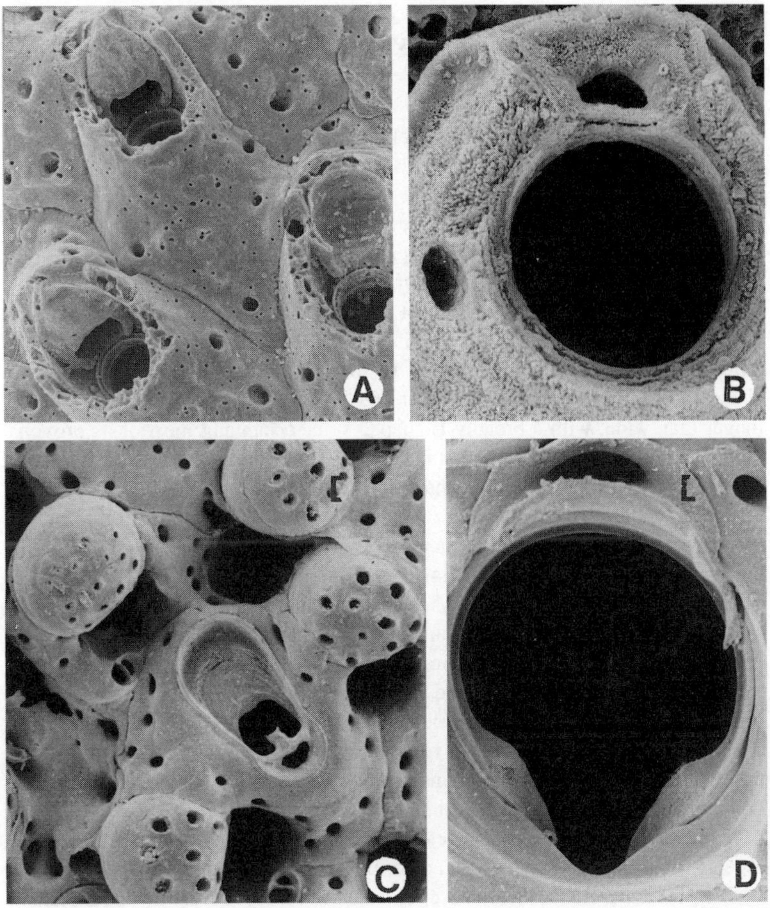

Fig. 155.
A, B, *Lagenipora lepralioides* (Faroes).
A, Ovicellate zooids; x90. B, The primary orifice; x400.

C, D, *Turbicellepora avicularis* (Guernsey).
A, Ovicellate autozooids and a vicarious avicularium; x80. B, The primary orifice; x400.

Turbicellepora avicularis (Hincks)
(Figs 155C, D; 156)

Cellepora avicularis Hincks 1860a: 278; 1880: 406, pl. 54, figs 4-6
Cellepora armata: Hincks 1880: 410, pl. 54, figs 10-13
Turbicellepora avicularis: Hayward & Ryland 1979: 284, fig. 124

Colonies domed, nodular or massive; often large, up to 60 x 30 mm, frequently as thick, multilaminar cylinders on erect substrata, especially large hydroids. Decumbent autozooids at colony margins oval to broadly hexagonal, convex, separated by distinct grooves; most erect or partly erect, with no common orientation, and with orifices at varying levels over the colony surface. Frontal shield smooth, finely granular, with distinct marginal pores, especially conspicuous around the distal ends of erect autozooids. Primary orifice slightly longer than wide, with a broadly U-shaped sinus occupying most of its proximal border; depth of sinus accentuated by thick, prominent condyles. No oral spines. Peristome developed laterally and proximally, incorporating a single lateral suboral avicularium, on the right or left; cystid low and rounded, or columnar, rostrum acute to frontal plane, facing laterally, elongate oval, hooked distally, with a complete crossbar and large palatal foramen. Similar adventitious avicularia sporadically present. Vicarious avicularia often abundant, on squat, swollen cystids; rostrum most frequently spatulate, 0.2-0.6 mm long, but often broadly triangular, with hooked tip; in all cases the palate has a short foramen, extending about half its length, and the crossbar is thickened with a stout, usually bifid, median columella. Polypide with 13-16 tentacles. Ovicell prominent, smooth, spherical, with about 10 round frontal pores. Embryos yellow, present throughout the year but most abundant in summer months. Ancestrula tatiform with about ten spines around a circular opesia.

From the lower shore, to at least 130 m. On a wide variety of substrata, especially shell, large hydroids and other bryozoans. Widespread and common off all British coasts. Ranges from Mediterranean to Faroe, N Norway and S Iceland. It probably does not occur in Arctic seas.

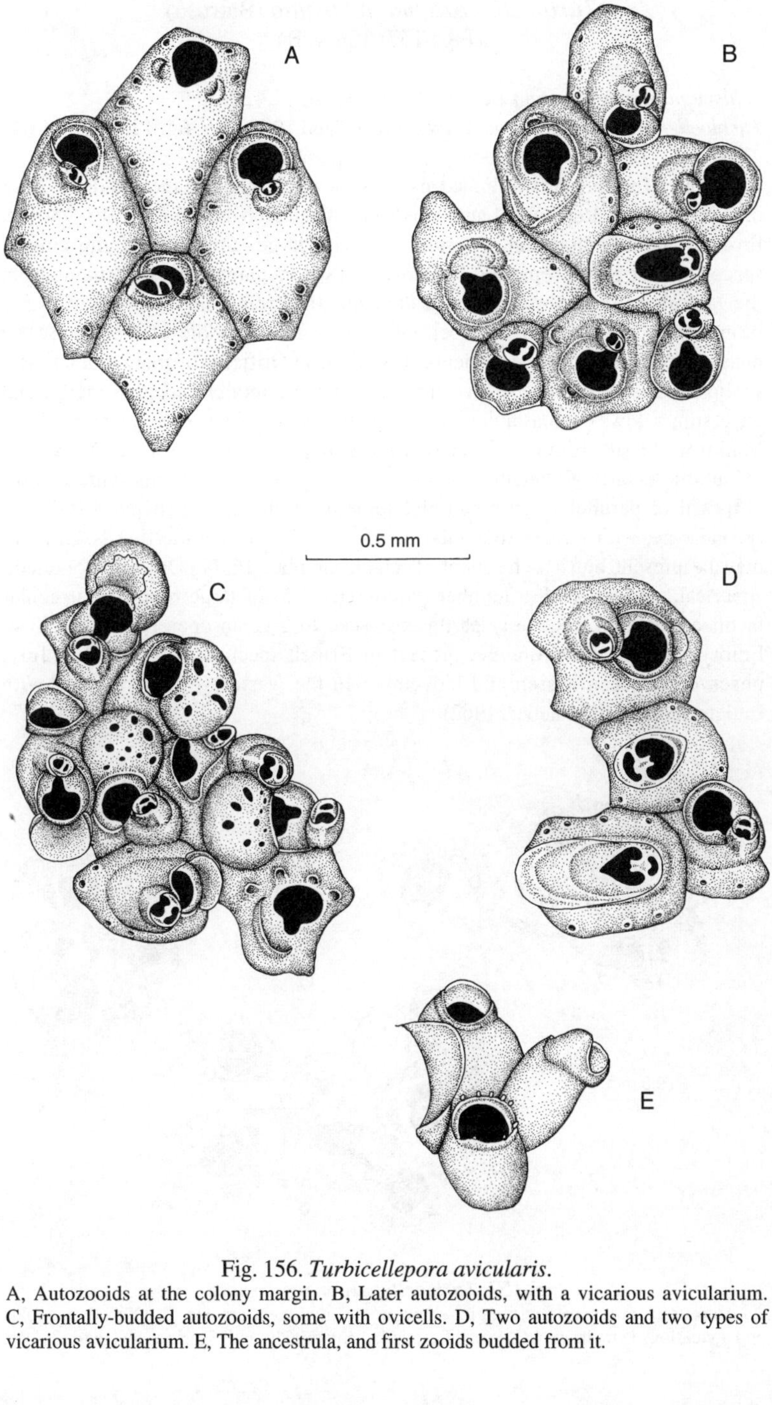

Fig. 156. *Turbicellepora avicularis*.
A, Autozooids at the colony margin. B, Later autozooids, with a vicarious avicularium. C, Frontally-budded autozooids, some with ovicells. D, Two autozooids and two types of vicarious avicularium. E, The ancestrula, and first zooids budded from it.

Turbicellepora magnicostata (Barroso)
(Figs 157; 158A, B)

Schismopora magnicostata Barroso 1919: 346, figs 23-32
Turbicellepora magnicostata: Hayward & Ryland 1979: 286, fig. 125

Colonies forming broad domed patches, or thick, nodular incrustations; deep orange in colour. Recumbent autozooids broad and convex, 0.5-0.6 x 0.3-0.4 mm. Frontal shield smooth and fine grained, with large, conspicuous and closely spaced areolar pores; separated by narrow, raised interareolar ribs extending from the lateral walls of the autozooid centripetally, frequently converging on a prominent suboral umbo. In erect autozooids in the middle of the colony the marginal pores and ribs are particularly prominent. Orifice orbicular, with a broad, shallow sinus. Peristome a low, thin rim, rising laterally and proximally, and enclosing a low, conical avicularium situated to the right or left of the sinus. Mandible broad, blunt and narrowed distally; apically directed. Vicarious avicularia usually abundant, 0.2-0.4 mm long; elongated, broad with a semi-elliptical or parallel-sided mandible, sometimes slightly narrowed distally. A second type, with a more triangular mandible, smaller than the first (<0.2 mm), may be present but less frequent. Tentacle number 16-17. Ovicell prominent, spherical, smooth, with a number (commonly 4-5) of large round, or irregular lacunae frontally, frequently partly coalesced to give an open reticulate effect. Embryos deep yellow-orange; present in British specimens collected in July; observed between March and November in the western Mediterranean, with settlement in April (Gautier, 1962).

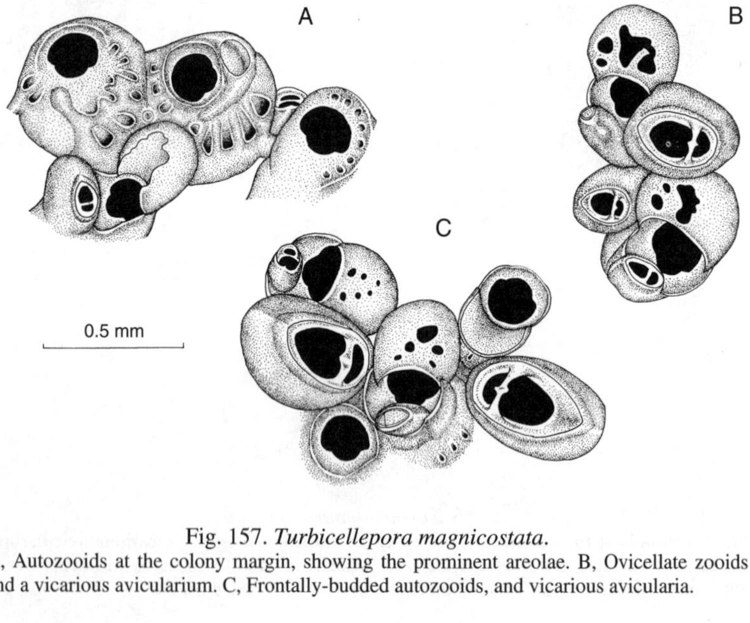

Fig. 157. *Turbicellepora magnicostata.*
A, Autozooids at the colony margin, showing the prominent areolae. B, Ovicellate zooids and a vicarious avicularium. C, Frontally-budded autozooids, and vicarious avicularia.

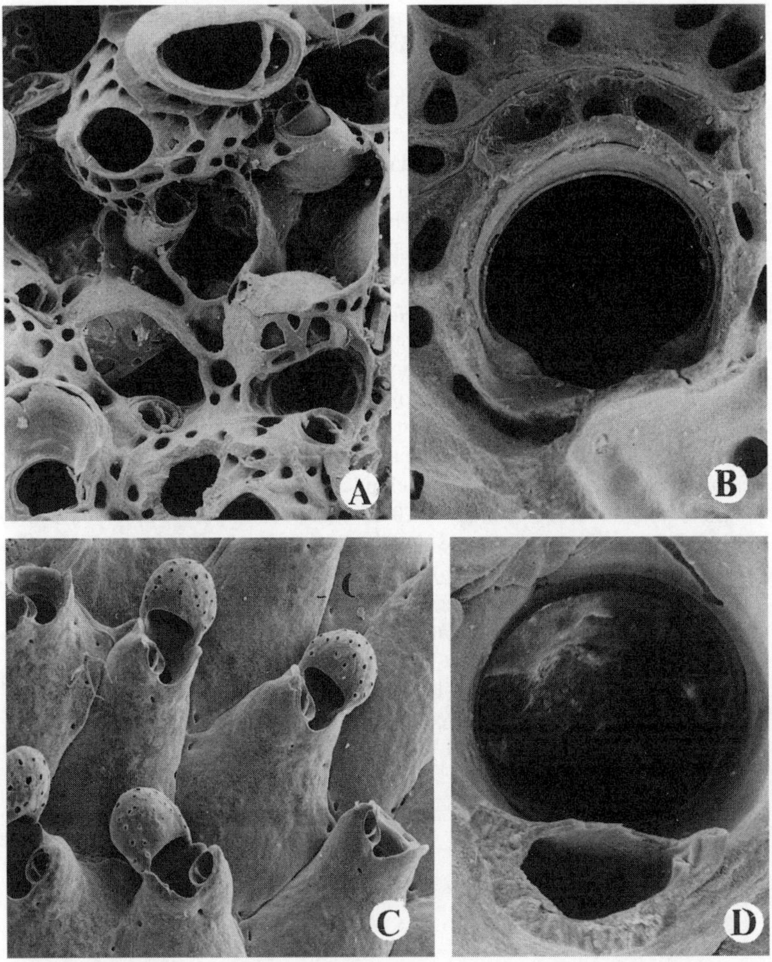

Fig. 158.
A, B, *Turbicellepora magnicostata*.
A, Group of autozooids, with ovicells and vicarious avicularia; x50. B, The primary orifice; x160.

C, D, *Turbicellepora boreale*.
C, Group of autozooids, some with ovicells; x40. D, The primary orifice; x200.

T. magnicostata is widespread throughout the Mediterranean, where it occurs typically from 0-50 m (Gautier, 1962), although specimens have been reported from 87 m off Oran. It encrusts a variety of hard substrata, forming large patches below boulders, and is often abundant on the stipes of algae. In British waters it has been recorded only from Scilly, where it constitutes a common, and obvious, member of the rocky shore community and is also frequent on sublittoral algae.

Turbicellepora boreale Hayward & Hansen
(Figs 158C, D; 159)

Turbicellepora smitti: Ryland 1963b: 34, fig. 11; Hayward & Ryland 1979: 288, fig. 126
Turbicellepora boreale Hayward & Hansen 1999:

Colony rising from an encrusting base to form massive or erect, branched growths, white. Autozooids large, visible to the naked eye, broad and convex, separated by distinct grooves; 0.8-1.1 x 0.6-0.7 mm. Frontal wall thick, smooth, fine grained; areolae small, widely spaced and inconspicuous. Primary orifice wider than long, a deep, broad U-shaped sinus occupying the majority of the proximal edge. Peristome low and thickened, not prominent, rising slightly one side of the orifice to meet the avicularian cystid. Oral avicularium squat, conical; rostrum elongate, triangular, hooked apically, directed laterally and steeply inclined to the frontal plane, mandible triangular. Vicarious avicularia infrequent, rostrum broadly spatulate or, rarely, elliptical. Ovicell prominent, broader than long, smooth-surfaced, with about 10-15 small round pores. Embryos white, appearing creamy in contrast to the calcification; present in August and September in the Bergen fjord (Ryland, 1963b). Ancestrula undescribed.

Turbicellepora boreale is distinguished from the Arctic *Cellepora smitti* Kluge by its deep peristome and elongate triangular suboral avicularium, and by its rather small, subspherical ovicell. Its distribution is incompletely known; it was recorded by Hayward & Ryland (1978) from the continental slope of the Bay of Biscay, and occurs infrequently around Faroe at depths greater than 30 m. Ryland (1963b) described specimens from West Norway as *T. smitti* (Kluge).

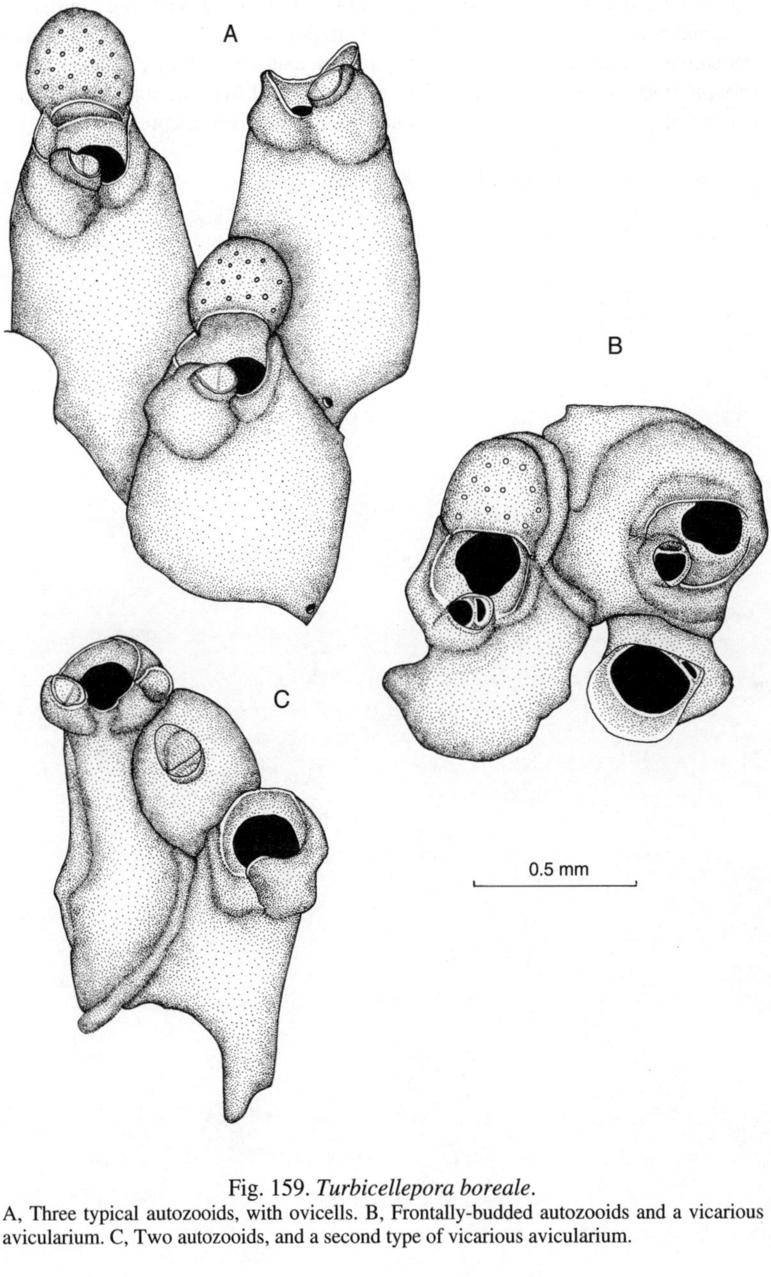

Fig. 159. *Turbicellepora boreale*.
A, Three typical autozooids, with ovicells. B, Frontally-budded autozooids and a vicarious avicularium. C, Two autozooids, and a second type of vicarious avicularium.

Genus OMALOSECOSA Canu & Bassler, 1925

Colony erect, branching. Frontal shield of autozooid with few, indistinct marginal pores. Primary orifice orbicular, with indistinct condyles. No oral spines. Peristome with lateral suboral pseudospiramen, and incorporating a single lateral suboral avicularium. No vicarious avicularia. Ovicell hyperstomial, imperforate, not closed by autozooid operculum. Small basal pore chambers present.

Type species: *Cellepora ramulosa* Linnaeus, 1767

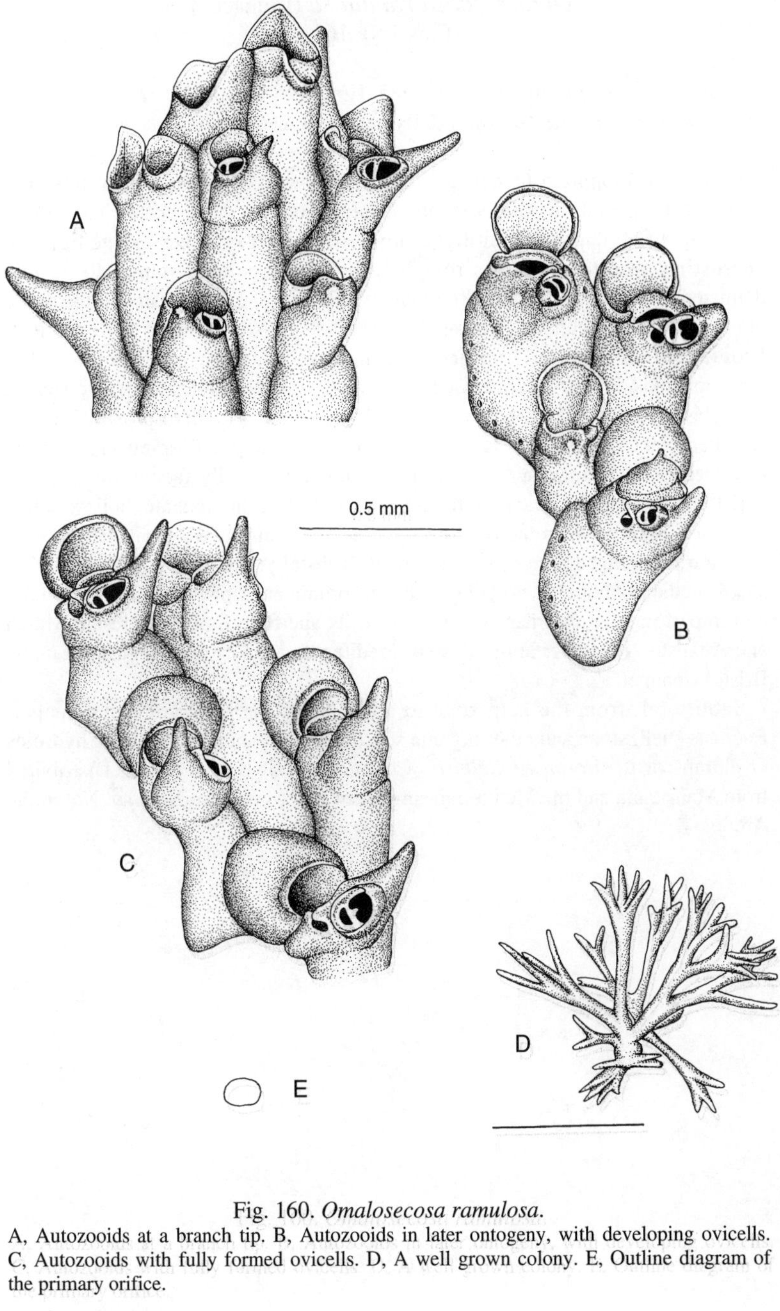

Fig. 160. *Omalosecosa ramulosa*.
A, Autozooids at a branch tip. B, Autozooids in later ontogeny, with developing ovicells. C, Autozooids with fully formed ovicells. D, A well grown colony. E, Outline diagram of the primary orifice.

Omalosecosa ramulosa (Linnaeus)
(Figs 160; 161)

Cellepora ramulosa Linnaeus 1767: 1285; Hincks 1880: 401, pl. 52, figs 7-9.
Omalosecosa ramulosa: Hayward & Ryland 1979: 290, fig. 127

Colony developing a branched, stagshorn growth, 30-40 mm high; branches cylindrical, tapered, dividing dichotomously in one or more planes; most often assuming a flat, fan shape, but frequently several erect branches radiate from the encrusting base to develop a rounded, three-dimensional shape, 40-50 mm diameter. Autozooids in alternating longitudinal series, disposed around the entire branch axis; oval, convex, separated by shallow grooves; 0.6-0.9 x 0.3-0.6 mm. Frontal shield smooth, fine grained, the few small marginal pores rather inconspicuous. Primary orifice wider than long, proximal edge shallowly concave; condyles indistinct, developed as small thickenings of the latero-proximal orifice rim. Peristome deep, with a large, round latero-suboral pseudospiramen, adjacent to a large avicularium, acute to frontal plane and laterally facing; distally, the peristome enfolds the avicularium and is produced as an elongate, hollow spike, with the suture between the two folds visible on its inner face. Avicularium with elongate triangular rostrum, strongly hooked distally; crossbar complete, with a small median columella; no palate. Ovicell prominent, spherical, finely granular and imperforate, with a narrow rim around its aperture. Embryos present March and August - November in the western Mediterranean, July and August at Lundy, Bristol Channel.

Sublittoral, from the kelp zone to about 80 m, perhaps not much deeper. Encrusts shell, stone and other organic substrata; often abundant on large hydroids or clumps of the bryozoan *Cellaria*. Common on all British coasts. Distributed from Mauritania and the Mediterranean northwards to western Norway. Not in the Arctic.

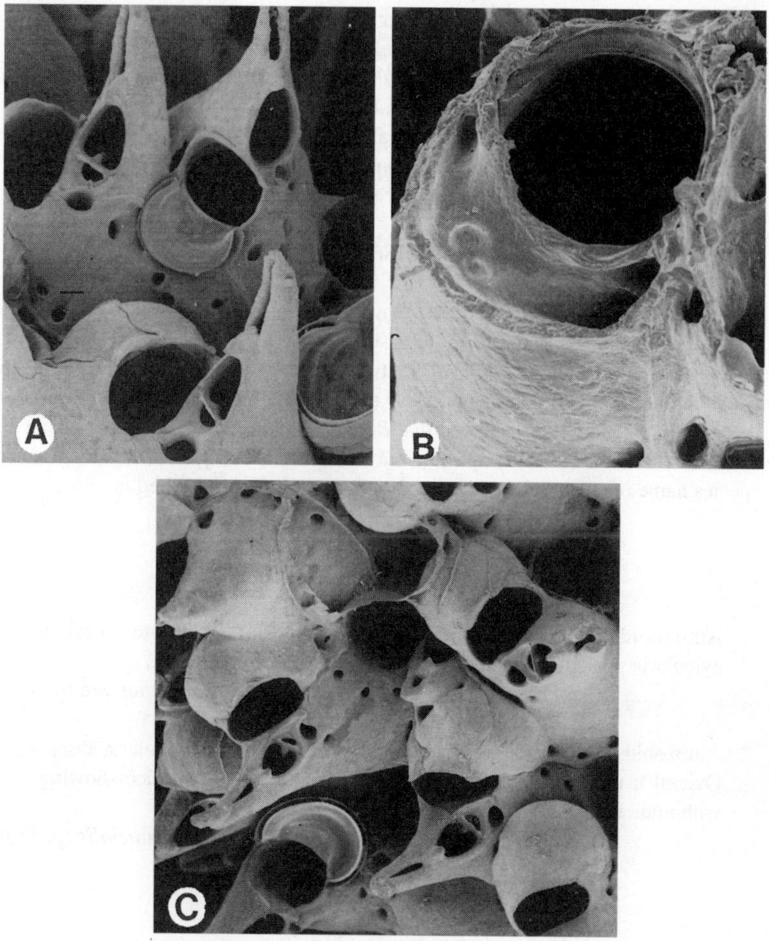

Fig. 161. *Omalosecosa ramulosa* (Scilly).
A, Autozooids with developing ovicells; x80. B, The primary orifice; x200. C, Ovicellate zooids in later ontogeny; x60.

Genus BUSKEA Heller, 1867

Colony erect. Frontal shield with small marginal pores. Primary orifice with a sinus; encircled by a peristome which encloses a suboral avicularium, with a median notch (pseudosinus). Additional adventitious avicularia, and spatulate vicarious avicularia present. Ovicell prominent, perforated by small frontal pores, not closed by zooidal operculum. Small, elongate pore chambers present.

Type species: *Buskea nitida* Heller, 1867

Only two species of *Buskea* are known to occur in British waters, *B. dichotoma* and *B. quincuncialis*. Both are found also in the western Mediterranean. A third, rarely reported species, *B. billardi* (Calvet), is known to be distributed in deep waters along the outer continental shelf of the Bay of Biscay, and southwards to the Azores. Zabala (1986) showed that *B. quincuncialis* and the Mediterranean *B. nitida* (Heller) are the same species, but (Zabala & Maluquer, 1988) the correct species name remains to be determined as both taxa were published in 1867.

Key to the British species of *Buskea*

1. Autozooids large (>0.5 mm), colony often stout. Spatulate vicarious avicularia present. Ovicell distinct, with irregular frontal pores
   .... ***Buskea dichotoma*** (p. 348)

   Autozooids small (<0.5 mm), colony delicate. Vicarious avicularia absent. Ovicell usually completely immersed, only the frontal surface showing, with numerous small pores
   .... ***Buskea quincuncialis*** (p. 350)

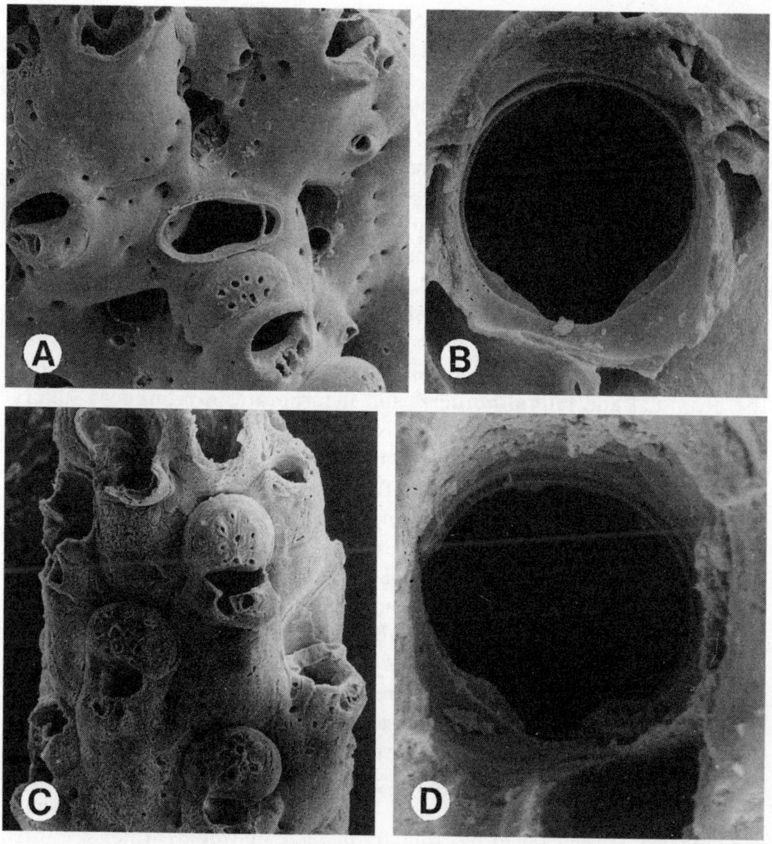

Fig. 162.
A, B, *Buskea dichotoma* (Shetland).
A, Group of autozooids, with an ovicell and a vicarious avicularium x60. B, Primary orifice; x260.

C, D, *Buskea quincuncialis* (English Channel).
C, Ovicellate zooids; x80. D, Primary orifice; x400.

Buskea dichotoma (Hincks)
(Figs 162A, B; 163)

Cellepora dichotoma Hincks 1862: 305, pl. 12, figs 7-8; 1880: 403, pl. 55, figs 1-10
Harmerella dichotoma: Gautier 1962: 248
Buskea dichotoma: Hayward & Ryland 1979: 294, figs 10D, 128

Colony slender, branching dichotomously, arising from a small encrusting base. Branches cylindrical, tapering, composed of alternating whorls of four to six autozooids. Height commonly 10-30 mm, rarely larger. Autozooids elongate, oval, 0.5-0.8 x 0.3-0.4 mm. Frontal shield convex, smooth, with a few indistinct marginal pores. Primary orifice semicircular, slightly wider than long, with a small U-shaped proximal sinus; rapidly obscured by a peristome which encircles the proximal and lateral borders, and incorporates a prominent suboral avicularium. Avicularian chamber situated to the right or left of the midline of the peristome; rostrum slanted upwards from the frontal plane of the autozooid and directed laterally; mandible triangular. The peristome rim with a small notch (or pseudosinus) adjacent to the proximal end of the avicularium. Up to four or more small adventitious avicularia frequently present on later autozooids, close to the lateral walls; mandible semi-elliptical or bluntly triangular, with no regular orientation. Larger vicarious avicularia infrequent, up to 0.3 mm long, with broad, spatulate mandible. Ovicell spherical, prominent, with up to a dozen small, round pores frontally, often linked by faint sutures - in older autozooids usually immersed in secondary calcification. Embryos orange, reported in all months of the year around the Isle of Man (Eggleston, 1969). Ancestrula tatiform with nine spines around an oval opesia.

Basally, the colony is thickened and strengthened by continual calcification, giving a smooth even stem with deeply immersed autozooidal orifices. Distally, the branches are thickened by frontally-budded autozooids which are often smaller than normal autozooids and without regular orientation.

Occurs offshore, from the lower levels of the kelp forest; usually attached to organic substrata and most typically on hydroids, *Cellaria*, or other bryozoans.

B. dichotoma occurs off all British coasts; it is distributed from Faroe and the North Sea to the western Mediterranean and Madeira, where it was recorded by Norman (1909) as *Cellepora margaritacea*.

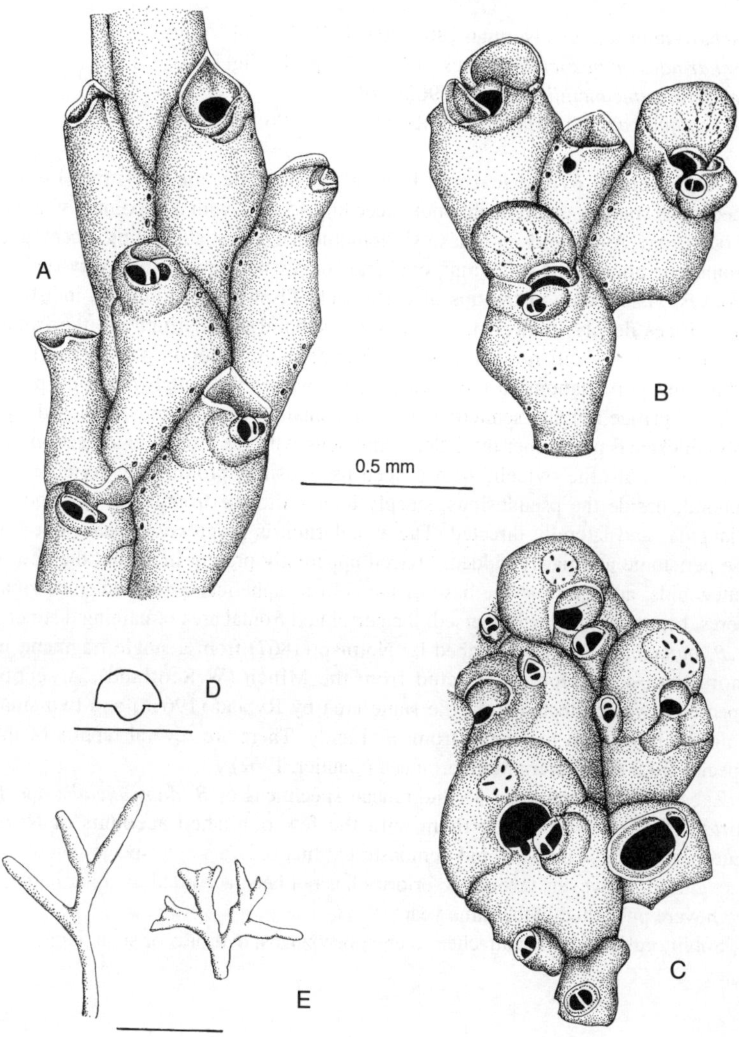

Fig. 163. *Buskea dichotoma.*
A, Autozooids at a branch tip. B, Ovicellate zooids. C, Autozooids in later ontogeny, with the ovicells largely covered by secondary calcification. D, Outline diagram of the primary orifice. E, Two different colony habits.

Buskea quincuncialis (Norman)
(Figs 162C, D; 164)

Eschara quincuncialis Norman 1867: 204; 1868: 221, pl. 7, figs 1-3
Escharoides quincuncialis: Hincks 1880: 339, pl. 15, fig. 7
Leiosella quincuncialis: Gautier 1962: 178
Buskea quincuncialis: Hayward & Ryland 1979: 296, fig. 129

Colony erect, cylindrical and branching; delicate, up to 10 mm high in specimens studied, and probably not exceeding 15 mm, with a maximum width of about 1 mm. Autozooids in four to six longitudinal series, alternating, opening all around the branch; 0.4 x 0.2 mm; oval and convex but separated by very shallow grooves, visible only at the tips of the branch. Elsewhere, boundaries indistinct, the orifices deeply immersed, and only the suboral avicularium and the frontal region of the ovicell remain distinct. Frontal shield smooth, becoming glossy white and porcellanous in older autozooids, with a few small marginal pores. Primary orifice roughly semicircular, with a small U-shaped sinus, enclosed by a low, thickened peristome; the latter continuous with the frontal calcification and extending onto the ovicell; with a deep pseudosinus proximally. Avicularium suboral, beside the pseudosinus, steeply inclined to the frontal plane; mandible triangular and laterally directed. The avicularium is progressively enveloped by the peristome and partly hidden. Ovicell apparently present in all but the newest autozooids, and those at the base of the colony; spherical, with numerous small pores, becoming deeply immersed; the perforated frontal area remaining distinct.

B. quincuncialis was described by Norman (1867) from a single fragment, no more than 4 mm long, collected from the Minch (W Scotland). A second specimen was collected from the same area by Ryland (1963a) and two small colonies have been recovered from off Lundy. There are several reports of this species from the western Mediterranean (Gautier, 1962).

Zabala (1986) examined Mediterranean specimens of *B. quincuncialis* and *B. nitida* (Heller), comparing them with the few published accounts of North European *B. quincuncialis* and demonstrated that only a single species should be recognised. However, taxonomic priority has not been accorded to either name as both were published in the same year.

Sublittoral, most often attached to erect bryozoans, hydroids or small algae.

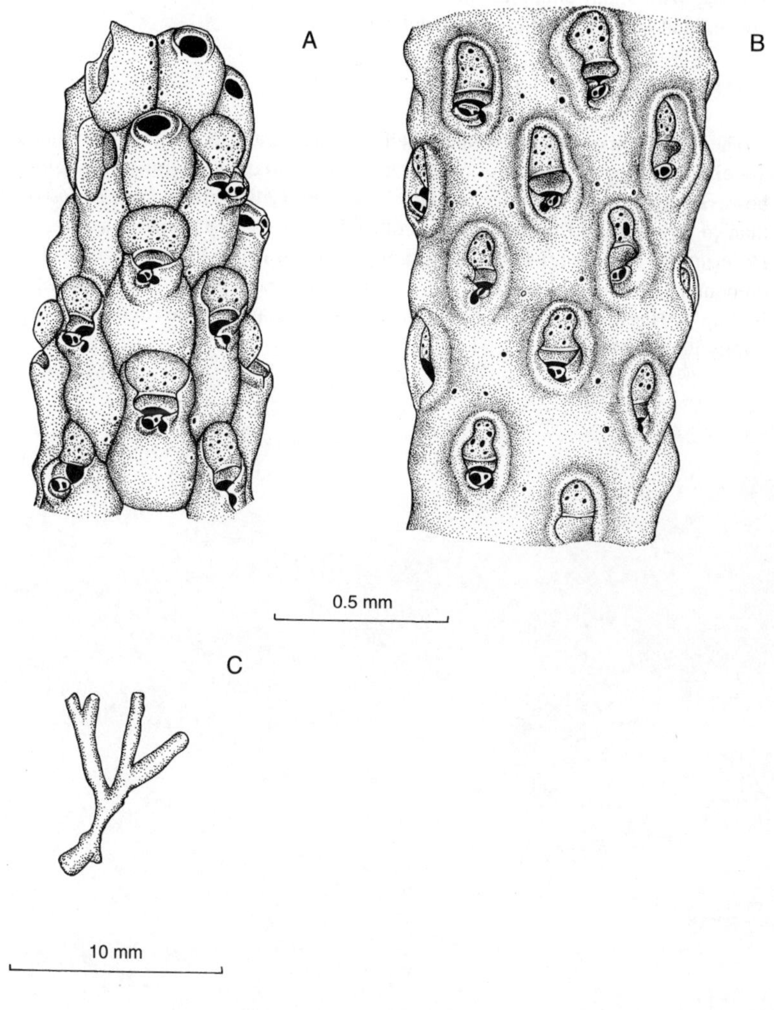

Fig. 164. *Buskea quincuncialis*.
A, Autozooids at a branch tip. B, Ovicellate zooids in later ontogeny. C, Diagram to show colony habit.

Genus PALMICELLARIA Alder, 1864

Colony erect, branching, attached by an encrusting sheet of autozooids. Frontal shield of autozooids cryptocystidean, with marginal perforations. Primary orifice sinuate, with small condyles; no oral spines. Avicularia adventitious, suboral. Ovicell undescribed. Vertical walls with small septula.

Type species: *Palmicellaria elegans* Alder, 1864

The placement of this genus in the Celleporidae must be considered provisional, pending the discovery of the ovicell and ancestrula. Its overall morphology, however, appears to be closer to that of the erect Celleporidae, such as *Buskea*, than to any smittinid genus. The two other British species formerly assigned to *Palmicellaria, skenei* Ellis & Solander, and *lorea* Alder, prove to have umbonuloid frontal shields and were assigned to a new genus, *Palmiskenea*, by Bishop & Hayward (1989), and are now accommodated in the umbonulomorph family Porellidae.

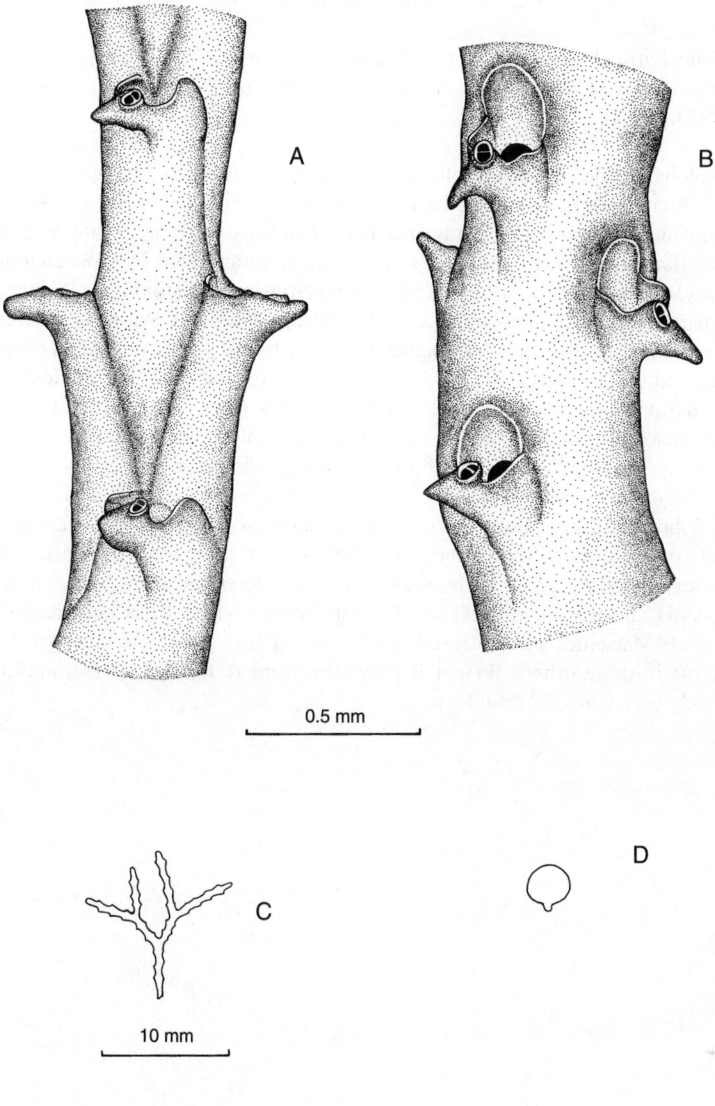

Fig. 165. *Palmicellaria elegans*.
A, Autozooids from the distal end of a branch. B, Autozooids in later ontogeny, with boundaries obscured by calcification. C, Diagram to show colony habit. D, Outline diagram of the primary orifice.

Palmicellaria elegans Alder
(Figs 165; 166)

Palmicellaria elegans Alder 1864: 100, pl. 2, figs 1-4
Palmicellaria elegans: Hincks 1880: 378, pl. 31, figs 7-9; Hayward & Ryland 1979: 130, fig. 50

Colony slender, branching dichotomously at regular intervals, up to 30 mm high. Autozooids in alternating back-to-back pairs, elongate, 0.9 - 1.0 x, about 0.3 mm; sutures between autozooids distinct only in early ontogeny. Primary orifice orbicular, with a very short, U-shaped sinus, emphasised by low, rounded condyles; encircled and obscured by deep, flaring peristome. Avicularium median-suboral, with a columnar cystid, produced apically, distal to the rostrum, as a stout spike; rostrum oval, perpendicular to frontal plane, orientated fronto-proximally. The lateral flaps of the peristome extend onto and enclose the avicularium. Frontal shield finely granular, with a single series of moderately-sized marginal pores visible in early ontogeny; calcification thickens in later ontogeny, becomes smooth and brilliantly white, obscuring boundaries between autozooids, and reducing the size of the marginal pores. Ovicell unknown.

Offshore, on hard substrata; usually recovered from dredgings of medium grade shell gravels. Little is known about the distribution or habitat requirements of this species, or indeed about its biology. Originally described from Shetland, and collected on subsequent occasions from the same region, it has been recorded from off Marseilles and the Balearic Islands (Gautier, 1962) and amongst shell gravels in the northern Bay of Biscay (Hayward & Ryland, 1978), and has recently been collected from Faroe.

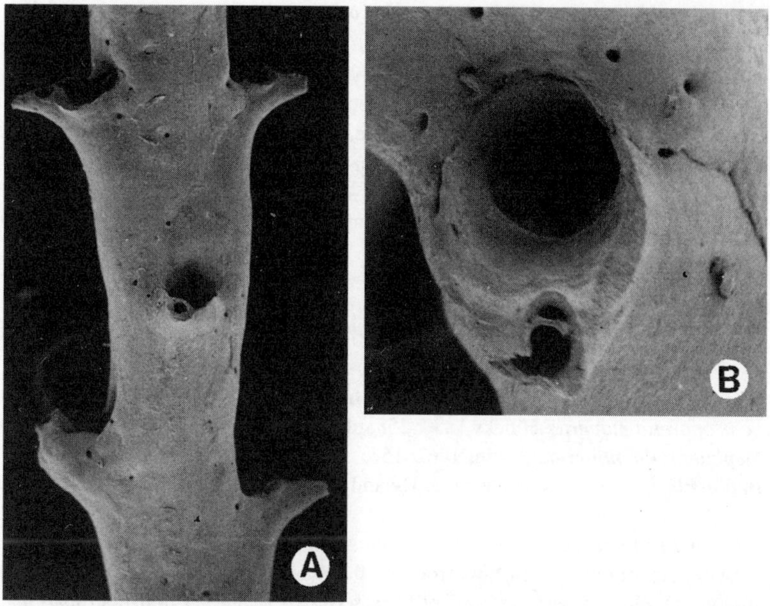

Fig. 166. *Palmicellaria elegans*.
A, Part of a colony; x50. B, The primary orifice; x170.

Genus BUFFONELLARIA Canu & Bassler, 1917

Frontal shield with marginal pores only. Primary orifice with a sinus. Oral spines present or absent. Avicularia adventitious, lateral to orifice, frequently enlarged, with acute or spatulate mandible. Ovicell prominent, imperforate, not closed by zooidal operculum. Basal pore chambers present.

Type species: *Buffonellaria divergens* (Smitt)

Buffonellaria divergens (Smitt)
(Fig. 167)

Hippothoa divergens Smitt 1873: 47, pl. 9, figs 177, 179
Schizoporella biaperta: Hincks 1880: 255, pl. 40, figs 7-9
Stephanosella biaperta: Gautier 1962: 155
Buffonellaria divergens: Hayward & Ryland 1979: 204, fig. 86

Colony forming extensive incrustations. Autozooids oval to hexagonal, slightly convex, separated by shallow grooves; 0.5-0.6 x 0.3-0.4 mm. Frontal shield smooth, hyaline, imperforate except for a series of small and indistinct marginal pores. Primary orifice longer than wide; anter semicircular, poster deep, U-shaped, with blunt lateral condyles. No oral spines. Avicularia situated lateral to the orifice, single or paired, the small chamber arising from one of the marginal pores; rostrum elliptical or triangular, directed proximally. Grossly enlarged avicularia may replace the lateral oral type, or arise elsewhere on the frontal surface of the autozooid, originating from one of the marginal pores; the inflated cystid supports an elongate, acutely triangular mandible, typically directed proximally. Of occasional occurrence on young parts of the colony, this second type is often abundant on older areas, frequently obscuring the borders of the autozooids. The vertical walls of the autozooid are deep, with small blister-like pore chambers basally. Ovicell prominent, spherical, opening above the autozooid orifice and not closed by the zooidal operculum; ectooecium smooth, an uncovered area of frontal entooecium marked with fine radiating striations. Embryos orange. Ancestrula about two-thirds the size of the later autozooids, with a smooth gymnocyst and a semi-orbicular aperture surrounded by a number of spines. Gautier (1962) reported embryos in June and November in the Mediterranean.

With increasing calcification the orifice of *B. divergens* becomes immersed and partially obscured, and the sutures between the autozooids are covered by coarser calcite. Further, the proliferation of avicularia and ovicells in older parts of the colony leads to a quite different appearance of these areas, compared with young, actively growing, areas. It seems to have been these features which convinced Gautier (1962) that the Mediterranean species of *Buffonellaria* (attributed to *Stephanosella biaperta* by Waters (1879: 37) and Canu & Bassler (1928: 32)) was different from that described by Hincks (1880: 225). However, comparison of old,

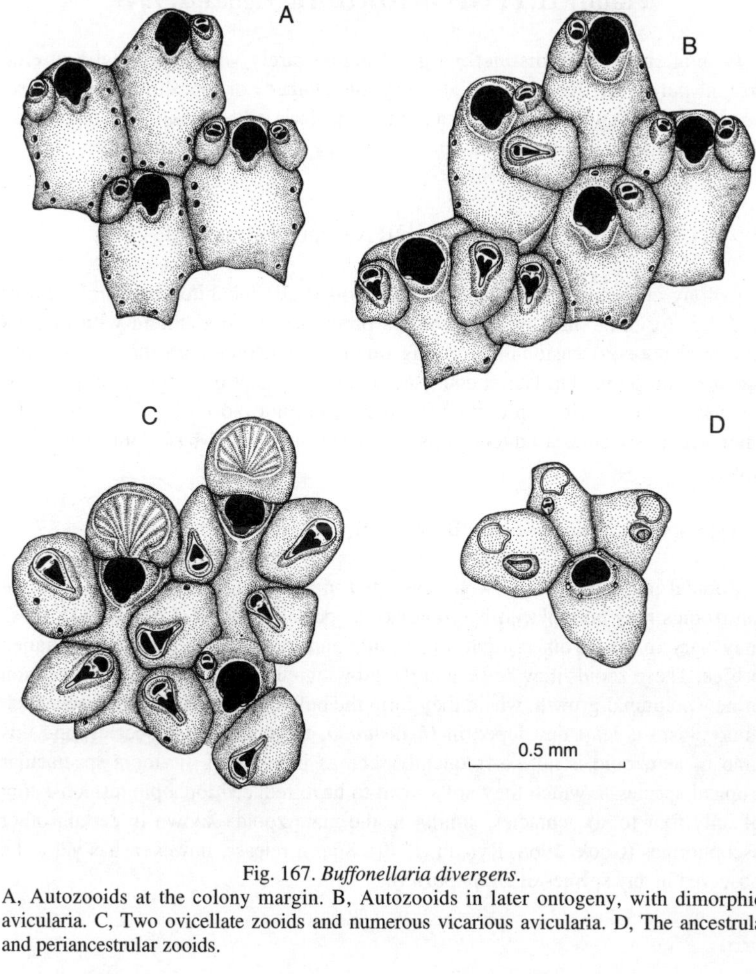

Fig. 167. *Buffonellaria divergens*.
A, Autozooids at the colony margin. B, Autozooids in later ontogeny, with dimorphic avicularia. C, Two ovicellate zooids and numerous vicarious avicularia. D, The ancestrula and periancestrular zooids.

ovicellate specimens of *B. divergens* from British waters with similar Mediterranean specimens will show no difference between them. The systematic status of the taxa *divergens* Smitt and *biaperta* Michelin have been clarified by Ryland (1969).

Little is known of the ecology of this species beyond that it occurs in the western Mediterranean (from Naples to Morocco), and in the Gulf of St. Malo, and has occurred in the Channel as far east as Hastings.

It was described by Smitt (1873) from Florida and is now also known from the Carolinas and Bermuda (Maturo, 1968).

Family HIPPOPORIDRIDAE Vigneaux, 1949

Frontal shield with distinct marginal pores; rarely with additional scattered frontal pores. Orifice cleithridiate (keyhole-shaped) or lepralioid, with lateral condyles. Oral spines present or absent. Avicularia adventitious or interzooidal. Ovicells present. Interzooidal communication by septula or pore chambers.

Genus HIPPOPORIDRA Canu & Bassler, 1927

Colony encrusting, typically on gastropod shells inhabited by hermit crabs; nodular, forming short or long conical processes. Frontal shield with distinct marginal pores, occasionally forming tubular extensions onto the frontal wall. Orifice with prominent lateral condyles, anter and poster of differing proportions. No oral spines. Ovicell prominent, imperforate, not closed by the autozooidal operculum. Avicularia adventitious and vicarious. Small basal pore chambers present.

Type species: *Cellepora edax* Busk, 1859

Zooidal polymorphism is well developed in all species of this genus. Normal autozooids may be replaced by zooids of a specialised kind (Cook, 1964); these may vary in size from normal autozooids and may have a differently shaped orifice. These zooids may be seen at the growing edge or at the point of inception of new columnar growth, where they form the bulk of the nodular processes. Such dimorphism is least developed in *H. lusitania*, the only British species, and this kind of autozooid is not so frequently seen as in some of the more spectacular tropical species in which they are known to have reduced lophophores consisting of only four to six tentacles, similar to the male zooids known in certain other ascophorines (Cook, 1968; Ryland, 1979). Sperm release, however, has yet to be observed in any species of *Hippoporidra*.

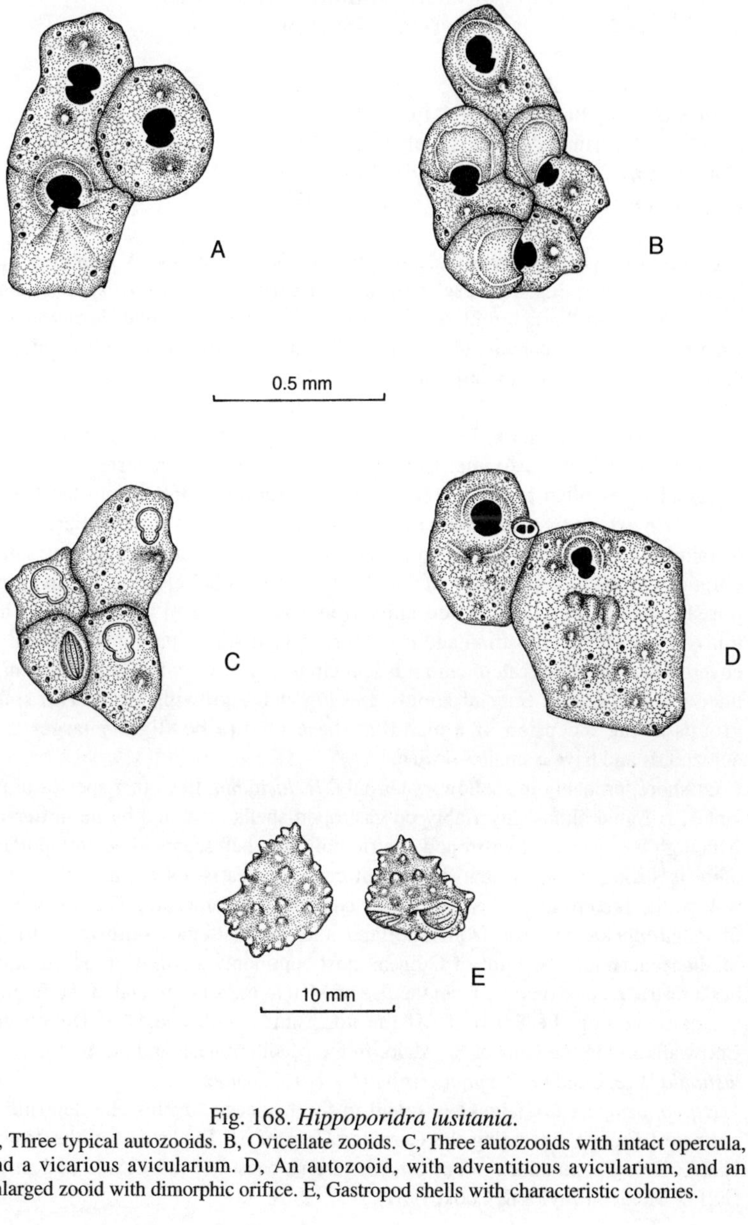

Fig. 168. *Hippoporidra lusitania.*
A, Three typical autozooids. B, Ovicellate zooids. C, Three autozooids with intact opercula, and a vicarious avicularium. D, An autozooid, with adventitious avicularium, and an enlarged zooid with dimorphic orifice. E, Gastropod shells with characteristic colonies.

Hippoporidra lusitania Taylor & Cook
(Figs 168; 169)

not *Cellepora edax* Busk 1859: 59, pl. 9, fig. 6; pl. 22, fig. 3
Cellepora edax: Busk 1861: 154, figs 3, 3a
Lepralia edax: Hincks 1880: 311, pl. 24, figs 7-8
Hippoporidra edax: Hayward & Ryland 1979: 214, fig. 91
Hippoporidra lusitania Taylor & Cook 1981: 247, fig. 6

Colony exclusively encrusting gastropod shells, almost always inhabited by a hermit crab; forming a coarse multilaminar growth raised into stout conical processes (mamillae); bright red when living. Autozooids oval, hexagonal or irregular in outline, convex, closely packed and not always easily differentiated; 0.4-0.5 x 0.2-0.3 mm, regularly orientated only in the first lamina of the colony, frontally budded in later stages of growth such that neighbouring orifices often have opposing orientation. Frontal shield coarsely granular, often raised into a stout, conical umbo proximal to the orifice; bordered by a series of distinct marginal pores often bridged over laterally, extending as tubes onto the frontal wall. Primary orifice approximately keyhole-shaped; anter orbicular, poster forming a slender ellipse below strong lateral condyles. Avicularia typically small, elliptical, developed on the frontal wall of the autozooid close to the marginal pores, of irregular occurrence and often absent; rarely vicarious, with a transversely elliptical rostrum and mandible. Ovicell at first prominent, becoming covered by secondary calcification; a semicircular area of entooecium remaining uncovered frontally. Special zooids develop at the growing edge or in small groups at the inception of a mamilla; these tend to be slightly larger than autozooids and have a smaller orifice.

Offshore, probably in shallow water only. *H. lusitania*, like other species of the genus, is found almost invariably on gastropod shells inhabited by hermit crabs. Although the species of gastropod contributing the shell seems to be unimportant, there is strong evidence that the hermit crab species is not so, and, in British waters, the preferred species is *Pagurus cuanensis* (Thompson). The distribution of *P. cuanensis* extends from Shetland and the Kattegat southwards to the Mediterranean and the Gulf of Guinea, most commonly at 10-30 m. *H. lusitania* has a restricted distribution, from the Isle of Man to the western end of the English Channel, among the Scillies, off the north and south coasts of Devon and Cornwall, and in the Gulf of St. Malo. In the Mediterranean and off W Africa *H. lusitania* is replaced on *P. cuanensis* by *H. picardi* Gautier.

Hippoporidra edax (Busk) is found in the Miocene to Pliocene deposits of northwest Europe (including the Coralline Crag of SE England) and the eastern U.S.A; living populations occur off the eastern coasts of the U.S.A. and in the Gulf of Mexico (Taylor & Cook, 1981).

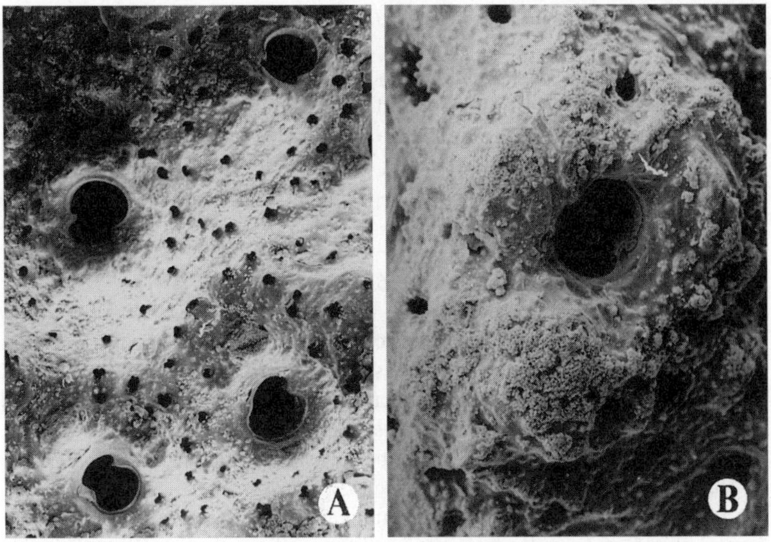

Fig. 169. *Hippoporidra lusitania*.
A, Group of autozooids; x80. B, A dimorphic male zooid; x130.

Genus HAGIOSYNODOS Bishop & Hayward 1989

Colony encrusting. Frontal shield with numerous, evenly spaced pores, or with marginal pores only. Orifice lepralioid, with strong lateral condyles. Ovicell partially immersed, sparsely perforated, closed by zooidal operculum. Numerous basal pore chambers present. Nodular umbones developed lateral to orifice, occasionally supporting small, elliptical avicularia. Other avicularia absent. Oral spines absent.

Type species: *Lepralia lata* Busk, 1856

Only one species, *H. latus* (Busk), occurs in British waters. It is also found in the Mediterranean together with another species, *H. kirchenpaueri* (Heller) and its variety *tregouboffi* Gautier. The distinction between the two nominate species remains unclear despite several attempts to define them (e.g., Canu & Bassler, 1930). Zabala & Maluquer (1988) describe both species, and var. *tregouboffi*.

Although Barroso (1924) selected *Lepralia adpressa* Busk as type species of his new genus *Hippopodinella*, his figures depicted *Lepralia lata* Busk. The frontal shield of *Lepralia lata* is quite different from that of *Hippopodinella adpressa*, which has marginal pores only.

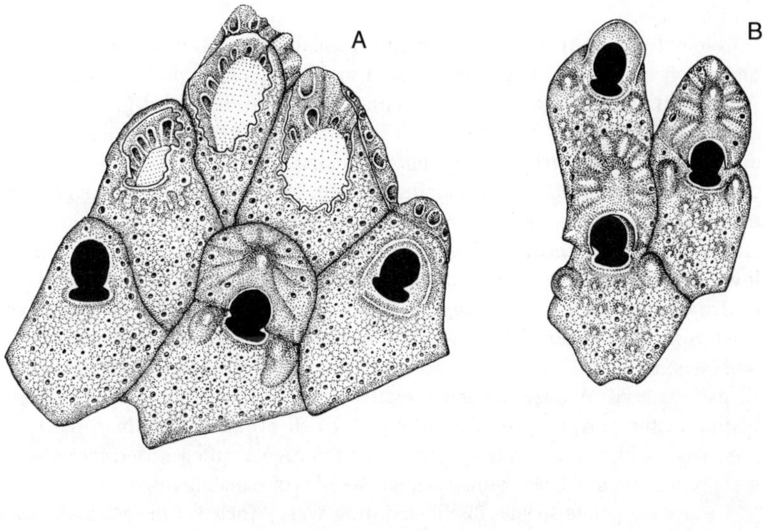

Fig. 170. *Hagiosynodos latus.*
A, The growing edge of the colony, showing developing autozooids. B, Autozooids in later ontogeny, two with ovicells.

Hagiosynodos latus (Busk)
(Figs 170; 171A, B)

Lepralia lata Busk 1856: 309, pl. 10, figs 1, 2
Lepralia adpressa: Hincks 1880: 307, pl. 33, figs 5-7
Hippopodinella lata: Hayward & Ryland 1979: 216, fig. 92
Hagiosynodos latus: Bishop & Hayward 1989: 46, figs 189-191

Colony forming irregular, often large incrustations, occasionally on dead shell, and rarely on stones, but typically and most frequently on live gastropods. Autozooids oval, hexagonal or irregular in outline, flat or slightly convex, separated by shallow grooves; 0.4-0.6 x 0.3-0.5 mm. Frontal shield rugose, closely and evenly perforated by numerous small, round pores, each becoming deeply immersed as the frontal calcification thickens. Primary orifice longer than broad, the anter horseshoe-shaped, poster shallowly curved below strong, blunt lateral condyles. Operculum brown, distinct, with strong marginal sclerites. A low, thickened peristome develops around the distal and lateral borders of the orifice and may develop into rugose umbones, one on each side of the orifice. According to Gautier (1962), minute avicularia may be developed on these umbones; he described them as rare and they have not been observed in any British material. A dozen or more small basal pore chambers present, clearly visible at the periphery of the colony. Ovicell broader than long, partially immersed, with a few scattered pores; surface rugose, often raised into one or more prominent umbones, with or without a series of radiating ridges.

A warm temperate species distributed from West Africa and the Mediterranean to the southwest British Isles. It is found in the Gulf of St. Malo, the English Channel as far east as Hastings, and as far north as the Isle of Man in the west.

Found in the shallow sublittoral, probably not deeper than 100 m; well characterised by its habitat, the shells of live gastropods - occasionally empty shells or those inhabited by hermit crabs.

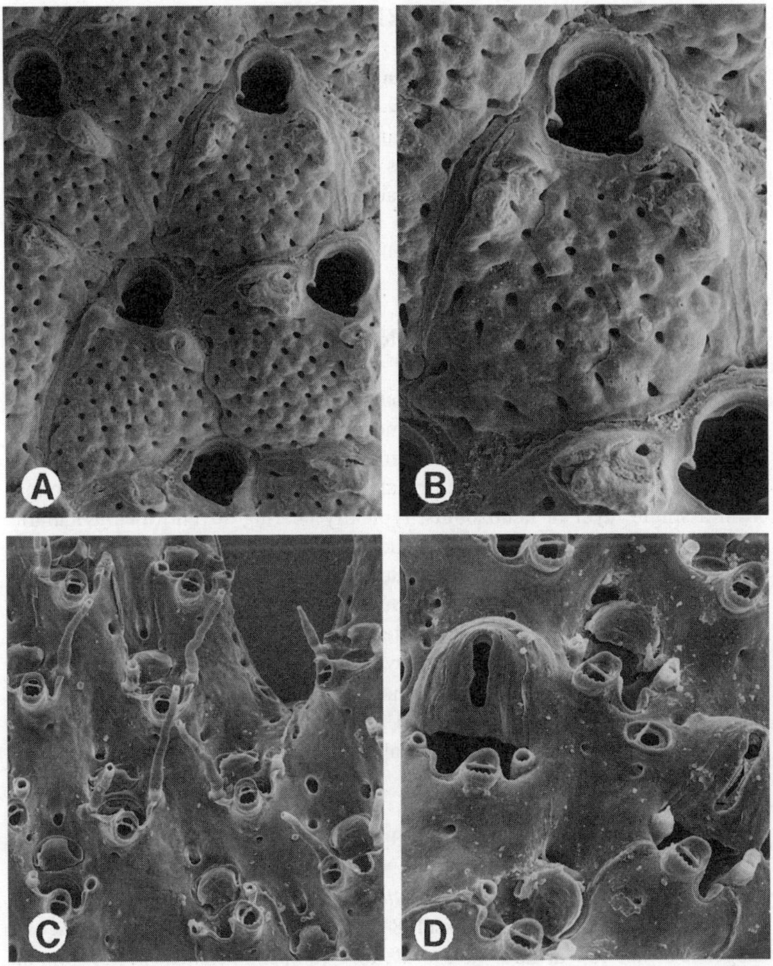

Fig. 171.
A, B, *Hagiosynodos latus*.
A, Group of autozooids; x80. B, Detail; x140.

C, D, *Reteporella beaniana* (specimen from Northumberland).
C, Group of autozooids below a fenestrula; x60. D, Ovicellate zooids; x170.

Family PHIDOLOPORIDAE Gabb & Horn, 1862

Colony encrusting; or erect, branching or reticulate. Autozooid frontal shield with few, small marginal pores. Primary orifice with toothed or beaded distal rim; condyles present; proximal border straight, concave or distinctly sinuate. Oral spines present. Avicularia adventitious and/or vicarious; polymorphic. Ovicell prominent, hyperstomial, not closed by autozooid operculum.

Gordon (1989) demonstrated that the family taxon Phidoloporidae should take precedence over the more familiar Sertellidae Jullien.

Genus RETEPORELLA Busk, 1884

Colony erect, branching dichotomously or irregularly, developing an open branched form, or with sporadic anastomoses, or forming a finely reticulate shape; attached by an encrusting sheet of kenozooids. Autozooids arranged in alternating, longitudinal series along a defined frontal face; abfrontal surface of colony formed from a single layer of kenozooids, typically bearing adventitious avicularia. Frontal shield of autozooids cryptocystidean, with a reduced umbonuloid shield, bounded by a clear ring scar, immediately proximal to orifice; characteristically with few, conspicuous marginal pores. Primary orifice with distinct distal denticulation and conspicuous condyles; concave proximally. Peristome well developed, asymmetrically notched, or with a larger fissure, which may close over to define a pseudospiramen. Ovicell smooth, imperforate, but with a longitudinal median fissure; labellum variably developed. Avicularia adventitious and vicarious, polymorphic; usually suboral and frontal. Vertical walls with small, multiporous septula.

Type species: *Reteporella flabellata* Busk, 1884

Gordon (1989) argued that *Reteporella* Busk should be regarded as the senior synonym of *Sertella* Jullien. Busk (1884) originally introduced *Reteporella* as a subgenus of the now invalid "*Retepora*", for two open-branched species from the southern Indian Ocean, and it has been used subsequently for mostly southern hemisphere, subtropical to temperate species with similar colony form. The northern hemisphere, and especially north Atlantic, species of *Sertella* are all characterised by closely reticulate growth fans, developing ordered, fan-like or bun-shaped colonies. However, as Gordon (1989) pointed out, in all other morphological features the two genera are identical, and as both loosely reticulate species of *Sertella* and partly reticulate species of *Reteporella* are now known, there is little utility in maintaining the distinction between the two genera.

Key to the British species of *Reteporella*

1. Proximal lip of peristome indistinctly notched adjacent to the avicularium, but not closing over to form a round foramen. Ovicell short and broad
.... ***Reteporella beaniana*** (p. 368)

Proximal lip of peristome notched in early development, the notch closing over to form a rounded foramen close to or distant from the peristome rim
.... **2**

2. Frontal shield, and abfrontal colony surface, finely or coarsely rugose. Peristomial foramen close to peristome rim, small
.... ***Reteporella watersi*** (p. 376)

Frontal shield and abfrontal surfaces smooth. Foramen large, or distant from peristome rim
.... **3**

3. Foramen linked to peristome rim by a fused fissure at least three times length of foramen; peristome deep. Large lingulate frontal avicularia present. Ovicell much longer than wide
.... ***Reteporella couchii*** (p. 370)

Peristome shallow, with only a short fissure between its rim and the foramen. Frontal avicularia triangular or short oval. Ovicell as wide as long
.... **4**

4. Peristomial foramen larger than the peristomial flap distal to it. Frontal avicularia small, oval or triangular. Ovicell with arched aperture, lacking a labellum
.... ***Reteporella incognita*** (p. 372)

Peristomial foramen much smaller than the peristomial flap distal to it. Frontal avicularia small, oval, or gigantic, triangular, with hooked rostrum. Ovicell with a well developed labellum
.... ***Reteporella septentrionalis*** (p. 374)

Reteporella beaniana (King)
(Figs 171C, D; 172)

Retepora beaniana King 1846: 237
Retepora beaniana: Hincks 1880: 391, pl. 53, figs 1-5; Kluge 1962: 528; 1975: 642
Sertella beaniana: Hayward & Ryland 1979: 262, figs 112A, 113
Reteporella beaniana: Hayward & Ryland 1996: 104, figs 1A-C

Colony 40-50 mm high, with an equivalent horizontal spread; forming an open fan or a shallow cup, or more usually enrolled or infolded to a varying degree. Fenestrulae oval, longer than wide, commonly 1.5 x 0.5 mm. Autozooids in three to five alternating longitudinal series, doubled at points of trabecular fusion; hexagonal to oval, convex, separated by raised sutures, about 0.5 x 0.2 mm; boundaries obscured by thickening calcification in later ontogeny. Frontal shield smooth, each autozooid with few, large conspicuous marginal pores. Primary orifice slightly wider than long, broadly bell-shaped, about 0.09 x 0.11 mm; distal border distinctly toothed, the denticulations not extending to the rounded proximo-lateral condyles. Typically with a single stout spine on each side of the orifice, level with the condyles; elongate, to 0.2 mm, cylindrical, blunt, usually persisting in late ontogeny. Peristome developed laterally and proximally, partly obscuring orifice in frontal view; on its medio-proximal rim a single oval avicularium, 0.6 mm long, projecting prominently, directed proximally or proximo-laterally, slightly acute to frontal plane. Distal rim of avicularium finely toothed, its proximal edge variably developed as a bifid mucro projecting above the orifice of the autozooid. Rim of peristome deeply notched adjacent to the avicularium, on the left or right, to form an open pseudosinus. Additional oval avicularia sporadically developed on frontal walls of autozooids, often frequent on older parts of colony, and on basal (abfrontal) surface of colony. Ovicell slightly wider than long, thinly calcified, flattened frontally with an elongate median fissure; aperture arched, without a labellum; conspicuous in early ontogeny but later submerged by thickening calcification and becoming obscured.

Offshore, attached to hard substrata. Distributed from western Norway, Faroe and Shetland, southwards along the western coasts of the British Isles, and on the east as far south at least as the coast of Norfolk. It has been found below 200 m in the Bay of Biscay. It does not occur in the Mediterranean.

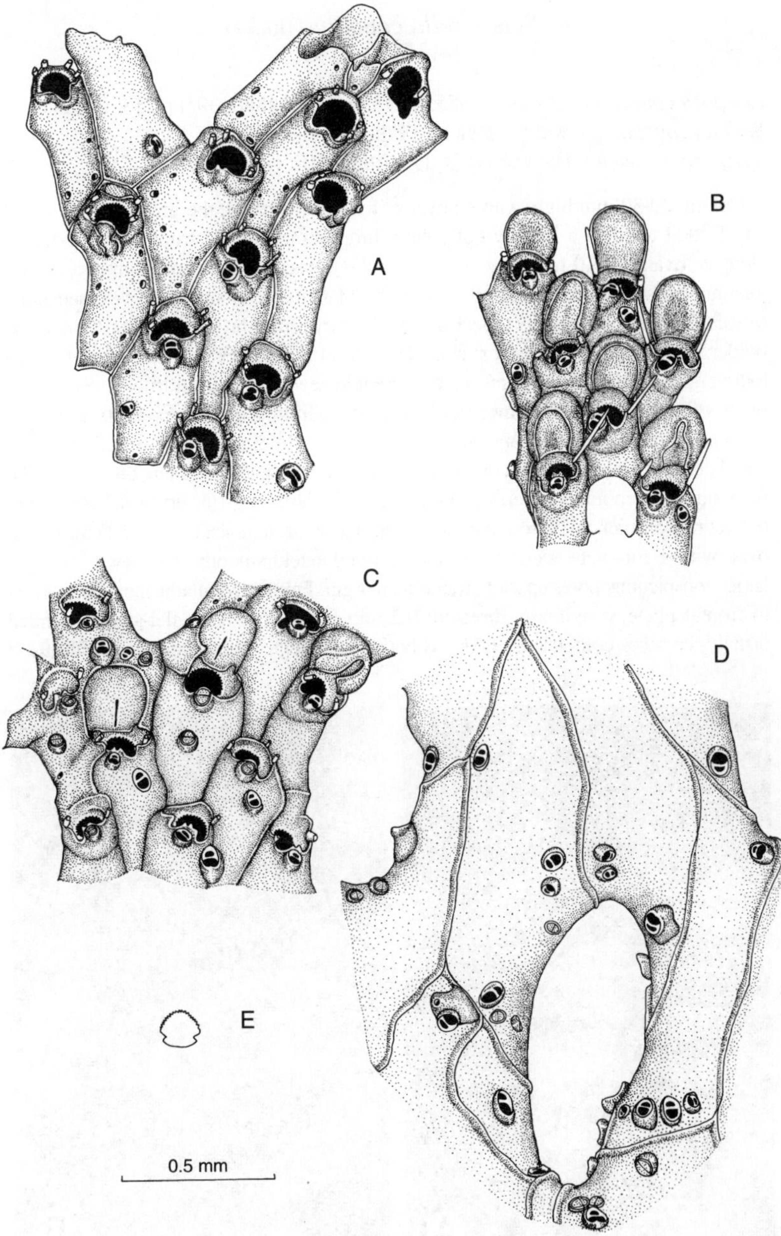

Fig. 172. *Reteporella beaniana.*
A, Autozooids at a colony margin. B, Young zooids with developing ovicells. C, Autozooids in later ontogeny, showing ovicells becoming immersed. D, Abfrontal surface of a portion of a colony. E, Outline diagram of the primary orifice.

Reteporella couchii (Hincks)
(Figs 173; 174)

Retepora couchii Hincks 1878: 355, pl. 18, figs 1-6; 1880: 395, pl. 53, figs 6-11
Sertella couchii: Hayward & Ryland 1979: 264, figs 112B, 114
Reteporella couchii: Hayward & Ryland 1996: 107, figs 1D, E

Colony 20-30 mm high, with equivalent horizontal spread; the growing edge lobed and folded to form a loosely convolute, three-dimensional structure. Fenestrulae elongate oval, 0.5-1.0 mm long, as wide as the trabeculae. Autozooids in two to four longitudinal series, doubled at points of trabecular fusion; elongate, convex, separated by raised sutures; commonly 0.5 x 0.2 mm. Primary orifice broader than long, proximal border straight or slightly convex; condyles indistinct, distal denticulations minute. Two to four delicate, disto-lateral oral spines present in newly budded autozooids, obliterated by development of peristome. Peristome deep, hiding orifice, consisting of two elongate proximo-lateral flaps, separated by a longitudinal fissure, fused along its length, except basally where it opens to define a round pseudospiramen; on one side the peristome incorporates an erect, cylindrical avicularian cystid, up to 0.3 mm long, projecting conspicuously from the colony surface; rostrum apical, about 0.05 mm long, oval, with a complete, slender crossbar. Frontal shield smooth, with fewer than ten large, conspicuous pores spaced around its margin. Frontal avicularia sporadic, normal to frontal plane, proximally directed; 0.2 mm long, almost parallel-sided, rounded distally; crossbar complete, slender, without a columella; palate extensive, with just a

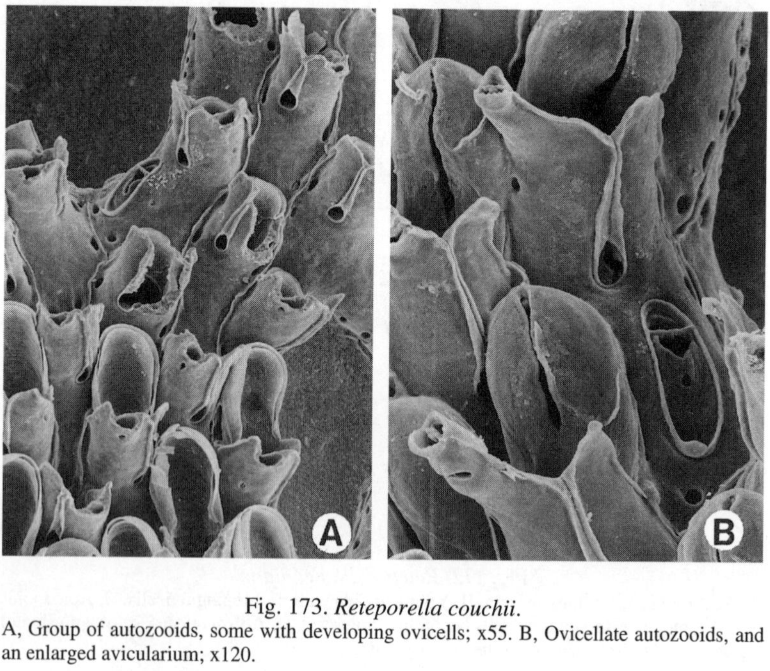

Fig. 173. *Reteporella couchii*.
A, Group of autozooids, some with developing ovicells; x55. B, Ovicellate autozooids, and an enlarged avicularium; x120.

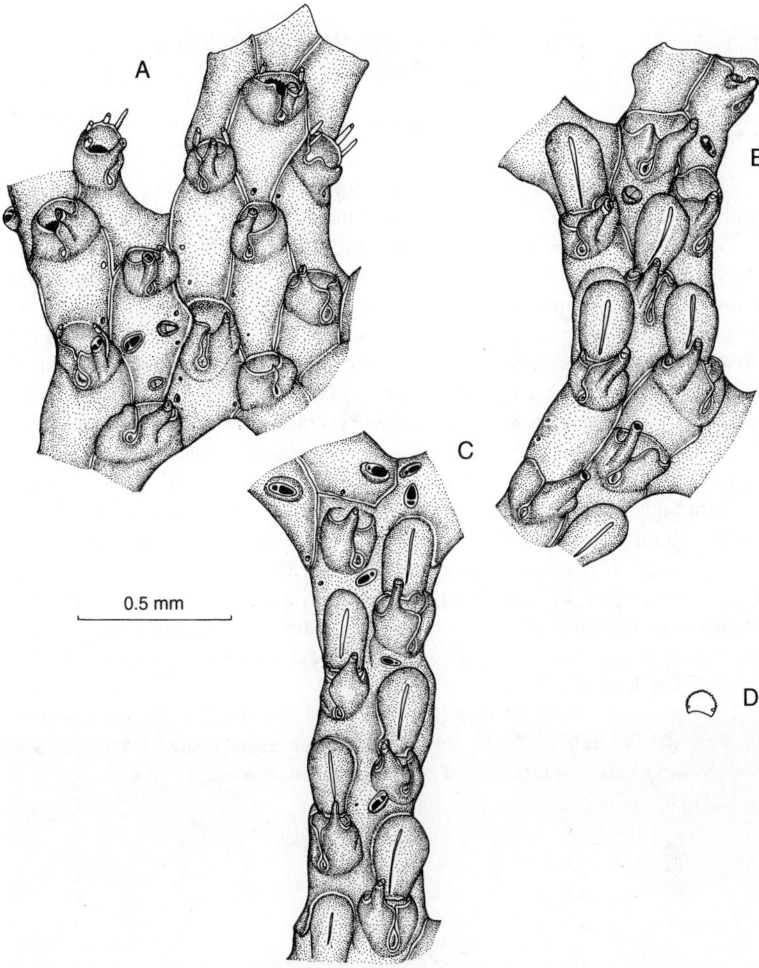

Fig. 174. *Reteporella couchii*.
A, Autozooids at a colony margin. B, Ovicellate zooids. C, Autozooids in later ontogeny. D, Outline diagram of the primary orifice.

small trilobed or triangular foramen distal to the crossbar. Numerous similar avicularia present on abfrontal surface of colony, mostly 0.1-0.2 mm long, and also a small, oval type, 0.05 mm long. Ovicell conspicuous, elongate, twice as long as wide, with a fissure extending the whole of its length.

Occurs offshore, on coarse gravels, attached to shell or stone. Recorded from 30-130 m in the western Mediterranean (Gautier, 1962). Predominantly southern in distribution, although it has been recorded from West Norway and Faroe (Hayward & Ryland, 1996). Present off the Channel Islands and South Cornwall, ranging south to the western Mediterranean and to the Gulf of Guinea (Cook, 1968).

Reteporella incognita Hayward & Ryland
(Fig. 175)

Reteporella incognita Hayward & Ryland 1996: 107, figs 2A-C

Colony broadly fan-shaped, with lobed, folded edges; exceeding 40 mm height, with equivalent horizontal spread. Fenestrulae irregularly oval, commonly 1.0 x 0.5 mm, wider than the trabeculae. Autozooids in three or four longitudinal series, doubled at points of trabecular fusion, elongate, convex, bounded by raised, sinuous sutures. Primary orifice longer than wide, proximal edge concave, condyles small and rounded, mid-distal border indistinctly toothed. Two widely spaced distal oral spines present in early ontogeny only. Peristome short, moderately developed, consisting of two proximo-lateral flaps, one of which is produced as a short lobe, extending across proximal border of orifice to fuse with avicularian cystid and defining a large, round foramen. Other flap of peristome fused with the lateral suboral avicularium, which is cylindrical, columnar, 0.15-0.3 mm high; rostrum terminal, oval, 0.08 mm long, with complete crossbar and small columella. Frontal shield smooth, with fewer than ten large, irregularly spaced marginal pores. Frontal avicularia often numerous, identical to suboral type, or larger, to 0.13 mm, with triangular rostrum. Both types of avicularia common around edges of fenestrulae, and on abfrontal surface of colony. Ovicell as wide as long; aperture highly arched, without a labellum; fissure extending most of length of ovicell.

This species has been reported from Faroe and Shetland, and from an unknown locality off SW Ireland. It has only recently been distinguished from *R. couchii* and its occurrence and distribution are thus scarcely known.

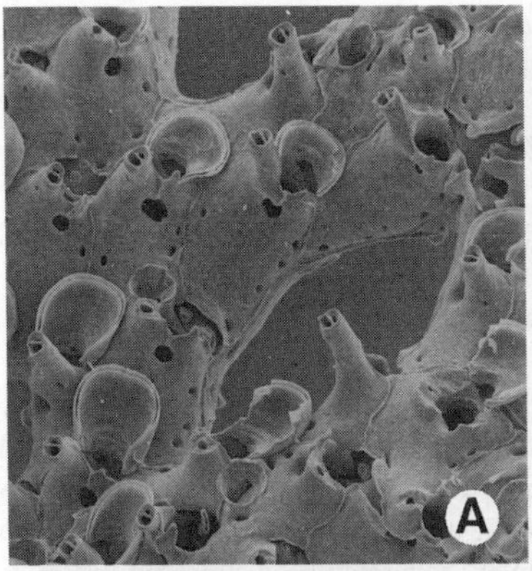

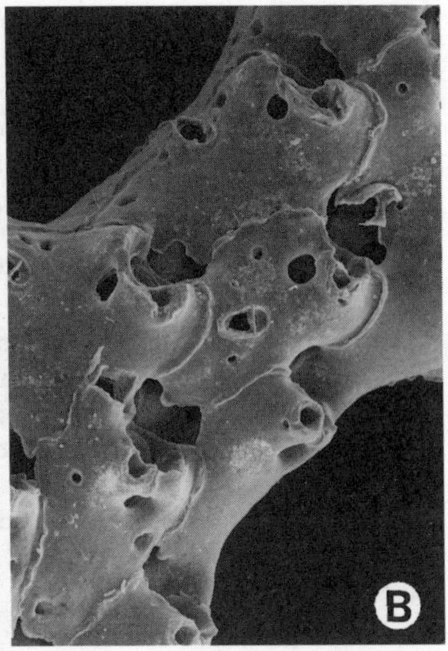

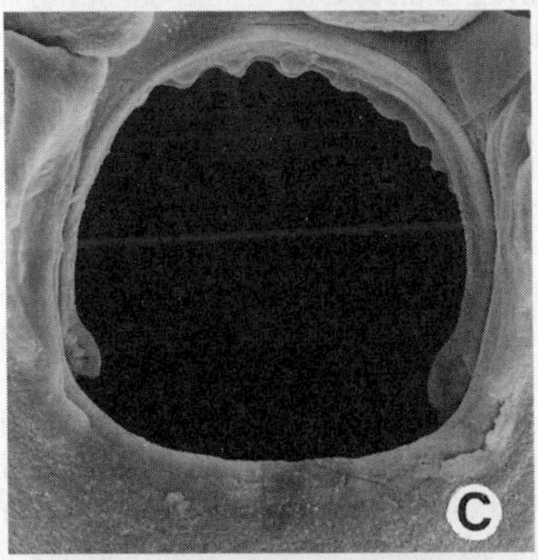

Fig. 175. *Reteporella incognita*.
A, Portion of the holotype specimen; x40. B, Group of autozooids showing characteristic peristome; x70. C, The primary orifice; x350.

Fig. 176. *Reteporella septentrionalis*.
C, Ovicellate autozooids; x55. D, Group of autozooids, and an enlarged avicularium; x90

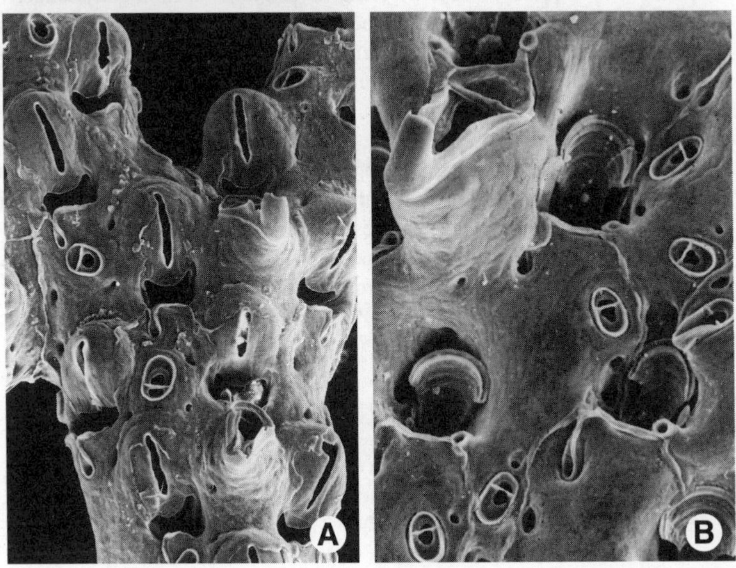

Reteporella septentrionalis (Harmer)
(Figs 176; 177)

Retepora cellulosa forma *cellulosa* Smitt 1868b: 35, pl. 28, figs 222-225
Retepora cellulosa: Kluge 1962: 526; 1975: 639
Sertella septentrionalis Harmer 1933: 620; Hayward & Ryland 1979: 266, figs 112C, 115
Reteporella septentrionalis: Hayward & Ryland 1996: 109, figs 2D-F, 3A

Colony 20-30 mm high, with equivalent horizontal spread, developing a shallow cup-shape. Fenestrulae elongate oval, 1.0-1.6 mm long, often wider than the relatively slender trabeculae, giving a widely open fenestration. Autozooids in two to four longitudinal series, doubled at points of trabecular fusion; elongate, commonly 0.4-0.6 x about 0.2 mm, convex, separated by distinct sutures in early ontogeny. Primary orifice about as wide as long, proximal edge straight; a pair of slender, lateral oral spines present in early ontogeny, obliterated by later calcification. Peristome deep, hiding orifice, consisting of two unequal-sized, proximo-lateral flaps, one appearing to overlie the other, with a short, closed fissure, and a small, basal pseudospiramen. Frontal shield finely granular, distinctly nodular, with three to five widely spaced and conspicuous marginal pores. Avicularia polymorphic, usually numerous, frontal only, not associated with peristome; most often oval, varying in length and width, typically 0.09-0.11 mm, with complete, slender crossbar, and a comparatively large oval, palatal foramen; enlarged avicularia frequent, with swollen cystid occupying most of frontal surface of bearing autozooid, rostrum almost perpendicular to frontal plane, 0.3-0.35 mm long, strongly hooked distally, and with conspicuous paired, tooth-like processes about halfway along its length. Ovicell slightly longer than wide, with a short, straight-edged labellum, and a widely open frontal fissure occupying about two-thirds of its length. Abfrontal surface of colony with numerous oval avicularia, similar to the frontal type. Encrusting base of colony consists of thickly calcified kenozooids, each with four large marginal pores and a small oval avicularium.

Offshore, on hard substrata, frequently attached to shell and other biogenic carbonates. An Arctic-boreal species, distributed in the NE Atlantic region from the White and the Kara Seas to western Greenland, south to the western continental shelf of the British Isles and into the Mediterranean, where it occurs commonly in shallow waters. Mediterranean specimens show slight differences from north European specimens, but appear to represent the same species (Hayward & Ryland, 1996).

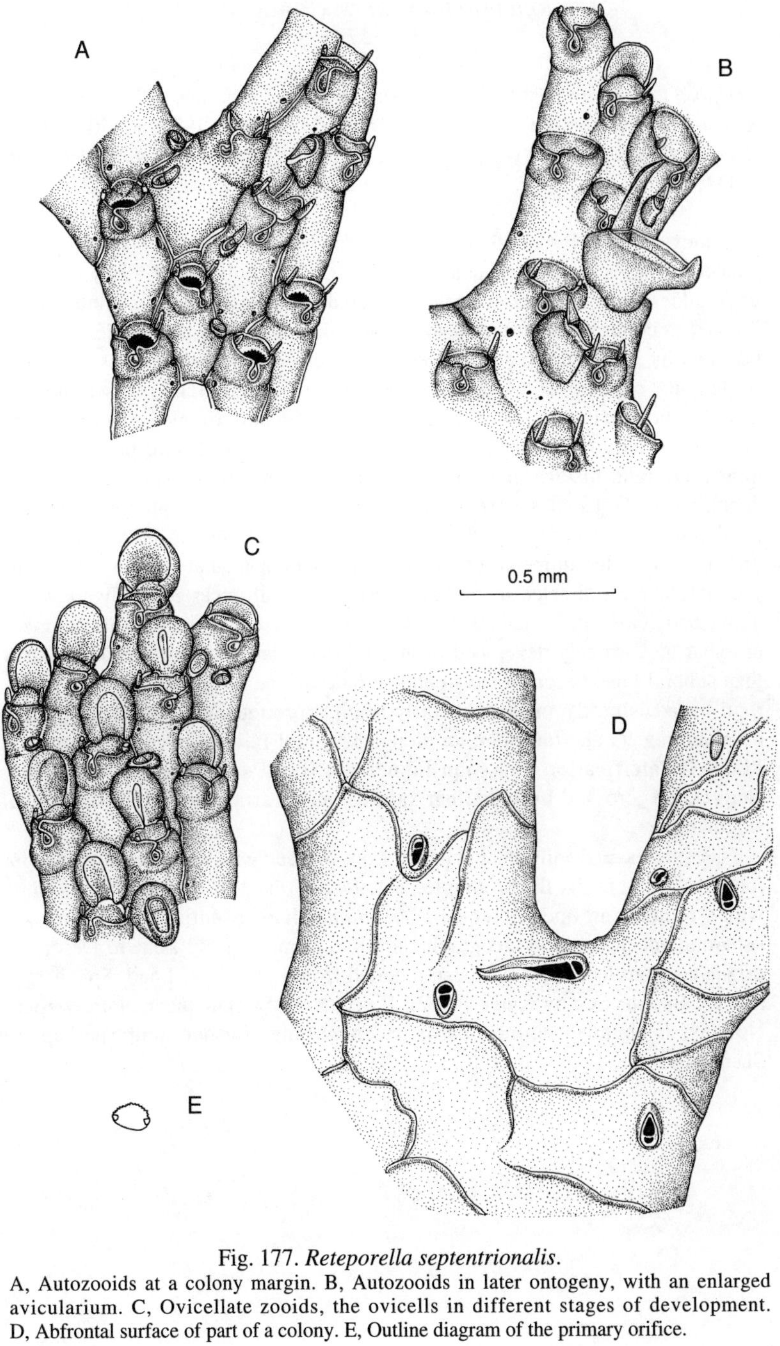

Fig. 177. *Reteporella septentrionalis*.
A, Autozooids at a colony margin. B, Autozooids in later ontogeny, with an enlarged avicularium. C, Ovicellate zooids, the ovicells in different stages of development. D, Abfrontal surface of part of a colony. E, Outline diagram of the primary orifice.

Reteporella watersi (Nordgaard)
(Fig. 178)

Retepora beaniana var. *watersi* Nordgaard 1907: 16
Retepora beaniana var. *watersi*: Kluge 1962: 529, fig. 370; 1975: 643, fig. 370
Reteporella watersi: Hayward 1994: 195, figs 8A, B; Hayward & Ryland 1996: 111, fig. 3B

Colony a broad, flat or slightly concave fan, thickly calcified, robust; up to or exceeding 40 x 40 mm. Fenestrulae small, oval, commonly 1.0 x 0.5 mm; trabeculae stout, consisting of three or four alternating series of autozooids, doubled at points of trabecular fusion. Autozooids 0.45-0.5 x about 0.25 mm, each bounded by raised, thickened sutures; frontal calcification distinctly shagreened, or dimpled, regularly so in newly budded autozooids, generally coarsely nodular in older, proximal parts of colony, a few large, thickly rimmed marginal pores present. Primary orifice wider than long, about 0.1 x 0.14 mm; proximal edge almost straight, distal rim with very indistinct denticulations, condyles large, angular. Four oral spines present in newly budded autozooids, only the lateral pair persisting in later ontogeny. Peristome low and thick, hiding proximal part of orifice, with a closed, medio-proximal pseudospiramen, adjacent to a short, oval, proximo-laterally directed suboral avicularium. Lateral spines generally projecting above peristome rim. Avicularium with finely toothed distal rostral rim, complete crossbar and scarcely developed palate; similar avicularia, as large as or larger than suboral type, occur on the frontal surfaces of the autozooids. Ovicell longer than broad, slightly inflated distally, with a pronounced, squared labellum overhanging its aperture, and an elongate frontal fissure; enveloped in thick, dimpled calcification and soon immersed. Basal surface of colony finely shagreened, crossed by indistinct sutures and bearing numerous small oval avicularia.

This species is readily distinguished from *S. beaniana*, with which it was for long confused, by its thick, dimpled calcification, and by the peristomial pore, which remains an open notch in *S. beaniana*. It is essentially an Arctic and subarctic species, but its geographical range is imprecisely known. Described originally from Iceland, it has been reported from Franz Josef Land, Spitzbergen and the Barents Sea by Kluge (1962). It occurs in the Hardanger Fjord, western Norway, is abundant around Faroe and may range further south, perhaps to Shetland.

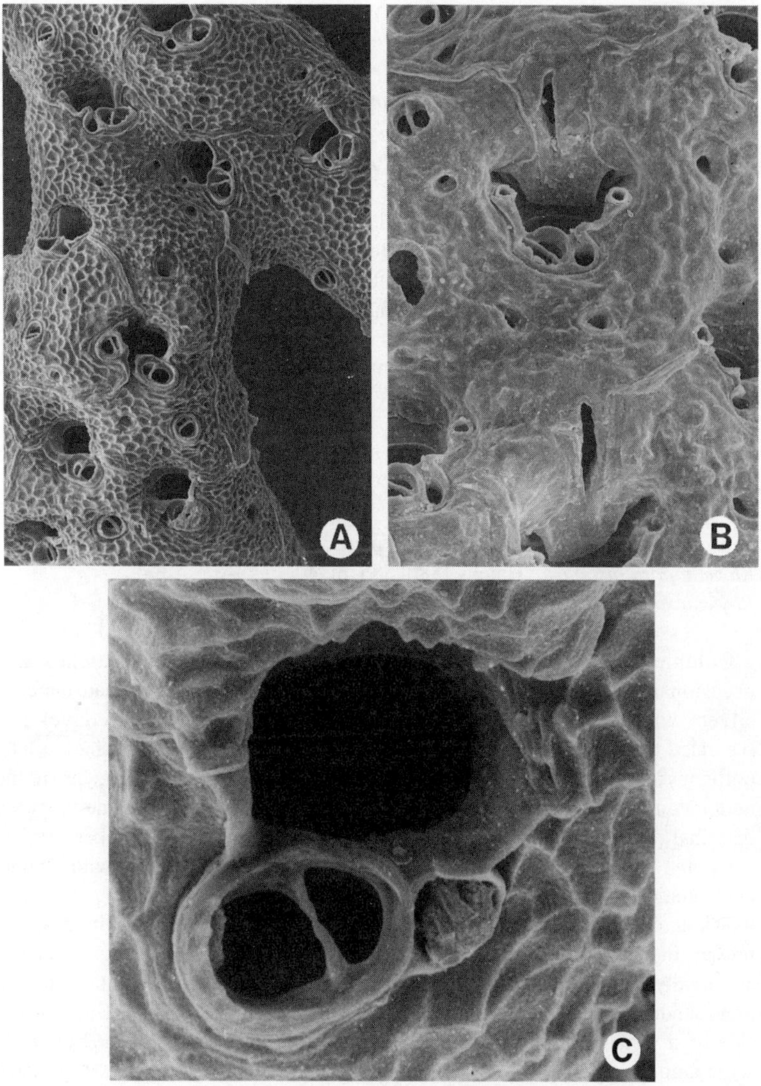

Fig. 178. *Reteporella watersi*.
A, Portion of a colony; x50. B, Ovicellate zooids; x120. C, The primary orifice; x280. (Specimen from Faroe).

Genus RHYNCHOZOON Hincks, 1895

Colony encrusting; unilaminar, to multilaminar or nodular. Frontal shield with marginal pores only. Primary orifice with denticulate distal margin, and proximal sinus; condyles present. Oral spines present. Peristome well developed, typically with an asymmetrical pseudosinus, a suboral avicularium, and variously developed mucrones; in many species a thick, blunt uncinate process projects above the proximal portion of the orifice from the inner face of the peristome. Avicularia adventitious, polymorphic, often abundant. Ovicell prominent, imperforate, with an uncovered area of endooecium frontally, and a variously developed labellum. Small basal pore chambers present.

Type species: *Lepralia bispinosa* Johnston, 1847

Rhynchozoon bispinosum (Johnston)
(Figs 179; 180A-C)

Lepralia bispinosa Johnston 1847: 326, pl. 57, fig. 10
Rhynchopora bispinosa: Hincks 1880: 385, pl. 40, figs 1-5
Rhynchozoon bispinosum: Hayward & Ryland 1979: 272, fig. 118

Colonies developing extensive unilaminar or multilaminar patches, and occasionally large nodular forms; living specimens pale lilac, dead material silvery white. Autozooids oval to hexagonal, initially strongly convex and separated by deep grooves, boundaries becoming progressively indistinct in later ontogeny; 0.35-0.5 x 0.2-0.35 mm. Frontal shield finely granular, becoming nodular in later ontogeny; each autozooid with up to 10 large, widely spaced marginal pores, particularly emphasised as frontal calcification thickens in late ontogeny. Primary orifice orbicular, as wide as long; distal border with distinct denticulation, proximal border with a broad V-shaped sinus occupying about half its width; condyles rounded, broad and short. Two widely spaced distal oral spines present in early ontogeny only. Peristome thick and deep, enclosing the orifice proximally and laterally, developed proximally as a stout, conical umbo, adjacent to a rounded pseudosinus; within the peristome the pseudosinus is defined on one side by a stout uncinate process, and on the other by a short, rounded knob. In some autozooids the proximal umbo is replaced by an avicularium, with inflated cystid, and broadly triangular, hooked rostrum, acute to plane of orifice and directed laterally. In late ontogeny additional, short knobbed processes develop on the lateral borders of the peristome, largely hiding the orifice. Small adventitious avicularia sporadically present on the frontal surfaces of autozooids, budded from the marginal pores; the rostrum slender, triangular, distally hooked, acute to frontal plane. Ovicell about as wide as long, flattened frontally, with an extensive area of endooecium uncovered but lacking a labellum; obscured by calcification originating from distally succeeding autozooids, and not readily visible in late ontogeny.

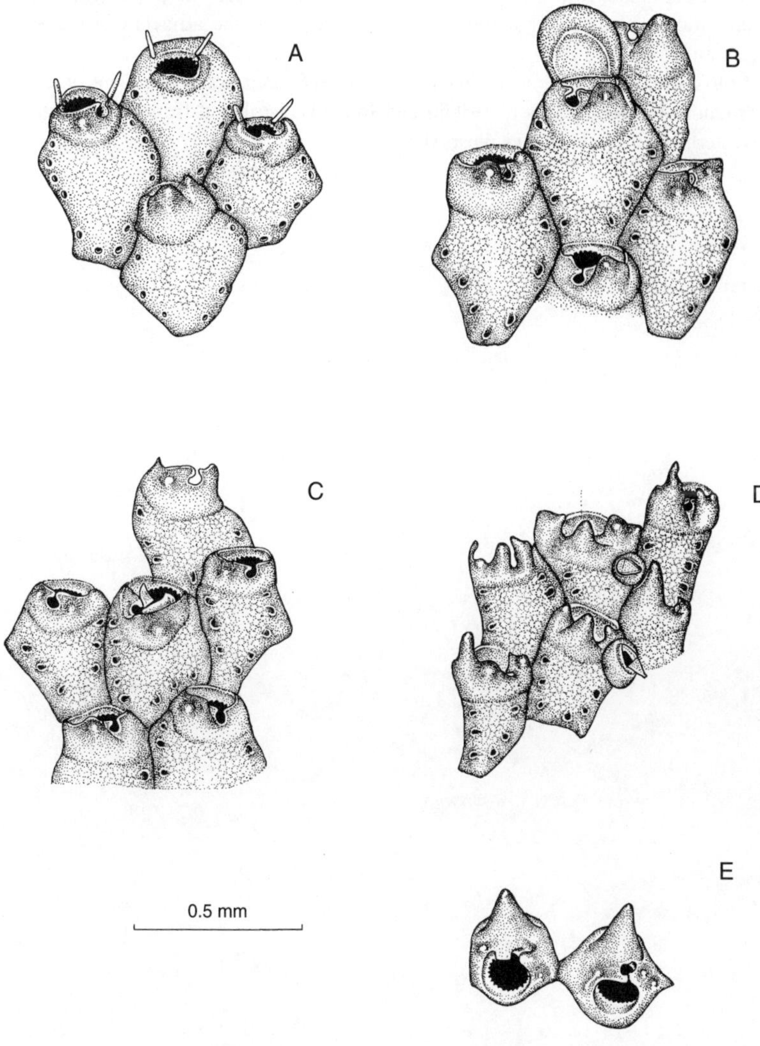

Fig. 179. *Rhynchozoon bispinosum.*
A, Autozooids at the colony margin, with oral spines and small peristomes. B, An ovicellate zooid. C, Autozooids in later ontogeny, with fully developed peristomes; avicularium well shown in the central one. D, Autozooids with mucronate peristomes. E, Distal view of two adjacent autozooids, showing the peristomial denticles.

The primary orifice and the characteristic, paired, distal oral spines are seen only in autozooids at the growing margin of the colony. The peristome deepens and thickens rapidly, but the uncinate process and the adjacent pseudosinus remain distinct.

Offshore, on stones, shells and other hard substrata, often abundant on coarse grounds. Western coasts of Britain and Ireland, distributed from Shetland to the western Mediterranean and West Africa.

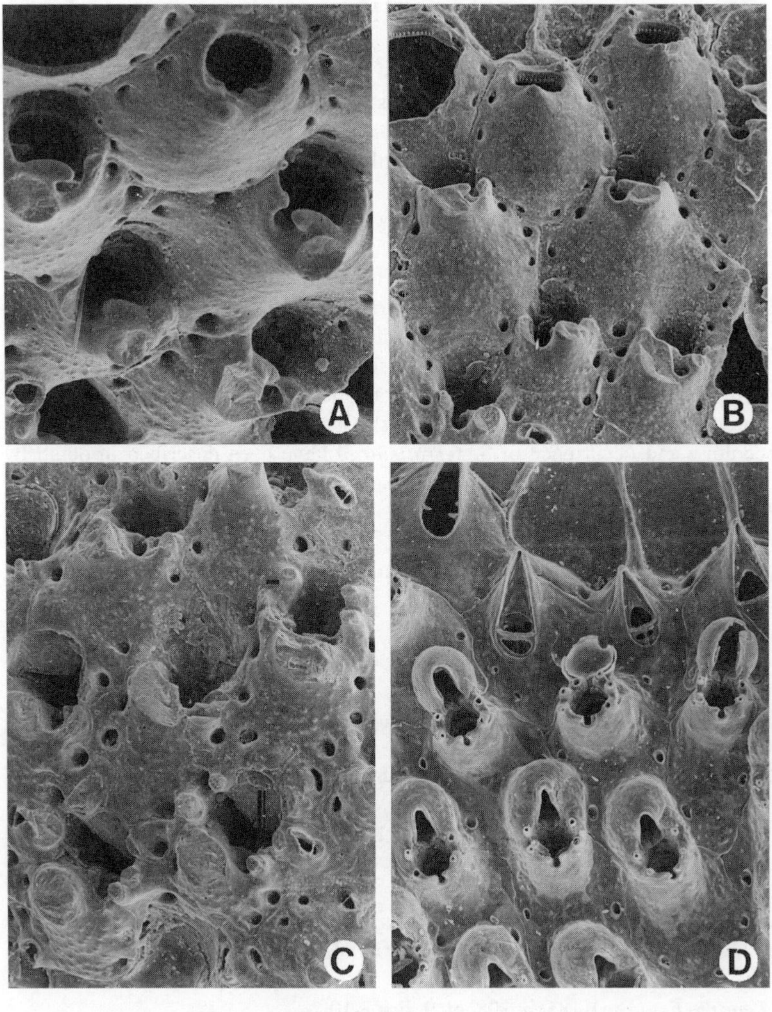

Fig. 180.
A-C, *Rhynchozoon bispinosum.*
A, Autozooids at the colony margin; x80. B, Colony margin in frontal view, showing development of peristome; x55. C, Autozooids in later ontogeny, some with ovicells; x60.
(Specimen from Eddystone)

D, *Schizotheca fissa.*
Colony margin, with vicarious avicularia and ovicellate zooids; x60.

Genus SCHIZOTHECA Hincks, 1877

Colony encrusting. Frontal shield of autozooids with few marginal pores. Primary orifice subcircular; border distal to blunt lateral condyles with fine denticulations. Avicularia vicarious. Ovicell with frontal fissure, imperforate, not closed by zooidal operculum. Large septula present in distal and lateral walls.

Type species: *Lepralia fissa* Busk, 1856

Two species of *Schizotheca* occur commonly in the British sea area, both develop small patch colonies. The Adriatic species *S. serratimargo* (Hincks) develops erect, three-dimensional, branching colonies which may exceed 15 cm diameter. *S. serratimargo* occurs also in the western Mediterranean and southwards to West Africa, and on rare occasions has been found off the southwestern approaches to the English Channel. The morphology, biology, ecology and occurrence of *S. serratimargo* have been described in detail by McKinney & McKinney (1993, 1994).

Key to the British species of *Schizotheca*

1. Ovicell with a triangular, open ended fissure. Avicularia frequent, with large pointed mandible
.... ***Schizotheca fissa*** (below)

Ovicell with a slender enclosed fissure. Avicularia rare, with small semi-elliptical mandible
.... ***Schizotheca divisa*** (p. 384)

Schizotheca fissa (Busk)
(Figs 180D; 181)

Lepralia fissa Busk 1856b: 308, pl. 9, figs 8-10
Schizotheca fissa: Hincks 1880: 284, pl. 41, figs 1-3; Hayward & Ryland 1979: 268, fig. 116

Colony forming small subcircular incrustations. Autozooids in linear series, alternating, oval to hexagonal, convex, separated by shallow grooves; 0.35-0.4 x 0.2-0.25 mm. Frontal shield initially smooth, fine grained and vitreous, becoming coarser with secondary calcification; imperforate except for a few indistinct marginal pores. Primary orifice suborbicular, distal border finely denticulate; a small U-shaped sinus proximally. Peristome well developed, thickened and erect; with a U-shaped notch medially in the proximal rim, extended as a groove down the inner side of the peristome. Six long, slender oral spines disto-laterally, four in

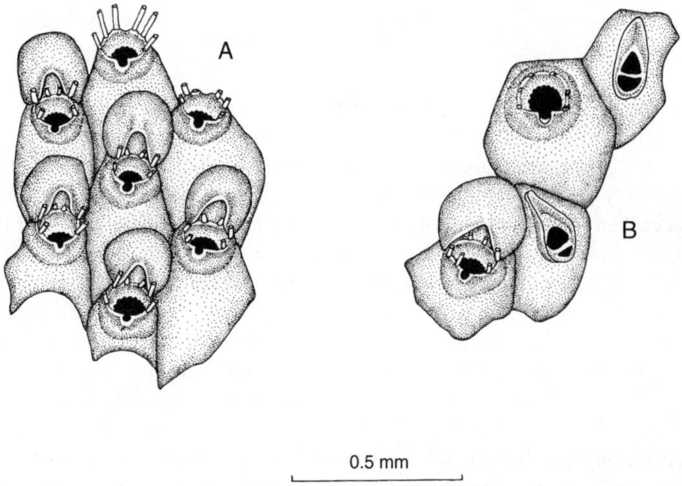

Fig. 181. *Schizotheca fissa*.
A, Group of autozooids, two with ovicells. B, Autozooids in later ontogeny, with well developed peristomes, and two vicarious avicularia.

ovicellate autozooids; delicate and frequently broken short. Avicularia vicarious; cystid oval or lozenge-shaped, as large as, and replacing, an autozooid; rostrum triangular, directed distally, with a slender pivotal bar. Avicularium rarely associated with an ovicell, the rostrum then inverted, directed proximally. Ovicell at first prominent, later immersed in secondary calcification extending from the distal autozooid; smooth surfaced and imperforate, with a broad, triangular, frontal fissure. Embryos red, present in February, June-October and December in Manx waters (Eggleston, 1969) and in the western Mediterranean (Gautier, 1962). Ancestrula undescribed.

Found predominantly encrusting shell, from below the kelp zone on the shallow areas of the continental shelf; 20-120 m in the western Mediterranean (Gautier, 1962). From the Cape Verde Islands (Cook, 1968) and the Mediterranean, *S. fissa* extends as far north as the British Isles. It has been recorded from numerous localities in the western end of the Channel but does not reach the North Sea; on western coasts it has been reported as far north as the Isle of Man and Antrim.

Schizotheca divisa (Norman)
(Figs 182; 183A, B)

Lepralia divisa Norman 1864: 86, pl. 10, fig. 6
Schizotheca divisa: Hincks 1880: 285, pl. 41, figs 4-6; Hayward & Ryland 1979: 270, fig. 117

Colony forming small round incrustations. Autozooids in regular, linear series, alternating, oval to hexagonal, convex and separated by deep grooves; 0.3-0.4 x 0.2 mm. Frontal shield smooth, fine grained and vitreous, imperforate except for a few widely spaced, small and inconspicuous, marginal pores. Primary orifice broader than long, suborbicular; distal border finely denticulate, proximal border straight; condyles blunt, downcurved. Peristome well developed with a deep U-shaped notch on the inner side of the proximal border, to the right or left of the midline. This notch may close over distally, forming a tube. As the peristome thickens an asymmetrical mucro may develop, with a number of denticles on the inner edge. Six long, delicate oral spines present disto-laterally, four in ovicellate autozooids; fragile and most usually lost in later autozooids. Vicarious avicularia rare, often absent; cystid as large as, and replacing, an autozooid; rostrum small, elliptical, directed distally or disto-laterally. Ovicell prominent, smooth surfaced and elongate, opening into the peristome; flattened frontally, with a long, narrow median fissure, which is closed proximally. Ancestrula tatiform, with about eight spines.

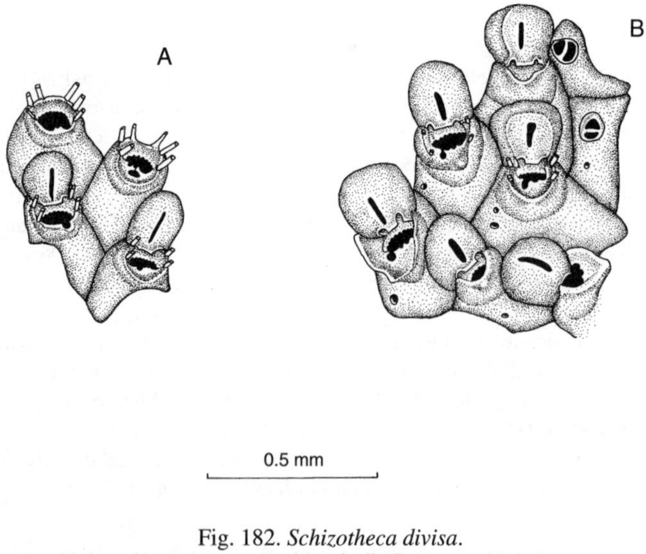

Fig. 182. *Schizotheca divisa*.
A, Autozooids in early ontogeny, two with ovicells. B, Autozooids in later ontogeny, with well developed peristomes, and two vicarious avicularia.

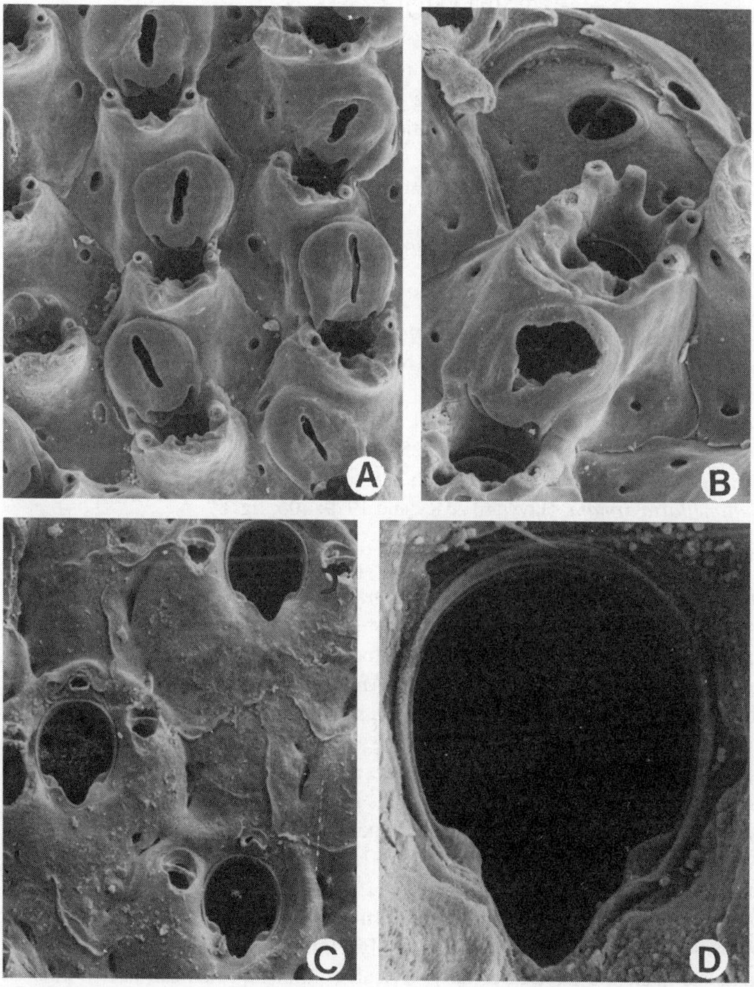

Fig. 183.
A, B, *Schizotheca divisa.*
A, Ovicellate zooids; x90. B, Colony margin, with a vicarious avicularium; x130.

C, D, *Stephanollona armata.*
C, Group of autozooids; x90. D, The primary orifice; x400.

Occurs offshore, mostly on shell. Nothing is known of its biology. A rarely-reported species of apparently very restricted distribution. It has been reported from the Antrim coast and the Channel Islands (Hincks, 1880), and has been discovered in dredgings from several localities in the Gulf of St. Malo and the western end of the Channel. There seem to be no records remote from the British Isles.

Genus STEPHANOLLONA Duvergier, 1920

Colony encrusting, unilaminar to multilaminar. Frontal shield of autozooids with marginal pores. Primary orifice sinuate, distal rim denticulate; oral spines present or absent. Avicularia adventitious, dimorphic: lateral to orifice, paired or single, with short oval or elongate, spatulate mandible. Ovicell hyperstomial, imperforate. Basal pore chambers present.

Type species *Stephanollona spinifera* Duvergier, 1920

Stephanollona armata (Hincks)
(Figs 183C, D; 184)

Lepralia armata Hincks 1862: 207, pl. 12, fig. 5
Schizoporella armata: Hincks 1880: 258, pl. 41, figs 7-8
Rhynchozoon armatum: Gautier 1962: 238
Buffonellaria armata: Hayward & Ryland 1979: 206, fig. 87

Colony encrusting. Autozooids oval to quadrate, slightly convex, separated by shallow grooves; 0.5-0.6 x about 0.4 mm. Frontal wall initially smooth; rapidly covered by secondary calcification, at first granular becoming smoother and quite vitreous as the process continues. Imperforate centrally; marginal pores distinct, most conspicuous in older autozooids. Anter of primary orifice orbicular, with finely denticulate rim, sinus deep, narrow, slit-like; operculum well chitinised and distinct, horn coloured. Four or five long, distal oral spines, distinctly jointed at the base, often cylindrical and slender, more usually flattened and oar-like, sometimes very long, attaining 0.6 mm. In older autozooids a low, thickened peristome may be present around the proximal and lateral borders of the orifice. Avicularia adventitious, single or paired, lateral to the orifice; not present on all autozooids and often missing from large areas of the colony. Cystid typically small, with an elliptical rostrum supporting a semicircular mandible, directed laterally or disto-laterally; occasionally elongate, enlarged, with a slender spatulate mandible directed distally. Ovicell prominent, spherical, later immersed in secondary calcification; finely granular and imperforate, with a short frontal lip.

The appearance of this species alters considerably as it ages; continuous calcification tends to obscure autozooid boundaries, which are indicated only by the conspicuous marginal pores. The orifice becomes deeply immersed and the peristome, incorporating the spine bases, is thickened, with an irregular indented rim. New autozooids, and additional avicularia, may be produced by frontal budding.

Gautier (1962) reported embryos in August and September in the western Mediterranean.

A warm temperate species, found on shell and stone offshore to 200 m, *S. armata* ranges from Madeira and the Mediterranean to the south-western shores of Britain.

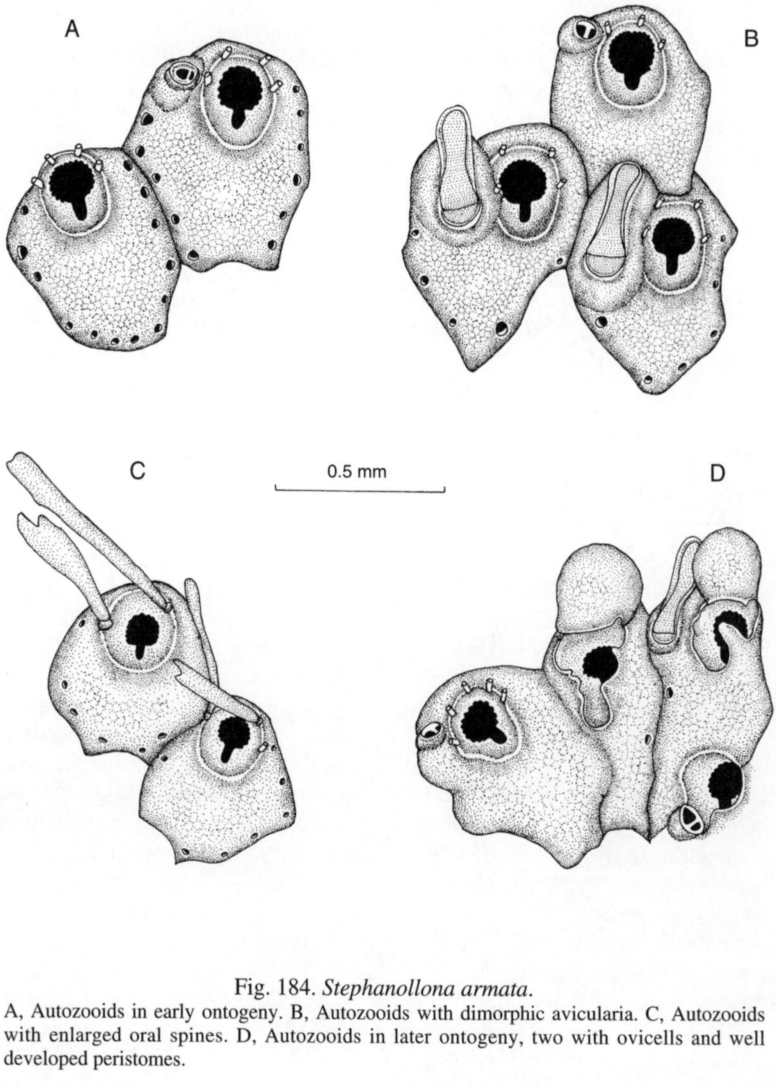

Fig. 184. *Stephanollona armata*.

A, Autozooids in early ontogeny. B, Autozooids with dimorphic avicularia. C, Autozooids with enlarged oral spines. D, Autozooids in later ontogeny, two with ovicells and well developed peristomes.

Glossary of Special Terms

Acleithral ooecium: One not closed by the parental operculum.

Adventitious avicularium: One occupying some position on the external wall of an autozooid (cf. vicarious).

Ancestrula: First-formed zooid of a colony, derived by metamorphosis of a free-swimming larva.

Androzooid: A male zooid, its skeletal morphology similar or dissimilar to other zooids, but developing testes only during the reproductive period (cf. gynozooid).

Anter: Part of the orifice distal to the condyles in ascophorine Cheilostomatida.

Apertural bar: Fused pair of costae immediately proximal to the orifice in cribriomorph Cheilostomatida.

Aperture: In ascophorines, an opening in the frontal wall not necessarily coextensive with the orifice.

Areola: Space, between two buttresses formed by secondary thickening, above an areolar pore in some Cheilostomatida.

Areolar pore: Marginal pore in the frontal wall of some Cheilostomatida.

Ascopore: Median frontal pore which serves as inlet of the ascus in some ascophorine Cheilostomatida.

Ascus: Sac-like hydrostatic organ in ascophorine Cheilostomatida (syn. compensation sac, compensatrix).

Astogenetic change: The progressive elaboration of zooidal morphology from the ancestrula through the first few generations of autozooids budded from it.

Astogeny: Development of a colony by budding.

Autozooid: Feeding zooid in Bryozoa.

Avicularium (pl. **Avicularia**): Specialised zooid in Cheilostomatida with reduced polypide but strong muscles which operate a mandible-like operculum (see adventitious, interzooidal, vicarious).

Basal: Under or reverse side of an encrusting or freely-growing colony (cf. proximal).

Brown body: Coloured spheroid in bryozoan autozooid resulting from aggregation of non-histolysable residue of a degenerated polypide.

Cardella: See condyle.

Cauda: Thread-like proximal portion of club-shaped autozooid.

Ciliated groove: The groove on the lower surface of a larva which enables it to grip the substratum when settling.

Cleithral ooecium: One closed by the operculum of the parent zooid.

Coalescent multiserial: A type of zooidal or intrazooidal budding in which new zooids form from the fusion of adjacent buds.

Coelomopore: A pore in the body wall connecting the coelom with the exterior. Applied specifically to the pore at the base of the tentacles through which ova are extruded (syn. supraneural pore).

Communication pore: Opening in interzooidal wall (syn. mural pore; cf. dietella, septulum).

Compound lineal: A type of zooidal or intrazooidal budding in which each new bud is divided by interior partition walls into two or more zooid series.

Compound non-lineal: A budding process in which a colony-wide body wall expands to create new extrazooidal coelomic space, which is then variously divided by interior calcification to define new zooids.

Condyle: One of a pair of oppositely placed protuberances on which the operculum pivots in the orifice of ascophorine Cheilostomatida (syn. cardella).

Costa: One of the modified spines overarching the frontal membrane in cribriomorph Cheilostomatida; usually united with neighbouring costae to form a frontal shield (syn. costula).

Costate shield: The frontal wall of cribriomorph ascophorines, formed from series of fused spines, or costae.

Cryptocyst: More or less horizontal calcareous lamina on the basal side of the frontal membrane in some Neocheilostomatina, developed from the vertical walls of the zooid but not completely dividing its body cavity.

Cryptocystidean: Ascophorines in which the frontal wall forms as an interior partition below an overlying hypostegal coelom.

Cyphonautes: Pelagic bivalved larva of some bryozoans.

Cystid: Cellular plus skeletal layers of the zooid wall (cf. polypide, zooecium).

Dietella: Small enclosed space near base of vertical walls in distal part of zooid in certain Cheilostomatida; its walls contain communication pores traversed by mesenchymatous fibres.

Discrete multiserial: Zooidal or multizooidal budding in which new zooids form in lineal progenitor-descendant series.

Distal: Pertaining to the side away from the ancestrula or origin of growth.

Divaricator muscles: Muscles employed in opening the operculum.

Ectooecium: Outer, generally calcified, layer of ooecial wall.

Endooecium: Inner, often membranous, layer of ooecial wall (= entooecium).

Endotoichal ooecium: One which appears immersed in the distal zooid, while opening independently to the exterior.

Endozooidal ooecium: One which appears immersed in the distal zooid, while opening below the operculum of the parent zooid (syn. entozooecial ooecium).

Entooecium: Inner, often membranous, layer of ooecial wall (= endooecium).

Epistege: In Cribrilinidae, the space between the frontal membrane and the overlying frontal shield (syn. epistegal space).

Exterior calcification: Formed by the deposition of calcite crystallites on the underside of outer, cuticular body wall.

Extrazooidal: Structures arising from, or shared by, more than a single zooid.

Fenestra: Uncalcified area in the ectooecium through which the endooecium is displayed.

Fenestrula: Open space in reticulate colony, e.g., in *Reteporella*.

Frontal: Pertaining to the exposed or orifice-bearing side of zooid or colony (syn. obverse).

Frontal membrane: Uncalcified part of frontal body wall in Cheilostomatida; may be exposed or covered by a frontal shield.

Frontal shield: The calcified frontal surface of ascophorine zooids, not necessarily homologous in all groups; coextensive with the outer body wall, formed as a partition below it, from the fusion of overarched spines or by the calcification of a proximally-derived fold.

Frontal wall: A calcareous frontal body wall (covering the ascus in ascophorine Cheilostomatida).

Funiculus: Mesenchymatous strands connecting the polypide with the zooidal wall (actually with the communication pores in Cheilostomatida and Ctenostomatida).

Gonozooid: Zooid modified as a brood chamber.

Gymnocyst: A calcified frontal shield formed of exterior calcification, constituting part or all of the frontal shield in cheilostomate zooids.

Gymnocystidean: Ascophorines in which the frontal wall forms by the calcification of the outer body wall.

Gynozooid: A female zooid, dimorphic or bearing an ovicell, but developing an ovary only during the reproductive period (cf. androzooid).

Heterozooid: Specialised zooid; one not an autozooid.

Hypostegal coelom: An extension of the main coelom overlying the frontal calcification in umbonuloid and cryptocystidean species.

Hyperstomial ooecium: One which rests on or is partly embedded in the distal zooid, and opens above the operculum of the parent zooid.

Independent ooecium: One which develops independently of the distal zooid.

Inner vesicle: Syn. of ooecial vesicle.

Intertentacular organ: Tubular extension of the coelomopore (used for the extrusion of ova) between the dorso-medial tentacles.

Interior calcification: Formed by the deposition of calcite crystallites on internal tissues, not adjacent to the outer body wall.

Interzooidal avicularium: One which extends to the basal surface of the colony, but is wedged in between autozooids rather than replacing one of them in a series.

Intrazooidal bud: One arising from a pre-existing space - a pore chamber, septulum or frontal coelomic space - within the parental zooid.

Introvert: See tentacle sheath.

Kenozooid: Heterozooid without a polypide, and usually without either orifice or muscles.

Lacuna (pl. **Lacunae**): Perforation (true pore) between costae making up the frontal shield in cribriomorph Cheilostomatida.

Lophophore: The ring of hollow, ciliated tentacles surrounding the mouth of the zooid.

Lucida: A transparent spot on the surface of a chitinous operculum.

Lyrula: Median tooth, often anvil-shaped, on the proximal side of the orifice in some Cheilostomatida.

Mandible: Articulated part of an avicularium, moved by muscles, and homologous with the operculum of an autozooid.

Marginal spine: Spine, often jointed at base, in a series surrounding the frontal area in Cheilostomatida.

Maternal zooid: The proximal reproductive zooid from which ova pass to an ooecium.

Mucro (pl. **Mucrones**): A blunt or spinous elevation of the (usually proximal) lip of the orifice in some Cheilostomatida.

Multiporous septulum: See septulum.

Multizooidal: See extrazooidal.

Multizooidal budding: A budding process in which continuously expanding giant buds are sequentially divided by interior, partition walls to define new autozooids.

Mural pore-chamber: Cavity enclosed by the rosette plate, in the vertical zooidal walls of some species.

Nanozooid: Dwarf zooid containing reduced polypide in some Cyclostomatida.

Occlusor muscles: Muscles closing the operculum.

Ontogenetic change: The development of zooidal morphology from undifferentiated buds at the growing edge of a colony, through the series of zooids preceding them.

Ooecial vesicle: The inner membrane and means of closure of an ooecium.

Ooecium (pl. **Ooecia**): Ovicell or brood-chamber in Cheilostomatida, excluding the inner vesicle.

Operculum (pl. **Opercula**): A generally uncalcified lamina, hinged or pivoting on condyles, which closes the zooidal orifice in Cheilostomatida.

Opesia: In zooids of Neocheilostomatina, the opening below the frontal membrane which remains after development of a cryptocyst.

Oral: Pertaining to the orifice of a bryozoan zooid (not to the mouth).

Orifice: Opening in the zooid wall through which the lophophore and tentacles are exserted. This primary orifice may, in Cheilostomatida, become surrounded by a peristome; the opening at the upper end is then known as the secondary orifice.

Ovicell: The globular brood chamber in Cheilostomatida.

Palate: Part of avicularium occupied by the mandible.

Peristome: A rim which may become elevated surrounding the primary orifice.

Plume cilia: Long sensory cilia present in most bryozoan larvae.

Polypide: Those organs and tissues in a bryozoan autozooid which undergo periodic replacement; namely, tentacles, tentacle sheath, alimentary canal, associated musculature and nerve ganglion.

Pore-chamber: See dietella.

Poster: Part of the orifice in ascophorine Cheilostomatida proximal to the condyles and leading to the ascus.

Primary orifice: The opening in the outer body wall through which the polypide is extruded, in cheilostomes coextensive with the operculum.

Proximal: Pertaining to the side toward the ancestrula or origin of growth.

Pseudopore: Tissue filled lacuna in the calcification of the outer zooid wall in many Bryozoa.

Pseudosinus: A notch or hole in the peristome of some ascophorans.

Quincuncial: Arrangement of five objects such that four are placed at the corners of a rectangle with the other in the centre.

Reteporiform: Having reticulate (*Retepora*-like) colonies, as in the Phidoloporidae (Cheilostomatida).

Retractor muscles: Muscles arising from the proximal end of the zooid, inserted on the pharynx and the base of the lophophore, serving to withdraw the polypide.

Rosette-plate: Multiporous subcircular area in vertical walls of Cheilostomatida for the passage of mesenchymatous fibres between zooids (syn. multiporous septulum).

Rostrum: (1) Spike-like prolongation of an avicularium; (2) distal part of avicularium occupied by mandible (syn. palate).

Sclerite: Thickened line in operculum, mandible or frontal membrane.

Secondary orifice: In those species with an immersed primary orifice, the outer opening of the peristome.

Septulum: Single (uniporous) or grouped (multiporous) perforations in the wall of Cheilostomatida for the passage of mesenchymatous fibres between zooids (see rosette-plate).

Septum: An interior wall or partition, not associated with cuticle; for example, transverse vertical walls of most species.

Sinus: Slit at proximal edge of orifice in some ascophorine Cheilostomatida.

Spiramen (pl. **Spiramina**): Median pore in the proximal wall of the peristome, into the cavity of which it leads.

Sulcus (pl. **Sulci**): A groove delineating the boundary between adjacent zooids.

Supraneural pore: See coelomopore.

Tatiform: An ancestrula with a membranous frontal wall, often surrounded by spines.

Tentacle sheath: Thin part of body wall introverted in retraction of tentacles and then enclosing them.

Trabecula: Branch separating the fenestrulae in reticulate Cheilostomatida (e.g., *Reteporella*).

Transverse parietal muscles: Short muscles inserted on flexible frontal wall which contract to depress the wall, thus raising coelomic pressure.

Umbo (pl. **Umbones**): A blunt prominence on the frontal wall or ovicell in some Cheilostomatida.

Umbonuloid: Ascophorines in which the frontal shield forms through the calcification of the lower side of a fold of body wall, derived from the proximal end of the zooid.

Vibraculum: Heterozooid in Cheilostomatida with operculum in form of a long seta slung between condyles.

Vicarious avicularium: One that replaces an autozooid in a series.

Zoarium: Syn. of colony.

Zooeciule: Small zooecium with orifice but not all parts of an autozooid.

Zooecium (pl. **Zooecia**): Skeleton of a bryozoan zooid.

Zooid: Single bryozoan individual. Various types are distinguished by prefixes.

Zooidal bud: One formed when the body wall of the parental zooid expands to create additional coelomic space, which is then partitioned from the parent to define a new zooid.

Acknowledgements

Sincere thanks are due to Mary E. Spencer-Jones (Natural History Museum, London) for responding cheerfully and swiftly to our constant demands for specimen loans, and to Dr. M. R. Fordy (School of Biological Sciences, University of Wales Swansea) for his unremitting labours with the scanning electron microscope. We are especially grateful to Jan Greengo for her patient, and valued, help in preparing the manuscript. The financial support of the Natural Environment Research Council is gratefully acknowledged.

References

Abildgaard, P. C., 1806. Descripsit et tabulas. Pp. 1-46 *in* Müller, O. F.: *Zoologica Danica, seu Animalium Daniae et Norvegiae rariorum ac minus notorum descriptiones et historia, etc.* 3rd Edition, Vol. 4. Hafniae.

Alder, J., 1856. A catalogue of the zoophytes of Northumberland and Durham. *Transactions of the Tyneside Naturalists' Field Club,* **3**, 93-160.

Alder, J., 1864. Descriptions of new British Polyzoa, with remarks on some imperfectly known species. *Quarterly Journal of Microscopical Science, new series,* **4**, 95-109.

Alvarez, J. A., 1992. *Fenestrulina asturiasensis* sp. nov. (Bryozoa: Cheilostomida) from the northern coast of the Iberian Peninsula. *Journal of the Marine Biological Association of the U.K.*, **72**, 727-730.

Alvarez, J. A., 1993. *Fenestrulina barrosoi* sp. nov. (Bryozoa Cheilostomida) with a review of the genus *Fenestrulina* on the Iberian Peninsula. *Journal of the Marine Biological Association of the U.K.,* **73**, 831-835.

Aristegui, J., 1984. *Briozoos Quilostomados (Ectoprocta, Cheilostomata) de Canarias: Estudio sistematico, Faunistico y Biogeografico.* Tesis Doctoral, Facultad de Biologia, Universidad de la Laguna. 524 pp.

Aristegui Ruiz, J., 1989. Consideraciones sobre el género *Celleporina* Gray, 1848 (Ectoprocta: Cheilostomata) en Canarias y descripcion de tres especies nuevas: *C. canariensis* sp. n., *C. fragilis* sp. n. y *C. labiata* sp. n. *Cahiers de Biologie Marine,* **30**, 143-165.

Atkins, D., 1955. The cyphonautes larvae of the Plymouth area and the metamorphosis of *Membranipora membranacea* (L.). *Journal of the Marine Biological Association of the U.K.,* **34**, 441-449.

Audouin, J. V., 1826. Explication sommaire des planches de polypes de l'Egypte et de la Syrie, publiées par Jules-César Savigny. *Description de l'Egypte,* **1**, 225-244.

Banta, W. C., 1969. The body wall of cheilostome Bryozoa. II. Interzooidal communication organs. *Journal of Morphology,* **129**, 149-170.

Barroso, M. G., 1919. Notas sobre briozoos espa–oles. *Boletin de la Real Sociedad española de historia natural,* **19**, 200-204, 340-347.

Barroso, M. G., 1920. Notas sobre briozoos españoles. 8. *Boletin de la Real Sociedad española de historia natural,* **20**, 353-362.

Barroso, M. G., 1924. Notas sobre briozoos de Tangier. *Boletin de la Real Sociedad española de historia natural,* **24**, 291-298.

Bassler, R. S., 1934. Notes on fossil and Recent Bryozoa. *Journal of the Washington Academy of Science,* **24**, 404-408.

Bassler, R. S., 1935. Bryozoa. Generum et Genotyporum. Index et Bibliographica. Part 67 *in* Quenstedt, W. (ed.): *Fossilium Catalogus 1: Animalia.* W. Junk: s'Gravenhage, 229 pp.

Beneden, P. J. van, 1845. Recherches sur l'anatomie, la physiologie et le development des Bryozoaires qui habitent la côte d'Ostende. *Mémoires de l'Academie royale des sciences, des lettres et des beaux-arts de Belgique,* **18**, 1-44.

Beneden, P. J. van, 1848. Recherches sur les polypes bryozoaires de la Mer du Nord. *Bulletin de l'Academie Royale de Belgique. Classe des Sciences,* **15**, 67-82.

Best, M. A. & Thorpe, J. P., 1983. Effects of particle concentration on clearance rate and feeding current velocity in the marine Bryozoan *Flustrellidra hispida. Marine Biology,* **77**, 85-92.

Best, M. A. & Thorpe, J. P., 1986a. Effects of food particle concentration on feeding current velocity in six species of marine Bryozoa. *Marine Biology,* **93**, 255-262.

Best, M. A. & Thorpe, J. P., 1986b. Feeding current interactions and competition for food among the bryozoan epiphytes of *Fucus serratus. Marine Biology,* **93**, 371-375.

Best, M. A. & Thorpe, J. P., 1994. An analysis of potential food sources available to intertidal bryozoans in Britain. pp. 1-7 *in* P. J. Hayward, J. S. Ryland and P. D. Taylor (eds.): *Biology and Palaeobiology of Bryozoans.* Olsen and Olsen, Fredensborg, 240 pp.

Bille Hansen, K., 1962. The Godthaab Expedition 1928. Bryozoa. *Meddelelser om Grønland,* **81**, 6, 1-74.

Bishop, J. D. D., 1987. Type and figured material from 'The Pliocene Bryozoa of the Low Countries' (Lagaaij, 1952) in the collection of the Royal Belgium Institute of natural Sciences. *Studiedocumenten, Koninklijk Belgisch Instituut voor Natuurwetenschappen,* **37**, 1-36.

Bishop, J. D. D., 1989. Colony form and the exploitation of spatial refuges by encrusting Bryozoa. *Biological Reviews,* **64**, 197-218.

Bishop, J. D. D. & Hayward, P. J., 1989. SEM Atlas of type and figured material from Robert Lagaaij's '*The Pliocene Bryozoa of the Low Countries*' (1952). *Mededelingen Rijks Geologische Dienst,* **43**, 1-64.

Blainville, H. M. D. de, 1830. Zoophytes. Vol. 60 *in* Cuvier, G. F. (ed.): *Dictionnaire des sciences naturelles, dans lequel on traite méthodiquement des différents êtres de la nature, par plusieurs professeurs du Muséum National d'Histoire Naturelle et des autres principales Écoles de Paris.* F. G. Levrault, Paris, 546 pp.

Bradstock, M. & Gordon, D. P., 1983. Coral-like bryozoan growths in Tasman Bay, and their protection to conserve commercial fish stocks. *New Zealand Journal of Marine and Freshwater Research,* **17**, 159-163.

Brock, F. J., 1985. South Australian fouling bryozoans. pp. 45-49 *in* Nielsen, C. and Larwood, G. P. (eds.): *Bryozoa: Ordovician to Recent.* Olsen and Olsen, Fredensborg. 364 pp.

Burgh, M. E. de, & Fankboner, P. V., 1978. A nutritional association between the bull kelp *Nereocystis luetkeana* and its epizootic bryozoan *Membranipora membranacea. Oikos,* **31**, 69-72.

Busk, G., 1851. Notices of three undescribed species of Polyzoa. *Annals and Magazine of Natural History, series 2,* **7**, 81-85.

Busk, G., 1852. *Catalogue of marine Polyzoa in the collection of the British Museum, 1. Cheilostomata (part).* Trustees of the British Museum, London. 54 pp.

Busk, G., 1854. *Catalogue of marine Polyzoa in the collection of the British Museum, 2. Cheilostomata (part).* Trustees of the British Museum, London. pp. 55-120.

Busk, G., 1856a. Zoophytology. *Quarterly Journal of Microscopical Science,* **4**, 93-96.

Busk, G., 1856b. Zoophytology. *Quarterly Journal of Microscopical Science,* **4**, 308-312.

Busk, G., 1856c. Polyzoa collected by Mr. M'Andrew on the coast of Norway and Finmark in 1856. *Annals and Magazine of Natural History, series 2,* **18**, 32-36.

Busk, G., 1859. *A monograph of the fossil Polyzoa of the Crag.* The Palaeontographical Society, London., 136 pp.

Busk, G., 1860a. Descriptions of new species of Polyzoa. Collected by George Barlee Esq., in Shetland. *Quarterly Journal of Microscopical Science,* **8**, 123-125.

Busk, G., 1860b. Zoophytology. Shetland Polyzoa. Collected by Mr. Barlee. (Continued). *Quarterly Journal of Microscopical Science,* **8**, 143-145.

Busk, G., 1860c. Zoophytology. Shetland Polyzoa. Collected by Mr. Barlee. (Continued and concluded). *Quarterly Journal of Microscopical Science,* **8**, 213-214.

Busk, G., 1861. Zoophytology. Description of new Polyzoa, collected by J. Y Johnson, Esq., at Madeira in the years 1859 and 1860. *Quarterly Journal of Microscopical Science, new series,* **1**, 77-80.

Busk, G., 1884. Report on the Polyzoa. The Cheilostomata. *Report of the Scientific Results of the Voyage of H.M.S. Challenger during the years 1873-1876. Zoology,* **10**, pt. 30, 1-216.

Buss, L. W., 1981. Mechanisms of competition between *Onychocella alula* (Hastings) and *Antropora tincta* (Hastings) on an eastern Pacific rocky shore. pp. 39-49 *in* Larwood, G. P. and Nielsen, C. (eds.): *Recent and Fossil Bryozoa.* Olsen and Olsen, Fredensborg. 334 pp.

Calvet, L., 1900. Contribution à l'histoire naturelle des bryozoaires ectoproctes marins. *Travaux de l'Institut de zoologie de l'Université de Montpellier et de la Station zoologique de Cette,* **8**, 1-488.

Cancino, J. M., Hughes, R. N. & Orellana, M. C., 1994. A comparative study of larval release in bryozoans. pp. 41-46 *in* P. J. Hayward, J. S. Ryland and P. D. Taylor (eds.): *Biology and Palaeobiology of Bryozoans.* Olsen and Olsen, Fredensborg. 240 pp.

Canu, F., 1900. Contribution à l'histoire naturelle des bryozoaires ectoproctes marins. *Travaux de l'Institut de zoologie de l'Université de Montpellier et de la Station zoologique de Cette,* **8**, 1-488.

Canu, F., 1904. Les Bryozoaires du Patagonien. Echelle des Bryozoaires pour les terrains tertiaires. *Mémoires de la Société Geologique de France, Paléontologie,* **12**, 1-30.

Canu, F. & Bassler, R. S., 1917. A synopsis of American Early Tertiary cheilostome Bryozoa. *Bulletin of the U.S. National Museum,* **96**, 1-87.

Canu, F. & Bassler, R. S., 1920. North American Early Tertiary Bryozoa. *Bulletin of the U.S. National Museum* **106**, 879 pp. (Vol. 1), 162 pls. (Vol. 2).

Canu, F. & Bassler, R. S., 1925. Les Bryozoaires du Maroc et de Mauretanie. *Mémoires de la Société des Sciences naturelles du Maroc,* **10**, 1-79.

Canu, F. & Bassler, R. S., 1927. Classification of the Cheilostomatous Bryozoa. *Proceedings of the U.S. National Museum,* **69**, 1-42.

Canu, F. & Bassler, R. S., 1928. Fossil and Recent Bryozoa of the Gulf of Mexico region. *Proceedings of the U.S. National Museum,* **72**(14), 1-199.

Canu, F. & Bassler, R. S., 1930. Bryozoaires marins de Tunisie. *Annales. Station océanographique de Salammbô,* **5**, 1-91.

Castric-Fey, A., 1971. Sur quelques Bryozoaires de l'Archipel de Glénan (Sud-Finistère). *Vie et Milieu,* **22**, 69-86.

Chadwick, S. R. & Thorpe, J. P., 1981. An investigation of some aspects of bryozoan predation by dorid nudibranchs (Mollusca: Opisthobranchia). pp. 51-58 *in* Larwood, G. P. and Nielsen, C. (eds.): *Recent and Fossil Bryozoa.* Olsen and Olsen, Fredensborg. 334 pp.

Cheetham, A., 1972. Cheilostome Bryozoa of Late Eocene Age from Eua, Tonga. *Professional Papers of the U.S. Geological Survey,* no. 640-**E**, 1-26.

Christopherson, C., 1985. Secondary metabolites from marine bryozoans. A review. *Acta Chemica Scandinavica,* B **39**, 517-529.

Chumley, J., 1918. *The fauna of the Clyde Sea area.* Glasgow University Press. 200 pp.

Cook, P. L., 1964. Polyzoa from West Africa. Notes on the genera *Hippoporina* Neviani, *Hippoporella* Canu, *Cleidochasma* Harmer and *Hippoporidra* Canu and Bassler. *Bulletin of the British Museum (Natural History), Zoology,* **12**(1), 1-35.

Cook, P. L., 1968. Bryozoa (Polyzoa) from the coast of tropical West Africa. *Atlantide Reports,* **10**, 115-262.

Cook, P. L., 1973. Preliminary notes on the ontogeny of the frontal body wall in the Adeonidae and the Adeonellidae (Bryozoa, Cheilostomata). *Bulletin of the British Museum (Natural History), Zoology,* **25**(6), 245-263.

Cook, P. L. & Chimonides, P. J., 1980. Further observations on water current patterns in living Bryozoa. *Cahiers de Biologie marine,* **21**, 393-402.

Couch, R. Q., 1844. *A Cornish Fauna Part III. The Zoophytes and calcareous corallines.* The Royal Institution of Cornwall, Truro., 164 pp.

Dick, M. H. & Ross, J. R. P., 1988. Intertidal Bryozoa (Cheilostomata) of the Kodiak vicinity, Alaska. *Center for Pacific Northwest studies, Western Washington University. Occasional Paper* no. 23, 1-133.

Duvergier, J., 1921. Note sur les Bryozoaires du Néogène de l'Aquitaine. *Actes de la Société linnéene de Bordeaux,* **72**, 145-181.

Dyrynda, P. E. J., 1994. Hydrodynamic gradients and bryozoan distributions within an estuarine basin (Poole Harbour, U.K.). pp. 57-63 *in* Hayward, P. J., Ryland, J. S. and Taylor, P. D. (eds.): *Biology and Palaeobiology of Bryozoans.* Olsen and Olsen, Fredensborg. 240 pp.

Dyrynda, P. E. J. & King, P. E., 1982. Sexual reproduction in *Epistomia bursaria* (Bryozoa: Cheilostomata), an endozooidal brooder without polypide recycling. *Journal of Zoology, London,* **198**, 337-352.

Dyrynda, P. E. J. & Ryland, J. S., 1982. Reproductive strategies and life histories in the cheilostome marine bryozoans *Chartella papyracea* and *Bugula flabellata. Marine Biology,* **71**, 241-256.

Eggleston, D., 1963. *The marine Polyzoa of the Isle of Man.* Thesis, University of Liverpool. 297 pp.

Eggleston, D., 1969. Marine fauna of the Isle of Man: revised lists of phylum Entoprocta (= Kamptozoa) and phylum Ectoprocta (= Bryozoa). *Reports of the marine biological station, Port Erin,* **81**, 57-80.

Eggleston, D., 1970. Embryo colour in Manx ectoprocts. *Reports of the marine biological station, Port Erin,* **82**, 39-42.

Eggleston, D., 1972a. Patterns of reproduction in marine Ectoprocta of the Isle of Man. *Journal of Natural History,* **6**, 31-38.

Eggleston, D., 1972b. Factors influencing the distribution of sublittoral ectoprocts off the south of the Isle of Man (Irish Sea). *Journal of Natural History,* **6**, 247-260.

Eggleston, D., 1975. The marine fauna of the Cullercoats district. Ectoprocta. *Reports of the Dove marine Laboratory, series 3,* no. 18, 1-30.

Ellis, J. & Solander, D., 1786. *The natural history of many curious and uncommon zoophytes, collected by the late John Ellis, systematically arranged and described by the late Daniel Solander.* Benjamin White and Son, London. 208 pp.

Esper, E. J. C., 1790. *Die Pflanzenthiere in Abbildungen nach der Natur mit Farbenerleuchtet nebst Beschreiben,* Nürnberg, **1**, (Parts 5-6) 193-320.

Esper, E. J. C., 1796. *Fortsetzungen der Pflanzenthiere in Abbildungen nach der Natur mit Farbenerleuchtet nebst Beschreibungen.* Nürnberg, Vol. 1, 117-168.

Fabricius, O., 1780. *Fauna Groenlandica.* J. G. Rothe, Hafniae and Lipsiae. 452 pp.

Fischer, G., 1807. *Muséum Démidoff (Moscou), III. Végétaux et Animaux.* Moscow. 330 pp.

Fleming, J., 1814. Contributions to the British fauna. *Memoirs of the Wernerian natural History Society,* **2**, 238-251.

Fleming, J., 1828. *A history of British animals, exhibiting their descriptive characters and systematical arrangement of the genera and species of quadrupeds, birds, reptiles, fishes, Mollusca, and Radiata of the United Kingdom.* Bell and Bradfute, Edinburgh, 565 pp.

Gabb, W. M. & Horn, G. H., 1862. The fossil Polyzoa of the Secondary and Tertiary Formations of North America. *Journal of the Academy of Natural Sciences of Philadelphia,* **5**(2), 111-179.

Gautier, Y. V., 1962. Recherches ƒcologiques sur les Bryozoaires chilostomes en Mediterranée occidentale. *Recueil des Travaux de la Station marine d'Endoume, Faculté des Sciences de Marseille,* **38**, 1-434.

Geraci, S., 1974. On a new species of Bryozoa from the Ligurian Sea: *Escharella rylandi* nov. sp. *Doriana,* **5**(214), 1-3.

Gordon, D. P., 1984. The marine fauna of New Zealand: Bryozoa: Gymnolaemata from the Kermadec Ridge. *New Zealand Oceanographic Institute Memoir,* **91**, 1-198.

Gordon, D. P., 1986. The marine fauna of New Zealand: Bryozoa: Gymnolaemata (Ctenostomata and Cheilostomata Anasca) from the western South Island continental shelf and slope. *New Zealand Oceanographic Institute Memoir,* **95**, 1-121.

Gordon, D. P., 1988. The bryozoan families Sclerodomidae, Bifaxariidae, and Urceoliporidae and a novel type of frontal wall. *New Zealand Journal of Zoology,* **15**, 249-290.

Gordon, D. P., 1989. The marine fauna of New Zealand: Bryozoa: Gymnolaemata (Cheilostomida Ascophorina) from the western South Island continental shelf and slope. *New Zealand Oceanographic Institute Memoir,* **97**, 1-158.

Gordon, D. P., 1993. Bryozoan frontal shields: studies on umbonulomorphs and impacts on classification. *Zoologica Scripta,* **22**, 203-221.

Gordon, D. P., & Grischenko, A. V., 1994. Bryozoan frontal shields: the type species of *Desmacystis, Rhamphostomella, Rhamphosmittina, Rhamphostomellina*, and new genus *Arctonula. Zoologica Scripta,* **23**, 61-72.

Gordon, D. P. & Hastings, A. B., 1979. The interzooidal communications of *Hippothoa sensu lato* (Bryozoa) and their value in classification. *Journal of natural History,* **13**, 561-579.

Gordon, D. P. & Mawatari, S. F., 1992. Atlas of marine-fouling Bryozoa of New Zealand Ports and Harbours. *Miscellaneous Publications New Zealand Oceanographic Institute,* **107**, 1-52.

Gray, J. E., 1848. *List of the specimens of British animals in the collection of the British Museum. Part 1. Centroniae or radiated animals.* Trustees of the British Museum (Natural History), London, 173 pp.

Gregory, J. W., 1893. On the British Palaeogene Bryozoa. *Transactions of the Zoological Society of London,* **13**, 219-279.

Hamond, R., 1973. The marine and brackish-water Bryozoa of Norfolk. *Transactions of the Norfolk and Norwich Naturalists' Society,* **22**, 406-423.

Harmelin, J. G., 1969. Bryozoaires récoltés au cours de la campagne du *Jean Charcot* en Mediterranée occidentale (août-septembre 1967)., 1. Dragages. *Bulletin du Muséum National d'Histoire Naturelle, Paris (2e serie),* **40**, 1179-1208.

Harmelin, J. G. & D'Hondt, J. L., 1992. Bryozoaires des parages de Gibraltar (campagne océanographique BALGIM 1984)., 1. Chéilostome. *Bulletin du Muséum national d'Histoire naturelle, Paris, (4e série)*, **14**, 23-67.

Harmer, S. F., 1933. The genera of Reteporidae. *Proceedings of the Zoological Society of London,* **1933**, 615-627.

Harmer, S. F., 1957. The Polyzoa of the Siboga Expedition. Part 4. Cheilostomata Ascophora. *Siboga Expedition Reports,* **28d**, 641-1147.

Harvell, C. D., 1986. The ecology and evolution of inducible defenses in a marine bryozoan: cues, costs, and consequences. *The American Naturalist,* **128**, 810-823.

Harvell, C. D., 1990. The ecology and evolution of inducible defenses. *Quarterly Review of Biology,* **65**, 323-340.

Harvell, C. D., 1991. Coloniality and inducible polymorphism. *The American Naturalist,* **138**, 1-14.

Hassall, A. H., 1840. Catalogue of Irish Zoophytes. *Annals and Magazine of Natural History,* **6**, 166-175.

Hassall, A. H., 1841. Supplement to a catalogue of Irish Zoophytes. *Annals and Magazine of Natural History,* **7**, 363-373.

Hassall, A. H., 1842. Remarks on the genus *Lepralia* of Dr. Johnston with descriptions of six undescribed species. *Annals and Magazine of Natural History,* **9**, 407-414.

Hastings, A. B., 1944. Notes on Polyzoa (Bryozoa), I. *Umbonula verrucosa* auctt.: *U. ovicellata*, sp. n. and *U. littoralis*, sp. n. *Annals and Magazine of Natural History, series 11*, **11**, 273-284.

Hastings, A. B., 1979. The genus *Hippothoa* Lamouroux Polyzoa (Bryozoa) Cheilostomata. *Journal of Natural History,* **13**, 535-560.

Hayward, P. J., 1974. Studies on the Cheilostome bryozoan fauna of the Aegean island of Chios. *Journal of Natural History,* **8**, 369-402.

Hayward, P. J., 1976. The marine fauna and flora of the Isles of Scilly. Bryozoa II. *Journal of Natural History,* **10**, 319-330.

Hayward, P. J., 1977. The marine fauna of Lundy: Bryozoa. *Reports of the Lundy Field Society,* **27**, 16-34.

Hayward, P. J., 1978. Bryozoa from the west European continental slope. *Journal of Zoology, London,* **184**, 207-224.

Hayward, P. J., 1979. Systematic notes on some British Ascophora (Bryozoa: Cheilostomata). *Zoological Journal of the Linnean Society,* **66**, 73-90.

Hayward, P. J., 1981. The Cheilostomata (Bryozoa) of the deep sea. *Galathea Report,* **15**, 21-68.

Hayward, P. J., 1994. New species and new records of cheilostomatous Bryozoa from the Faroe Islands, collected by BIOFAR. *Sarsia,* **79**, 181-206.

Hayward, P. J., 1995. *Antarctic Cheilostomatous Bryozoa.* Oxford University Press. 355pp.

Hayward, P. J., & Hansen, K. B., 1999. Three newly recognized cheilostomate bryozoans from the British sea area. *Journal of the Marine Biological Association, UK,* **79**,

Hayward, P. J. & Ryland, J. S., 1978. Bryozoa from the Bay of Biscay and western approaches. *Journal of the Marine Biological Association of the U.K.*, **58**, 143-159.

Hayward, P. J. & Ryland, J. S., 1979. *British Ascophoran Bryozoans.* Synopses of the British fauna (new series) **14**, published for the Linnean Society of London and the Estuarine and Brackish Water Sciences Association by Academic Press. 312 pp.

Hayward, P. J. & Ryland, J. S., 1995. The British species of *Schizoporella* (Bryozoa Cheilostomatida). *Journal of Zoology, London,* **237**, 37-47.

Hayward, P. J. & Ryland, J. S., 1996. Some British Phidoloporidae (Bryozoa: Cheilostomatida). *Zoological Journal of the Linnean Society,* **117**, 103-112.

Hayward, P. J. & Ryland, J. S., 1998. Cheilostomatous *Bryozoa Part I. Aeteoidea - Cribrilinoidea* . Synopses of the British fauna (new series) **10**, published for the Linnean Society of London and the Estuarine and Coastal Water Sciences Association by Field Studies Council, Shrewsbury.

Hayward, P. J. & Thorpe, J. P., 1989. Membraniporoidea, Microporoidea and Cellarioidea

(Bryozoa, Cheilostomata) collected by Discovery Investigations. *Journal of Natural History,* **23**, 913-959.

Hayward, P. J. & Thorpe, J. P., 1995. Some British species of *Schizomavella* (Bryozoa: Cheilostomatida). *Journal of Zoology, London,* **235**, 661-676.

Heller, C., 1867. Die Bryozoën des adriatischen Meeres. *Verhandlungen der Zoologisch-botanischen Gesellschaft in Wien,* **17**, 77-136.

Hincks, T., 1857. On some new British Polyzoa. *Quarterly Journal of Microscopical Science,* **5**, 175-176.

Hincks, T., 1860a. Zoophytology. Descriptions of new Polyzoa from Ireland. *Quarterly Journal of Microscopical Science,* **8**, 275-285.

Hincks, T., 1860b. Supplement to the list of zoophytes in the "Natural History of Ireland, by W. Thompson, Esq." (Vol. 4, 1856); with descriptions of new species. *Proceedings of Dublin University Zoological and Botanical Association,* **2**, 67-78.

Hincks, T., 1862. A catalogue of the zoophytes of south Devon and south Cornwall. *Annals and Magazine of Natural History, series 3,* **9**, 22-30, 200-207, 303-310, 467-475.

Hincks, T., 1877a. On British Polyzoa, Part 1. *Annals and Magazine of Natural History, series 4,* **20**, 212-218.

Hincks, T., 1877b. On British Polyzoa, Part 2. *Annals and Magazine of Natural History, series 4,* **20**, 520-532.

Hincks, T., 1878. Notes on the genus *Retepora*, with descriptions of new species. *Annals and Magazine of Natural History, series 5,* **1**, 353-365.

Hincks, T., 1879. On the classification of the British Polyzoa. *Annals and Magazine of Natural History, series 5,* **3**, 153-164.

Hincks, T., 1880. *A history of the British marine Polyzoa.* John Van Voorst, London, 2 vols.

Hincks, T., 1886. The Polyzoa of the Adriatic: a Supplement to Prof. Heller's "Die Bryozoën des Adriatischen Meeres, 1867". *Annals and Magazine of Natural History, series 5,* **17**, 254-271.

Hincks, T., 1888. The Polyzoa of the St. Lawrence: a study of arctic forms. *Annals and Magazine of Natural History, series 6,* **1**, 214-227.

Hincks, T., 1895. Index (to "Marine Polyzoa: contributions towards a general history"). Privately published, London. 6 pp.

Hondt, J.-L. d' 1981. Bryozoaires Cheilostomes bathyaux et abyssaux provenant des compagnes océanographiques américaines (1969-1972) de "l'Atlantis II", du "Chain" et du "Knorr" (Woods Hole Oceanographic Institution). *Bulletin du Muséum National d'Histoire Naturelle, Paris, serie 4,* **3A**, 5-71.

Hondt, J.-L. d' 1983. Nouvelle contribution à l'étude des Bryozoaires Eurystomes bathyaux et abyssaux de l'Océan Atlantique. *Bulletin du Muséum National d'Histoire Naturelle, Paris, serie 4,* **5A**, 73-99.

Hondt, J.-L. d' 1985. Contribution à la Systématique des Bryozoaires Eurystomes. Apports récents et nouvelles propositions. *Annales des Sciences Naturelles, Zoologie, série 13,* **7**, 1-12.

Hondt, J.-L. d' & Schopf, T. J. M., 1984. Bryozoaires des grandes profondeurs recueillis lors des campagnes océanographiques de la Woods Hole Oceanographic Institution de 1961 à 1968. *Bulletin du Muséum National d'Histoire Naturelle, Paris, série 4,* **6A**, 907-973.

Hughes, D. J., 1987. Gametogenesis and embryonic brooding in the cheilostome bryozoan *Celleporella hyalina. Journal of Zoology, London,* **212**, 691-711.

Hunter, E. & Hughes, R. N., 1993a. Self-fertilization in *Celleporella hyalina. Marine Biology,* **115**, 495-500.

Hunter, E. & Hughes, R. N., 1993b. Effects of diet on life-history parameters of the marine bryozoan, *Celleporella hyalina* (L.). *Journal of Experimental Marine Biology and Ecology,* **167**, 163-177.

Jackson, J. B. C., 1979. Overgrowth competition between encrusting cheilostome ectoprocts in a Jamaican cryptic reef environment. *Journal of Animal Ecology,* **48**, 805-823.

Jeffreys, J. G., 1876. Preliminary report of the biological results of a cruise in H.M.S. "Valorous" to Davis Strait in 1875. *Proceedings of the Royal Society,* **25**, 177-237.

Johnston, G., 1832. A descriptive catalogue of the recent zoophytes found on the coast of north Durham. *Transactions of the natural History Society, Newcastle,* **2**, 239-272.

Johnston, G., 1838. *A history of the British zoophytes.* W. H. Lizars, Edinburgh, 341 pp.

Johnston, G., 1840. Miscellanea Zoologica. Description of a new genus of British zoophyte. *Annals and Magazine of Natural History,* **5**, 272-274.

Johnston, G., 1847. *A history of the British zoophytes.* 2nd edn. John Van Voorst, London, 2 vols.

Jullien, J., 1882. Note sur une nouvelle division des Bryozoaires cheilostomiens. *Bulletin de la Société Zoologique de France,* **76**, 271-285.

Jullien, J., 1883. Bryozoaires. Espèces draguées dans l'océan Atlantique en 1881. *Bulletin de la Société Zoologique de France,* **7**, 497-529.

Jullien, J., 1888. Bryozoaires. *Mission scientifique du Cap Horn 1882-1883, 6 (Zoologie Part 3),* 1-92.

Jullien, J. & Calvet, L., 1903. Bryozoaires provenent des campagnes de l'Hirondelle. *Résultats des campagnes scientifiques accompli par le Prince Albert 1,* **23**, **1-188.**

King, W., 1846. An account of some shells and other invertebrate forms found on the coasts of Northumberland and Durham. *Annals and Magazine of Natural History,* **18**, 233-251.

Kluge, G. A., 1946. New and little known species of Bryozoa from the Arctic Ocean. *Trudy dreifuyushchei Ekspeditsiya "G. Sedov" 1937-1940,* **3**, 194-223.

Kluge, G. A., 1955. New and little known Bryozoa from the northern Ocean, 2. *Trudy Zoologicheskogo Instituta. Akademiya Nauk SSSR. Leningrad,* **18**, 63-99.

Kluge, G. A., 1962. Mshanki severnykh morei SSSR. *Opredeliteli po faune SSSR, Izdavaemye Zoologicheskim Muzeem Akademii Nauk,* **76**, 1-584.

Kluge, G. A., 1975. *Bryozoa of the northern seas of the U.S.S.R.* Amerind Publishing Company Pvt. Ltd., New Delhi, 711 pp.

Lagaaij, R., 1952. The Pliocene Bryozoa of the Low Countries and their bearing on the marine stratigraphy of the North Sea region. *Mededelingen van de Geologische Stichting,* series C, **5**, 1-233.

Lamouroux, J. V. F., 1812. Extrait d'un mémoire sur la classification des Polypiers coralligènes non entièrement pierreux. *Nouvelle Bulletin Scientifique de la Société Philosophique,* **3**, 181-188.

Lamouroux, J. V. F., 1816. *Histoire des Polypiers Coralligènes Flexibles, vulgairement nommés Zoophytes.* F. Poisson, Caen, 559 pp.

Lamouroux, J. V. F., 1821. *Exposition méthodique des genres de l'ordre des polypiers, avec leur description et celles des principales espèces figurées dans 84 planches; les 63 premières apartenant a l'Histoire naturelle des Zoophytes d'Ellis et Solander.* V. Agasse, Paris, 115 pp.

Landsborough, D., 1852. *A. popular history of British zoophytes, or corallines.* Reeve and Co., London. 404 pp.

Lanza Suarez, N. & Fernandez Pulpeiro, E., 1984. Briozoos infralitorales de Galicia: Queilostomados. *Investigacion Pesquera* **48**, 269-284.

Levinsen, G. M. R., 1902. Studies on Bryozoa. *Videnskabelige Meddelelser fra den naturhistoriske Foreningi Kjbenhavn,* **54**, 1-31.

Levinsen, G. M. R., 1909. *Morphological and systematic studies on the cheilostomatous Bryozoa.* Nationale Forfatteres Forlag, Copenhagen.

Levinsen, G. M. R., 1916. Danmark-Ekspeditionen til Grønlands Nordstkyst 1906-1908. Bind III, nr. 16, Bryozoa. *Meddelelser om Grønland,* **43**, 431-472.

Lidgard, S., 1981. Water flow, feeding, and colony form in an encrusting cheilostome. pp. 135-142 *in* Larwood, G. P. and Nielsen, C. (eds.): *Recent and Fossil Bryozoa.* Olsen and Olsen, Fredensborg. 334 pp.

Lidgard, S., 1985. Zooid and colony growth in encrusting cheilostome bryozoans. *Palaeontology,* **28**, 255-291.

Linnaeus, C., 1758. *Systema naturae per regna tria naturae, secundum classes, ordines, genera, species cum characteribus, differentiis, synonymis, locis.*, 10th Edn. Laurentii Salvii, Holmiae, Vol. 1, 824 pp.

Linnaeus, C., 1761. *Fauna Suecica.* Laurentii Salvii, Holmiae. 576 pp.

Linnaeus, C., 1767. *Systema Naturae.*, 12th Edn. Vol. 1 Regnum Animale. Laurentii Salvii, Holmiae, 1327 pp.

Lopez de la Cuadra, C. M. & Garcia Gomez, J. C., 1988. Briozoos queilostomados del Estrecho de Gibraltar y areas proximas. *Cahiers de Biologie Marine,* **29**, 21-36.

Lopez de la Cuadra, C. M. & Garcia Gomez, J. C., 1993. Little known Atlantic Cheilostome Bryozoans at the entrance to the Mediterranean. *Journal of Natural History,* **27**, 457-469.

Macgillivray, J., 1842. Catalogue of the marine zoophytes of the neighbourhood of Aberdeen. *Annals and Magazine of Natural History* **9**, 462-469.

Macgillivray, P. H., 1895. A monograph of the Tertiary Polyzoa of Victoria. *Transactions of the Royal Society of Victoria, new series,* **4**, 1-166.

Manzoni, A., 1870a. Bryozoi fossili Italiani. Terza contribuzione. *Sitzungsberichte der Akademie der Wissenschaften in Wien (Abt. 1),* **60**, 930-944.

Manzoni, A., 1870b. Bryozoi fossili Italiani. Quarta contribuzione. *Sitzungsberichte der Akademie der Wissenschaften in Wien (Abt. 1),* **61**, 323-349.

Marcus, E., 1940. Mosdyr (Bryozoa eller Polyzoa). *Danmarks Fauna,* **46**, 1-401.

Maturo, F. J., 1968. The distributional pattern of the Bryozoa of the east coast of the United States exclusive of New England. *Atti della Società Italiana di Scienze Naturali e del Museo Civico di Storia Naturale di Milano* **108**, 261-284.

McKinney, F. K., 1981. Planar branch systems in colonial suspension feeders. *Paleobiology,* **7**, 344-354.

McKinney, F. K., 1986. Evolution of erect marine bryozoan faunas: repeated success of unilaminate species. *The American Naturalist,* **128**, 795-809.

McKinney, F. K., 1989. Historical record of erect bryozoan growth forms. *Proceedings of the Royal Society, London,* **B228**, 133-149.

McKinney, F. K. & Jackson, J. B. C., 1989. *Bryozoan Evolution.* The University of Chicago Press, Chicago. 238 pp.

Milne Edwards, H., 1836. Histoire des polypes. Vol. 2, *in* Deshayes, G. P. and Milne Edwards, H. (eds.): *Histoire naturelle des animaux sans vertèbres par J.-B. P. A. Lamarck,* deuxième Édition revue et augmentée. J. B. Baillière, London, 684 pp.

Moll, J. P. C. von 1803. *Eschara, ex zoophytorum, seu, phytozoorum ordine pulcherrimum ac notatu dignissimum genus novis speciebus auctum, methodice descriptum et iconibus ad naturam delineatis illustratum.* Vindobonae, 70 pp.

Moore, P. G., 1973. Bryozoa as a community component on the northeast coast of Britain. pp. 21-36 *in* Larwood, G. P. (ed), *Living and Fossil Bryozoa.* Academic Press: London.

Moyano G. H. I., 1986. Bryozoa Marinos Chilenos. VI. Cheilostomata Hippothoidae: South Eastern Pacific species. *Boletin de la Sociedad de Biologia de Concepcion,* **57**, 89-135.

Moyano G. H. I. & Gordon D. P., 1980. New species of Hippothoidae (Bryozoa) from Chile, Antarctica and New Zealand. *Journal of the Royal Society of New Zealand,* **10**(1), 75-95.

Neviani, A., 1895. Briozoi neozoici di alcune località d'Italia. Part 1. *Bolletina della societ^ Romana per gli studi Zoologici,* **4**, 109-123.

Nichols, A. R., 1911. Polyzoa from the coasts of Ireland. *Scientific Investigations, Fisheries Branch, Ireland 1910*, 1-37.

Nielsen, C., 1981. Ovicell formation in *Tegella* and four cellularioids (Bryozoa, Cheilostomata). pp. 213-220 *in* Nielsen, C. and Larwood, G. P. (eds.): *Bryozoa Ordovician to Recent.* Olsen and Olsen, Fredensborg. 364 pp.

Nordgaard, O., 1900. Polyzoa. *The Norwegian North-Atlantic Expedition, 1876-1878. Zoology*, 1-30.

Nordgaard, O., 1906. Die Bryozoen des westlichen Norwegens. *Meeresfauna,* **2**, 76-107.

Nordgaard, O., 1907. Bryozoen von dem norwegischen Fischereidampfer "Michael Sars" in den Jahren 1900-1904 gesammelt. *Bergens Museums Aarbog*, **1907** (no. 2), 1-20.

Norman, A. M., 1864. On undescribed British Hydrozoa, Actinozoa and Polyzoa. *Annals and Magazine of Natural History, Series 3,* **13**, 82-90.

Norman, A. M., 1867. Report of the committee appointed for the purpose of exploring the coasts of the Hebrides by means of the dredge. Part 2. On the Crustacea, Echinodermata, Polyzoa, Actinozoa and Hydrozoa. *Report of the British Association for the Advancement of Science,* **1866**, 193-206.

Norman, A. M., 1868. Notes on some rare British Polyzoa, with descriptions of new species. *Quarterly Journal of Microscopical Science, new series,* **8**, 212-222.

Norman, A. M., 1869. Last report on dredging among the Shetland Islands. Polyzoa. *Report of the British Association for the Advancement of Science,* **1868**, 303-312.

Norman, A. M., 1894. A month on the Trondheim Fjord. *Annals and Magazine of Natural History, series 6,* **13**, 112-133.

Norman, A. M., 1903. Notes on the natural History of East Finmark. Polyzoa. *Annals and Magazine of Natural History, series 7,* **11**, 567-598.

Norman, A. M., 1909. The Polyzoa of Madeira and neighbouring islands. *Journal of the Linnean Society, Zoology,* **30**, 275-314.

Norton, T. A., Ebling, F. J. & Kitching, J. A., 1971. Light and the distribution of organisms in a sea cave. pp. 409-432 *in* Crisp, D. J. (ed), *Fourth European Marine Biology Symposium.* Cambridge University Press.

Occhipinti Ambrogi, A., 1981. *Guide per il riconoscimento delle specie animali delle acque lagunari e costiere italiane AQ/1/1267. Briozoi Lagunari.* Consiglio Nazionale delle Richerche, Genova, 145 pp.

Occhipinti Ambrogi, A., 1985. The zonation of bryozoans along salinity gradients in the Venice Lagoon (northern Adriatic). pp. 221-231 *in* Nielsen, C. and Larwood, G. P. (eds.): *Bryozoa: Ordovician to Recent.* Olsen and Olsen, Fredensborg. 364 pp.

Okamura, B., 1992. Microhabitat variation and patterns of colony growth and feeding in a marine bryozoan. *Ecology,* **73**, 1502-1513.

Oken, L., 1815. *Lehrbuch der Naturgeschichte, III. Zoologie*, Abteilung 1. Fleischlose Thiere. Leipzig. 842 pp.

Orbigny, A. D' 1841. *Voyage dans l'Amérique méridionale,* **5**, Pt. 4 Zoophytes, 7-28 (1847), Pl. 1, 3, 5 (1841), 2,4, 6-13 (1842). Paris and Strasbourg.

Orbigny, A. D' 1851-4. *Paléontologie Francaise, Terrains Crétacés,* **5**, 1-1192. Paris.

Osburn, R. C., 1952. Bryozoa of the Pacific Coast of America, 2. Cheilostomata Ascophora. *Allan Hancock Pacific Expeditions,* **14**, 276-611.

Packard, A. S., 1863. A list of animals dredged near Caribou Island, Southern Labrador, during July and August, 1860. *Canadian Naturalist and Geologist,* **8**, 401-429.

Pallas, P. S., 1766. *Elenchus zoophytorum sistens generum adumbrationes generaliores et specierum cognitarum succintas descriptiones cum selectis auctorus synonymis*. Petrum van Cleef, Hagae-Comitum, 451 pp.

Patzold, J., Ristedt, H. & Wefer, G., 1987. Rate of growth and longevity of a large colony of *Pentapora foliacea* (Bryozoa) recorded in their oxygen isotope profiles. *Marine Biology,* **96**(4), 535-538.

Peach, C. W., 1868. On a new British *Eschara. Journal of the Royal Institution of Cornwall,* **3**, 116-117.

Picton, B. E. & Morrow, C. C., 1994. *A field guide to the nudibranchs of the British Isles.* Immel Publishing Ltd., London., 143 pp.

Powell, N. A., 1968. Bryozoa (Polyzoa) of Arctic Canada. *Journal of the Fisheries Research Board of Canada,* **25**(11), 269-320.

Powell, N. A. & Crowell, G. D., 1967. Studies on Bryozoa (Polyzoa) of the Bay of Fundy Region., 1. Bryozoa from the intertidal zone of Minas Basin and Bay of Fundy. *Cahiers de Biologie Marine,* **8**, 331-347.

Prenant, M. & Bobin, G., 1966. Bryozoaires, 2[e] partie Chilostomes Anasca. *Faune de France,* **68**, 1-647.

Quelch, J. J., 1884. On *Schizoporella Ridleyi*, MacG., and *Schizoporella simplex*, D'Orbigny and Johnston. *Annals and Magazine of Natural History, series 5,* **13**, 215-217.

Reuss, A. E. von 1848. Die fossilen Polyparien des Wiener Tertiärbeckens. *Naturwissenschaftliche Abhandlungen, Wien,* **11**, 1-109.

Reverter Gil, O., 1994. *Briozoos de la Ria de Ferrol.* Thesis, Universidad de Santiago de Compostela. 386 pp.

Reverter Gil, O. & Fernandez Pulpeiro, E., 1996. Some species of *Schizomavella* (Bryozoa, Cheilostomida) from the Atlanto-Mediterranean Region.

Reverter Gil, O., Fernandez Pulpeiro, E. & Ramil Blanco, F., 1992. Briozoos marinos de Galicia: Queilostomados. *Boletin de la Real Sociedad Espa–ola de Historia Natural (Biol.),* **88**, 99-115.

Ristedt, H., 1991. Ancestrula and early astogeny of some anascan Bryozoa: their taxonomic importance and possible phylogenetic implications. pp. 371-382 *in* Bigey, F. P. (ed.): *Bryozoaires actuels et fossiles: Bryozoa living and fossil. Bulletin de la Société des Sciences Naturelles de l'Ouest de la France. Mémoire H.S.1.* 599 pp.

Rogick, M. D. & Croasdale, H., 1949. Studies on marine Bryozoa III. Woods Hole region Bryozoa associated with algae. *Biological Bulletin. Marine Biology Laboratory, Woods Hole, Massachusetts,* **96**, 32-69.

Rubin, J. A., 1982. The degree of intransitivity and its measurement in an assemblage of encrusting cheilostome Bryozoa. *Journal of Experimental Marine Biology and Ecology,* **60**, 119-128.

Rubin, J. A., 1985. Mortality and avoidance of competitive overgrowth in encrusting Bryozoa. *Marine Ecology Progress Series,* **23**, 291-299.

Ryland, J. S., 1963a. A collection of Polyzoa from the west of Scotland. *Scottish Naturalist,* **71**, 13-22.

Ryland, J. S., 1963b. Systematic and biological studies on Polyzoa (Bryozoa) from western Norway. *Sarsia,* **14**, 1-59.

Ryland, J. S., 1963c. The species of *Haplopoma* (Polyzoa). *Sarsia,* **10**, 9-18.

Ryland, J. S., 1967. Polyzoa. *Oceanography and Marine Biology Annual Review,* **5**, 343-369.

Ryland, J. S., 1968. On marine Polyzoa. III. *Schizoporella ansata* auctt. *Journal of Natural History,* **2**, 535-546.

Ryland, J. S., 1969. A nomenclatural index to "A History of the British Marine Polyzoa" by T. Hincks (1880). *Bulletin of the British Museum (Natural History), Zoology,* **17**, 6, 207-260.

Ryland, J. S., 1975. Parameters of the lophophore in relation to population structure in a bryozoan community. pp. 363-393 *in* Barnes, H. (ed.): *Proceedings of the 9th European Marine Biology Symposium.* Aberdeen University Press, Aberdeen, 760 pp.

Ryland, J. S., & Bishop, J. D. D., 1993. Internal fertilization in hermaphroditic colonial invertebrates. *Oceanography and Marine Biology Annual Review,* **31**, 445-477.

Ryland, J. S., & Gordon, D. P., 1977. Some New Zealand and British species of *Hippothoa* (Bryozoa: Cheilostomata). *Journal of the Royal Society of New Zealand,* **7**, 17-49.

Ryland, J. S., & Hayward, P. J., 1977. *British Anascan Bryozoans.* Synopses of the British fauna, new series, no. **10**, published for the Linnean Society of London Academic Press. 188 pp.

Sandberg, P. A., 1977. Ultrastructure, mineralogy, and development of bryozoan skeletons. pp. 143-181 *in* Woollacott, R. M. and Zimmer, R. L. (eds.): *Biology of Bryozoans.* Academic Press, New York, 566 pp.

Sars, G. O., 1872. On some remarkable forms of animal life from the great deeps off the Norwegian coast. I. Partly from the posthumous manuscripts of the late Professor Dr. Michael Sars. *Univ. Prog., 1,* **869**. Christiana. 82 pp.

Sars, M., 1851. Beretning om en i Sommeren 1849 foretagen zoologisk Reise i Lofoten og Finmarken. *Nyt Magazin for Naturvidenskaberne,* **6**, 121-211.

Silén, L., 1972. Fertilization in the Bryozoa. *Ophelia,* **10**, 27-34.

Silén, L. & Harmelin, J. G., 1976. *Haplopoma sciaphilum* sp. n., a cave-living bryozoan from the Skagerrak and the Mediterranean. *Zoologica Scripta,* **5**, 61-66.

Smitt, F. A., 1867. Kritisk förteckning öfver Skandinaviens Hafs-Bryozoer, II. *...versigt Kungliga Vetenskapsakademiens Förhandlingar,* **23**, 395-534.

Smitt, F. A., 1868a. Kritisk förteckning öfver Skandinaviens Hafs-Bryozoer, III. *...versigt Kungliga Vetenskapsakademiens Förhandlingar,* **24**, 279-429.

Smitt, F. A., 1868b. Kritisk förteckning öfver Skandinaviens Hafs-Bryozoer, IV. *...versigt Kungliga Vetenskapsakademiens Förhandlingar,* **25**, 3-230.

Smitt, F. A., 1873. Floridan Bryozoa, collected by Count L. F. de Pourtales, Part II. *Kongliga. Svenska Vetenskaps-Akademiens Fπrhandlingar,* **11**, 1-83.

Sowerby, J., 1805. The British Miscellany (No. 8), **1**, 1-131.

Taylor, P. D., 1981. Bryozoa from the Jurassic Portland Beds of England. *Palaeontology,* **24**, 863-875.

Taylor, P. D., 1986. The ancestrula and early growth pattern in two primitive cheilostome bryozoans: *Pyripora catenularia* (Fleming) and *Pyriporopsis portlandensis* Pokowsky. *Journal of Natural History,* **20**, 101-110.

Taylor, P. D., & Cook, P. L., 1981. *Hippoporidra edax* (Busk, 1859) and a revision of some fossil and living *Hippoporidra* (Bryozoa). *Bulletin of the British Museum (Natural History), Geology,* **35**, 243-251.

Thompson, T. E. & Brown, G. H., 1984. *Biology of Opisthobranch Molluscs.* Ray Society, London. 229 pp.

Todd, J. A., 1994. The role of bioimmuration in the exceptional preservation of fossil ctenostomates, including a new Jurassic species of *Buskia.* pp. 187-196 *in* Hayward, P. J., Ryland, J. S. and Taylor, P. D.: *Biology and Palaeobiology of Bryozoans.* Olsen and Olsen, Fredensborg. 240 pp.

Todd, C. D. & Havenhand, J. N., 1989. Nudibranch-bryozoan associations: the quantification of ingestion and some observations on partial predation among Doridoidea. *Journal of Molluscan Studies,* **55**, 245-249.

Vigneaux, M., 1949. Révision de Bryozoaires néogènes du Bassin d'Aquitaine et essai de classification. *Mémoires de la Société geologique de France, n.s.*, **28**, 1-153.

Ward, M. A. & Thorpe, J. P., 1989. Assessment of space utilization in a subtidal temperate bryozoan community. *Marine Biology,* **103**, 215-224.

Waters, A. W., 1878. The use of opercula in the determination of the cheilostomatous Bryozoa. *Proceedings of the Manchester Literary and Philosophical Society,* **18**, 8-11.

Waters, A. W., 1879. On the Bryozoa (Polyzoa) of the Bay of Naples. *Annals and Magazine of Natural History* (5), **3**, 27-43, 114-126, 192-202, 267-281.

Waters, A. W., 1899. Bryozoa from Madeira. *Journal of the Royal Microscopical Society,* **1899**, 6-16

Waters, A. W., 1900. Bryozoa from Franz-Josef Land collected by the Jackson-Harmsworth Expedition, 1896-1897. Pt. 1, Cheilostomata. *Journal of the Linnean Society, London (Zoology),* **28**, 43-105.

Winston, J. E., 1978. Polypide morphology and feeding behavior in marine ectoprocts. *Bulletin of Marine Science,* **28**, 1-31.

Winston, J. E., 1984. Why bryozoans have avicularia: a review of the evidence. *American Museum Novitates no.* **2789**, 1-26.

Winston, J. E., 1986. Victims of avicularia. *Pubblicazione della Stazione Zoologica di Napoli (Marine Ecology),* **7**, 193-199.

Winston, J. E., 1991. Avicularian behavior – a progress report. pp. 531-540 *in* Bigey, F. P. (ed.): *Bryozoaires actuels et fossiles: Bryozoa living and fossil. Bulletin de la Société des Sciences Naturelles de l'Ouest de la France, Mémoire H. S.*, 1, 599 pp.

Wood, S. V., 1844. Descriptive catalogue of the zoophytes of the Crag. *Annals and Magazine of Natural History,* **13**, 10-21.

Wyer, D. W. & King, P. E., 1973. Relationships between some British littoral and sublittoral bryozoans and pycnogonids. pp. 198-207 *in* Larwood, G. P. (ed.): *Living and fossil Bryozoa.* Academic Press, London. 634 pp.

Zabala, M., 1986. *Fauna dels Briozous dels Paãsos Catalans.* Institut d'Estudis Catalans, Barcelona. 833 pp.

Zabala, M. & Maluquer, P., 1988. Illustrated Keys for the classification of Mediterranean Bryozoa. *Treballs del Museu de Zoologia Barcelona,* **4**, 1-294.

Zabala, M., Maluquer, P. & Harmelin, J. G., 1993. Epibiotic bryozoans on deep-water scleractinian corals from the Catalonia Slope (western Mediterranean, Spain, France). *Scientia Marina,* **57**, 65-78.

Zimmer, R. L. & Woollacott, R. M., 1977. Structure and classification of gymnolaemate larvae. pp. 57-89 *in* Woollacott, R. M. and Zimmer, R. L. (eds.) *Biology of Bryozoans.* Academic Press, London and New York.

Index

to the scientific names of Superfamilies, families, genera and species included in the Systematic Part of this Synopsis

The names of families and superfamilies are entered in BOLD TYPE. Species are listed under both their generic name and their specific epithet: valid species in ***bold italic type***, synonyms in *italic type.* Only the principal page reference appears after the specific epithet entry. *Page references in italics* after the generic name entries indicate (additional) illustrations.